DYNAMICS OF FLIGHT— STABILITY AND CONTROL

SECOND EDITION

BERNARD ETKIN

Institute for Aerospace Studies
Faculty of Applied Science and Engineering
University of Toronto

JOHN WILEY & SONS

New York Chichester Brisbane Toronto Singapore

Library of Congress Cataloging in Publication Data:

Etkin, Bernard.
Dynamics of flight.

Includes biliographies and index.
1. Aerodynamics. 2. Stability of airplanes.
I. Title.
TL570.E75 1982 629.132'36 81-13058
ISBN 0-471-08936-2 AACR2

Printed in the United States of America

13 14 15 16 17 18 19 20

To the men and women of science and engineering whose contributions to aviation have made it a dominant force in shaping the destiny of mankind, and who, with sensitivity and concern, develop and apply their technological arts toward bettering the future.

PREFACE TO THE SECOND EDITION

Since the publication of *Dynamics of Atmospheric Flight* in 1972, many teachers, students, and engineers have expressed disappointment that my 1959 book, *Dynamics of Flight—Stability and Control* was out of print. It was still regarded, even after twenty years, as a useful text and reference book by many. As a result, we decided to reprint it. Inevitably, simple reprinting was not in order. Some of the material was too dated, and some I regarded as not appropriate for a book intended primarily as an undergraduate textbook. Hence I have made some changes in the first eight chapters—the core of the book—and deleted the original Chapters 9 through 14 (which treated control response, flight in turbulence, inverse problems, automatic controls, missiles, and machine computation). These five chapters have been replaced by two new chapters on open-loop and closed-loop control, drawn largely from Chapters 10 and 11 of my second book. Thus the book has been shortened somewhat and is more "single-minded" in its purpose—that is, to serve as a textbook for undergraduates.

All but one of the original appendixes—that giving the properties of the atmosphere—are included in this volume. I have learned that the extensive data on stability and control derivatives in Appendix B was one of the most useful features of the original book. This material has been reproduced without change. It still serves its original purpose, to show orders of magnitude and trends; it is not intended as accurate design data.

The mathematical techniques used in the first eight chapters remain, of course, essentially unchanged. In the two new chapters, however, I have taken advantage of the fact that virtually all engineering students today are skilled in matrix algebra and computing and know about eigenvalues and eigenvectors. I would expect teachers to comment in class on how Eqs. 4.15,7 and 4.15,8 can be put into the canonical first-order form $\dot{\mathbf{x}} = \mathbf{A}\mathbf{x} + \mathbf{B}\mathbf{u}$ to facilitate both analysis and computation.

The relationship between this book and *Dynamics of Atmospheric Flight* warrants some comment. Simply put, the latter is a deeper and more comprehensive treatment of the same subject. It allows for the fact that the Earth is a rotating sphere, instead of a stationary plane, and treats such matters as variation of the atmosphere with altitude and its turbulent nature. All reference to random processes has now been eliminated from this book. The practicing engineer or graduate student who has studied this book will find much new matter of interest in the other. I reiterate the thought expressed in the concluding sentence of the 1958 preface, which follows.

BERNARD ETKIN

PREFACE TO THE FIRST EDITION

Of all aspects of aeronautical engineering, the study of dynamics of flight is to me one of the most fascinating. I think that this is because it is an elegant blending of a trinity of classical disciplines; mechanics of solids, mechanics of fluids, and applied mathematics; and because it has at the same time important practical ramifications in design. The problems of airplane design are centered in three main areas: performance, structural integrity, and flying qualities. Dynamics of flight bears strongly on all of these. It is from studies of its dynamics that we learn about the flying qualities of an airplane and about the loads imposed on its structure when flying in rough air or in maneuvers. The importance of such studies is universally recognized in the aircraft and missile industry, and they form an integral part of all programs from preliminary project work to flight test.

For the past fifteen years, I have had a continuing interest in this subject. During that time I have taught courses in mechanics, fluid mechanics, applied aerodynamics, and airplane dynamics at the University of Toronto, while at the same time participating in analysis programs in the aircraft industry. This book is a result of those experiences, and of the urging of my colleagues at the Institute of Aerophysics to write it. I wish to thank them for their encouragement. Especially would I like to thank the director of the Institute, Dr. G. N. Patterson, whose support was practical as well as moral.

In my approach to the subject I have tried to emphasize the fundamental aspects, while at the same time bringing in some applications and typical results. The word "typical" is used with reservations. The art of airplane and missile design is progressing so rapidly, and the configurations and flight regimes of interest are so varied, that one is scarcely justified in speaking of typical configurations and typical results. Each new departure brings with it its own special problems. Engineers in this field must always be alert to discover these, and be adequately prepared to tackle them—they must be ready to discard long-accepted methods and assumptions and to venture in new directions with confidence. The proper background for such ventures is a thorough understanding of the underlying principles and the essential techniques. My primary aim in writing this book has been to set down these principles and techniques as clearly as I could, and to incorporate the most important recent developments.

The body of the book does not contain any numerical data, of the kind used in estimation of aerodynamic parameters. An appendix of such data has been prepared in cooperation with Mr. G. K. Dimock, of Avro Aircraft, Toronto, and is included as a convenience. This material is intended to show orders of magnitude and trends, and should not be interpreted as accurate design information.

I should like to express my indebtedness to the following, for reading portions of the manuscript, and for many helpful discussions: Dr. H. S. Ribner, Mr. E. D. Poppleton, Dr. J. M. Ham, Mr. G. K. Dimock, and Dr. G. V. Bull. Thanks are also due to Dr. C. C. Gottlieb and the staff of the University of Toronto Computing Centre, for making available time on the digital computer FERUT.

Prerequisites for reading this book are a knowledge of basic mechanics, aerodynamics, and mathematics including complex numbers and differential equations. It is my hope that it will be useful both as a textbook for senior and graduate students and as a reference book for practicing engineers.

BERNARD ETKIN

CONTENTS

CHAPTER 8
SOME MATHEMATICAL AIDS 211

CHAPTER 9
RESPONSE TO ACTUATION OF THE CONTROLS (OPEN LOOP) 239

CHAPTER 10
CLOSED-LOOP CONTROL 271

APPENDIX A
RÉSUMÉ OF VECTOR ANALYSIS 293

INTRODUCTION

CHAPTER 1

1.1 THE SUBJECT MATTER OF DYNAMICS OF FLIGHT

Dynamics of flight is an applied engineering subject, not one of the fundamentals. The latter subjects, for example, *mechanics of solids*, *thermodynamics*, or *electricity and magnetism*, constitute studies of the basic principles of certain branches of science; these principles find application or are manifested in a wide variety of machines and natural phenomena. In dynamics of flight, on the other hand, our interest is confined to certain features of a particular class of machines, namely airborne vehicles. To carry out the necessary studies, we find that we must apply information and methods drawn from several fundamental subjects, principally mechanics of solids, mechanics of fluids, and applied mathematics. Not until these fundamental subjects were sufficiently advanced was it possible to solve the main problems of flight mechanics.

The general mathematical formulation of the flight of an elastic body subject to aerodynamic (including propulsive) and gravitational forces through non-stationary air involves three sets of equations:

1. The *force equations*, which relate the motion of the mass-center to the external forces.
2. The *moment equations*, which relate the rotation about the mass center to the external moments.
3. The *elastic equations*, which relate the deformations of the structure to the loading imposed on it.

Likewise the problems that have to be solved fall into three general categories, as illustrated in the block diagram below:

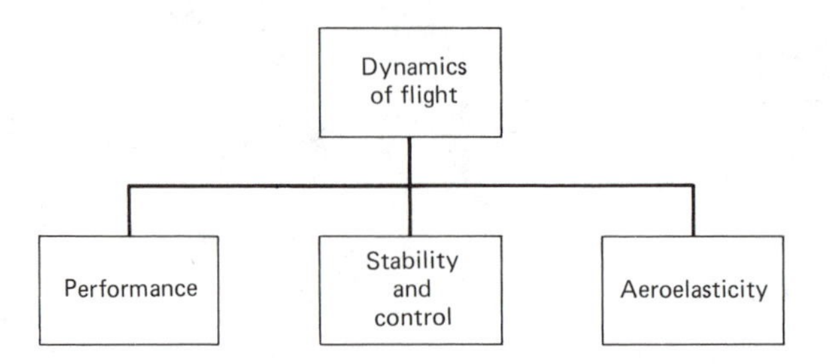

The problems of *performance*, e.g., the calculation of maximum speed, ceiling, and range, involve primarily the application of the force equations. The moment and elastic equations play secondary roles. The times that are characteristic of this class of problems are relatively long (minutes or hours).

The problems of *stability and control* involve motions with smaller characteristic times (seconds), and employ all three sets of equations, although the moment equations and the rotations of the body tend to be dominant.

The problems of *aeroelasticity* are dominated by the elastic equations, although they may involve the force and moment equations as well. The characteristic times of the dynamic aeroelastic phenomena (vibration and flutter) are still smaller, with typical periods less than 1 sec.

As might be expected, the boundaries between the three areas indicated are not well defined. For example, steady turning flight might be treated under either performance or control, and response to gusts under either stability or aeroelasticity.

A comprehensive treatment of the entire broad subject in one volume seems to be neither feasible nor desirable, and hence this book concentrates on the area of the center block, i.e., stability and control.[1] In this branch of the subject, we are concerned mainly (but not entirely) with (1) the tendency, or lack of it, of an airplane to fly straight with wings level (i.e., stability), and (2) steering an airplane on an arbitrary flight path (i.e., control). The term *flying qualities* is used to designate the airplane characteristics that are relevant to these two aspects of its behavior.

The subject of airplane stability and control appeared as a recognizable entity at the turn of the century, largely owing to the pioneering theoretical work of Lanchester (ref. 1.1) and Bryan (refs. 1.2, 1.3). Its experimental beginnings, of course, go back much farther. A long list of inventors and scientists made contributions—from Sir George Cayley, who experimented with flying models in 1843, to the Wright brothers, who finally made successful flights in 1903. The most difficult problem that these experimenters encountered, and the last to be surmounted,

[1] For treatments of the other two branches of the subject, the reader is referred to Miele (ref. 1.11), Fung (ref. 1.18) and Bisplinghoff, Ashley, and Halfman (ref. 1.19).

was the achievement of satisfactory flying qualities. It was here that the Wright brothers made their most notable contribution.

In contrast with other vehicles of transportation, such as ships, trains, and automobiles, the problems associated with stability and controlability of an airplane take a position of preeminence in its design. The reasons for this are mainly the following: (1) An airplane is supported in flight by a dynamic reaction of the air against its wings. This is in contrast to the simpler mechanisms of buoyancy and road reaction that support sea-going and land-going vehicles. (2) Airplanes fly at very high speed, and changes in their attitude can take place very quickly; this places severe demands on the human or automatic pilot. (3) Aircraft must be steered in a three-dimensional space, whereas ships require steering in only two dimensions, and trains are not steered at all. (It is noteworthy, however, that underwater vehicles such as submarines and torpedoes share many of the dynamic problems of aircraft.)

The effort devoted during design and development to ensuring that a finished airplane will have adequate flying qualities is made in several directions—mathematical analysis, machine computation, wind-tunnel testing, and flight testing. Mathematical analysis and machine computation are directed at setting up and solving the complex equations that describe the behavior of the airplane in flight. Wind-tunnel testing is conducted to obtain accurate information on the aerodynamic forces, required as input data for the equations of motion. Finally, flight testing is required to verify that the flying qualities actually achieved are satisfactory, or to find and eliminate imperfections in them.

Stability, Control, and Equilibrium

It is appropriate here to define what is meant by the terms *stability* and *control*. To do so requires that we begin with the concept of equilibrium.

A body is in equilibrium when it is at rest or in uniform motion (i.e., has constant linear and angular momenta). The most familiar examples of equilibrium are the static ones; that is, bodies at rest. The equilibrium of an airplane in flight, however, is of the second kind; i.e., uniform motion. Because the aerodynamic forces are dependent on the angular orientation of the airplane relative to its flight path, and because the resultant of them must exactly balance its weight, the equilibrium state is without rotation; i.e., it is a motion of rectilinear translation.

Stability, or the lack of it, is a property of an equilibrium state. The equilibrium is stable if, when the body is slightly disturbed in any of its degrees of freedom, it returns ultimately to its initial state. This is illustrated in Fig. 1.1*a*. The remaining sketches of Fig. 1.1 show neutral and unstable equilibrium. That in Fig. 1.1*d* is a more complex kind than that in Fig. 1.1*b* in that the ball is stable with respect to displacement in the y direction, but unstable with respect to x displacements. This has its counterpart in the airplane, which may be stable with respect to one degree of freedom and unstable with respect to another. Two kinds of instability

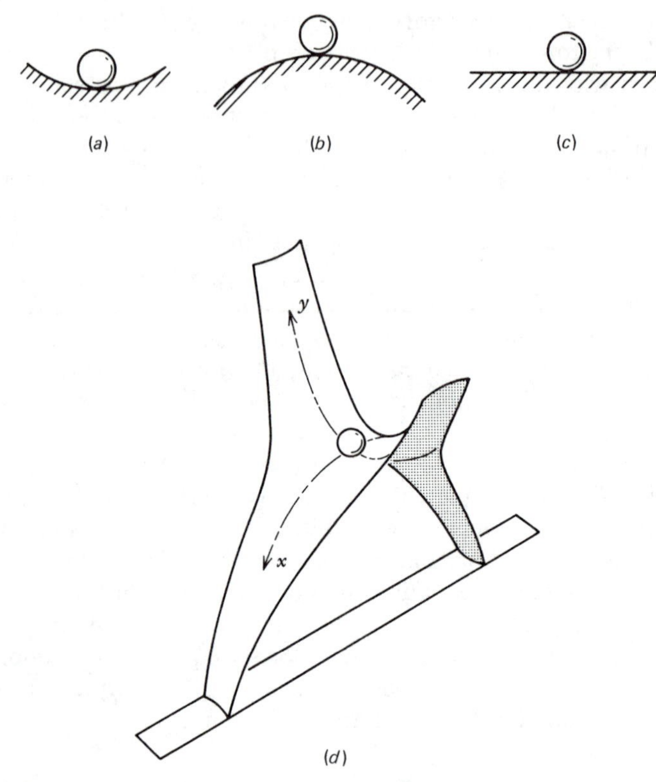

FIG. 1.1 (*a*) Ball in a bowl—stable equilibrium. (*b*) Ball on a hill—unstable equilibrium. (*c*) Ball on a plane—neutral equilibrium. (*d*) Ball on a saddle surface—unstable equilibrium.

are of interest in airplane dynamics. In the first, called *static instability*, the body departs continuously from its equilibrium condition. That is how the ball in Fig. 1.1*b* would behave if disturbed. The second, called *dynamic instability*, is a more complicated phenomenon in which the body oscillates about its equilibrium condition with ever-increasing amplitude.

The primary functions of control are twofold. The first is to fix or to change the equilibrium condition (speed or angle of climb). An adequate control must be powerful enough to produce the whole range of equilibrium states of which the airplane is capable from a performance standpoint. The dynamics of the transition from one equilibrium state to another are of interest and are closely related to stability. The second function of the control is to produce nonequilibrium, or accelerated motions; i.e., maneuvers. These may be steady states in which the forces and accelerations are constant when viewed from a reference frame fixed to the airplane (e.g., a steady turn), or they may be transient states. Investigations of the transition from equilibrium to a nonequilibrium steady state, or from one maneuvering steady state to another, form part of the subject matter of airplane control. Very large aerodynamic forces may act on the airplane when it maneuvers—a knowledge of these forces is required for the proper design of the structure.

Response to Atmosphere Turbulence

A topic that belongs in dynamics of flight and that is closely related to stability is the response of the airplane to atmospheric turbulence. This response is important from several points of view. It has a strong bearing on the adequacy of the structure, it influences the acceptability of the airplane as a passenger transport, and it limits its accuracy as a gun or bombing platform.

1.2 THE HUMAN PILOT

In dealing with airplanes that are to be flown by a human pilot, it is obvious, that to some extent the machine must be designed to match the pilot. The two are parts of a single system, the pilot–machine combination. Some aspects of this matching are so self-evident that they scarcely need comment. For example, the forces and movements required of the pilot must be within the ranges of which he or she is capable. Some data on forces and motions (from ref. 1.4) are given in Tables 1.1 through 1.3.

Less obvious are the dynamic characteristics of human pilots. Pilots may be regarded as links in a closed-loop system, sensing the motion and position of the airplane, and actuating the controls in response. Their sensitivity and dynamic behavior are central to the problem of manually controlled flight; the ideal dynamic design of an airplane requires dynamic studies of the pilot-plus-airplane system. The importance of pilot dynamics may be seen by considering the following fact. Many airplanes have a spiral instability, such that, if left alone, they would slowly proceed into an ever steeper turn. Yet such airplanes are no more difficult to fly than airplanes that are spirally stable—pilots are virtually unaware of this instability. The reason is that the time of growth of the spiral mode is so long that the pilot unconsciously compensates for it. The combination pilot-plus-machine is stable even though the machine alone is not. To carry out analyses of pilot–machine combinations requires the development of an adequate mathematical model of the pilot (e.g., a transfer function, see Chap. 8). This model should ideally allow for the effects of psychological stress, fatigue, noise and vibration in the environment, etc. The difficulty of deriving such a model is aggravated by the fact that it is not constant, either in form or in the values of parameters. It changes with the nature of the task (i.e., pursuit, landing, cruise) and with the circumstances under which it is performed (i.e., daylight or night, weather, stress, and fatigue). Much progress has nevertheless been made in recent years, and satisfactory "describing functions" are available for some applications.

TABLE 1.1
Estimates of the Maximum Rudder Forces that Can Be Exerted for Various Positions of the Rudder Pedal (Ref. 1.4)

Rudder Pedal Position	*Distance from Back of Seat, in.*	*Pedal Force, lb*
Back	31	246
Neutral	$34\frac{3}{4}$	424
Forward	$38\frac{1}{2}$	334

TABLE 1.2
Hand-Operated Control Forces (From Flight Safety Foundation Human Engineering Bulletin 56-5H) (see figure on page 7)

Direction of Movement		180°	150°	120°	90°	60°	
Pull	Rt. hand	52	56	42	37	24	Values given represent maximum exertable force in pounds by the 5 percentile man
	Lft. hand	50	42	34	32	26	
Push	Rt. hand	50	42	36	36	34	
	Lft. hand	42	30	26	22	22	
Up	Rt. hand	14	18	24	20	20	
	Lft. hand	9	15	17	17	15	
Down	Rt. hand	17	20	26	26	20	
	Lft. hand	13	18	21	21	18	
Outboard	Rt. hand	14	15	15	16	17	
	Lft. hand	8	8	10	10°	12	
Inboard	Rt. hand	20	20	22	18	20	
	Lft. hand	13	15	20	16	17	

Note: The above results are those obtained from unrestricted movement of the subject. Any force required to overcome garment restriction would reduce the effective forces by the same amount.

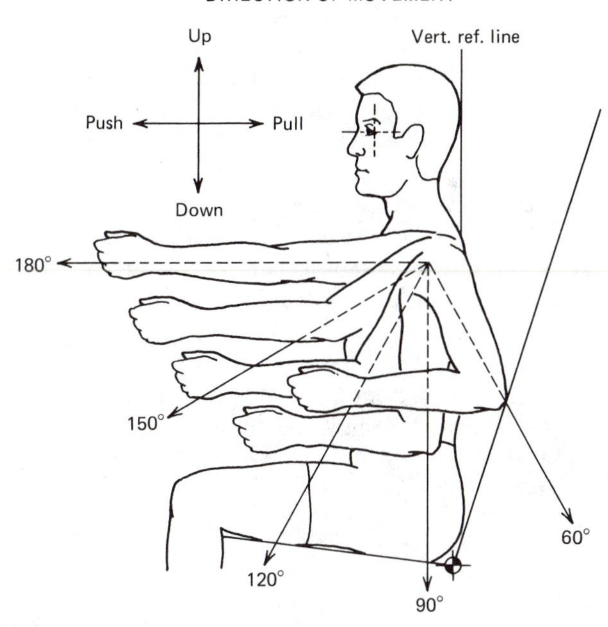

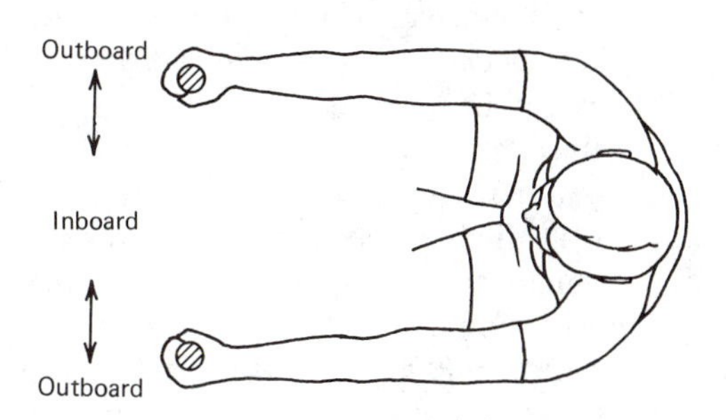

TABLE 1.3
Rates of Stick Movement in Flight Test Pull-ups Under Various Loads (Ref. 1.4)

Pull-up	*Maximum Stick load, lb*	*Average Rate of Stick Motion, in./sec*	*Time for Full Deflection, sec*
1	35	51.85	0.162
2	74	15.58	0.475
3	77	11.00	0.600
4	97	10.27	0.750

1.3 FLYING QUALITIES REQUIREMENTS

As a result of the inability to carry out completely rational design of the pilot–machine combination, it is customary for the government agencies responsible for the procurement of military airplanes, or for licensing civil airplanes, to specify compliance with certain "flying qualities requirements" (e.g., refs. 1.5 and 1.6).

These requirements have been developed from extensive and continuing flight research. In the final analysis they are based on the opinions of research test pilots, substantiated by careful instrumentation. They vary from country to country and from agency to agency, and, of course, are different for different types of aircraft. They are subject to continuous study and modification in order to keep them abreast of the latest research and design information. Because of these circumstances, it is not feasible to present a detailed description of such requirements here. The following is intended to show the nature, not the detail, of typical flying qualities requirements.[2] Most of the specific requirements can be classified under one of the following headings.

Control Power

The term control power is used to describe the efficacy of a control in producing a range of steady equilibrium or maneuvering states. For example, an elevator control, which by taking positions between full up and full down can hold the airplane in equilibrium at all speeds in its speed range, for all configurations[3] and C.G. positions, is a powerful control. On the other hand, a rudder that is not capable at full deflection of maintaining equilibrium of yawing moments in a condition of one engine out and negligible sideslip is not powerful enough. The flying qualities requirements normally specify the specific speed ranges that must be achievable with full elevator deflection in the various important configurations, and the asymmetric power condition that the rudder must balance. They may also contain references to the elevator angles required to achieve positive load factors, as in steady turns and pull-up maneuvers ("elevator angle per *g*," Sec. 3.1).

Control Forces

The requirements invariably specify limits on the control forces that must be exerted by the pilot in order to effect specific changes from a given trimmed condition, or to maintain the trim speed following a sudden change in configuration or throttle setting. They frequently also include requirements on the control forces in pull-up maneuvers ("stick force per *g*," Sec. 3.2).

[2] For a more complete discussion, see ref. 1.7.

[3] This word describes the position of movable elements of the airplane—e.g., landing configuration means that landing flaps and undercarriage are down, climb configuration means that landing gear is up, and flaps are at take-off position, etc.

Static Stability

The requirement for static longitudinal stability (see Chap. 2) is usually stated in terms of the *neutral point* (defined in Sec. 2.3). It is usually required that the relevant neutral point (stick free or stick fixed) shall lie some distance (e.g., 5% of the mean aerodynamic chord) behind the most aft position of the C.G. This ensures that the airplane will tend to fly at a constant speed and angle of attack as long as the controls are not moved.

The requirement on static lateral stability is usually mild. It is simply that the spiral mode (see Chap. 7) if divergent shall have a time to double greater than some stated minimum (e.g., 4 sec).

Dynamic Stability

The requirement on dynamic stability is typically expressed in terms of the damping and frequency of a natural mode. Thus the USAF (ref. 1.6) requires the damping and frequency of the lateral oscillation for various flight phases and stability levels to conform to the values in Table 1.4.

Stalling and Spinning

Finally, most requirements specify that the airplane's behavior following a stall or in a spin shall not include any dangerous characteristics, and that the controls must retain enough effectiveness to ensure a safe recovery to normal flight.

TABLE 1.4[1]
Minimum Dutch Roll Frequency and Damping

Level	*Flight Phase Category*	*Class*	*Min* ζ_d[2]	*Min* $\zeta_d\omega_{n_d}$,[2] *rad/sec*	*Min* ω_{n_d}, *rad/sec*
1	A	I, IV	0.19	0.35	1.0
		II, III	0.19	0.35	0.4
	B	All	0.08	0.15	0.4
	C	I, II-C, IV	0.08	0.15	1.0
		II-L, III	0.08	0.15	0.4
2	All	All	0.02	0.05	0.4
3	All	All	0.02	—	0.4

[1] *Level*, *Phase* and *Class* are defined in Ref. 1.6.
[2] *Note:* The damping coefficient ζ, and the undamped natural frequency ω_n, are defined in Chap. 8.

1.4 AXES AND NOTATION

Two frames of reference are used herein in describing the motion of an airplane. One of these is fixed to the earth (the rotation of which is assumed to be negligible). The other is fixed to the airplane with origin at the mass center and moves with it. The position of the mass center of the airplane is given by its Cartesian coordinates relative to the fixed frame of reference, and its angular orientation is given by three rotations relative to the same frame of reference (see Chap. 4). The instantaneous motion *relative to the fixed axes* is described by the six projections of the linear and angular velocity vectors *on the moving axes*. The directions of the latter axes are shown in Fig. 1.2. The xz plane is the plane of symmetry. The velocity of C, denoted $\mathbf{v}_c$, is in general inclined to the x axis. This inclination is defined by the two angles (see Fig. 1.4):

$$\text{Angle of attack,} \quad \alpha = \tan^{-1}\frac{W}{U}$$

$$\text{Angle of sideslip,} \quad \beta = \sin^{-1}\frac{V}{v_c} \tag{1.4,1}$$

With these definitions, the sideslip angle β is not dependent on the direction of Cx in the plane of symmetry.

The symbols used throughout the text correspond generally to current usage, and are mainly used in a consistent manner. The reader's attention is drawn to

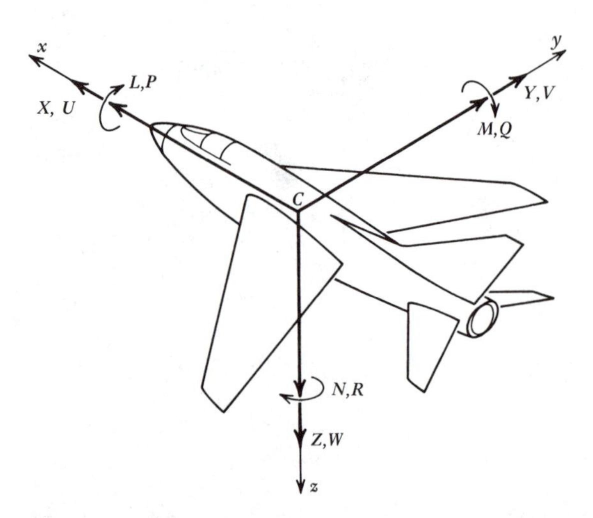

FIG. 1.2 Notation. L = rolling moment; M = pitching moment; N = yawing moment; P = rolling velocity; Q = pitching velocity; R = yawing velocity; (X, Y, Z) = components of resultant aerodynamic force; (U, V, W) = components of velocity of C.

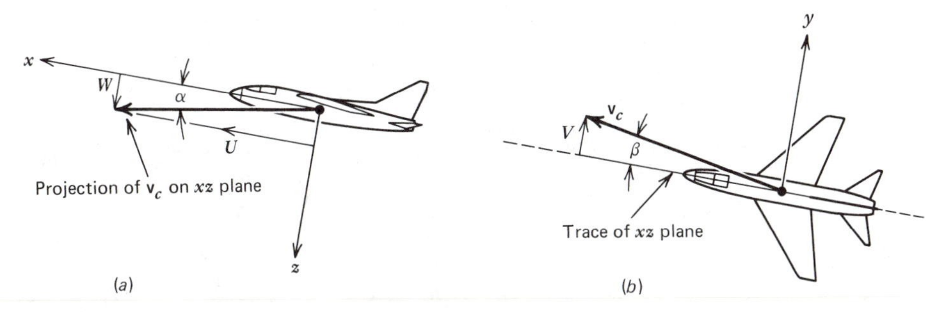

FIG. 1.3 (*a*) Definition of α. (*b*) View in plane of y and $\mathbf{v}_c$, definition of β.

two exceptions. In Chaps. 2 and 3, α denotes the angle of attack of the zero lift line of the airplane (with elevator angle zero). In the succeeding chapters it denotes the angle of attack of the x axis, as required by Eq. 1.4,1. In Chaps. 2 and 3 the symbol V denotes the resultant velocity of the airplane mass center, whereas in the remaining chapters it denotes only the y component.

1.5 BIBLIOGRAPHY

1.1 F. W. Lanchester. *Aerodonetics*. A. Constable & Co. Ltd., London, 1908.

1.2 G. H. Bryan and W. E. Williams. The Longitudinal Stability of Aerial Gliders. *Proc. Roy. Soc. London*, ser. A, vol. 73, pp. 100–116, 1904.

1.3 G. H. Bryan. *Stability in Aviation*. Macmillan Co., London, 1911.

1.4 ———. The Human Pilot. Vol. III of *Bur. Aero. Rept. AF-6-4*, 1954.

1.5 ———. *Airworthiness of Aircraft*, Annex 8, Chap. 2.4—Flying Qualities, International Civil Aviation Organization (ICAO) Montreal, Canada.

1.6 ———. Flying Qualities of Piloted Airplanes. *USAF Spec.* MIL-F-8785B, 1969.

1.7 ———. *Flight Test Manual*, Vol. II, Stability and Control. AGARD, 1955.

1.8 W. H. Phillips. Appreciation and Prediction of Flying Qualities. *NACA TN. 1670*, 1948.

1.9 W. J. Duncan. *Control and Stability of Aircraft*. Cambridge University Press, Cambridge, 1952.

1.10 B. Etkin. *Dynamics of Atmospheric Flight*, John Wiley & Sons, New York, 1972.

1.11 A. Miele. *Flight Mechanics*, Addison Wesley Pub. Co., Reading, Mass., 1962.

1.12 T. Hacker. *Flight Stability and Control*, Elsevier Pub. Co., New York, 1970.

1.13 D. McRuer, I. Ashkenas, and D. Graham. *Aircraft Dynamics and Automatic Control*, Princeton Univ. Press, Princeton N.J., 1973.

1.14 H. Kwakernak and R. Sivan. *Linear Optimal Control Systems*. Wiley-Interscience, New York, 1972.

1.15 E. L. Houghton and R. P. Boswell. *Further Aerodynamics for Engineering Students*. Edward Arnold Ltd., London, 1969.

1.16 B. W. McCormick. *Aerodynamics, Aeronautics, & Fight Mechanics*, John Wiley & Sons, New York, 1979.

1.17 A. M. Kuethe, and C. Y. Chow. *Foundations of Aerodynamics*. Third Ed. John Wiley & Sons, New York, 1976.

1.18 Y. C. Fung. *The Theory of Aeroelasticity*. John Wiley & Sons, New York, 1955.

1.19 R. L. Bisplinghoff, H. Ashley, and R. L. Halfman. *Aeroelasticity*. Addison-Wesley Publishing Co., Cambridge, 1955.

STATIC STABILITY AND CONTROL—PART 1

CHAPTER 2

2.1 GENERAL REMARKS

A general treatment of the stability and control of airplanes requires a study of the dynamics of flight, and this approach is taken in later chapters. Much useful information can be obtained, however, from a more limited view, in which we consider not the *motion* of the airplane, but only its *equilibrium states*. This is the approach in what is commonly known as *static* stability and control analysis.

The unsteady motions of an airplane can frequently be separated for convenience into two parts. One of these consists of the *longitudinal* or *symmetric* motions; i.e., those in which the wings remain level, and in which the center of gravity moves in a vertical plane. The other consists of the *lateral* or *asymmetric* motions; i.e., rolling, yawing, and sideslipping, while the angle of attack, the speed, and the angle of elevation of the x axis remain constant.

This separation can be made for both dynamic and static analyses. However, the results of greatest importance are those associated with the longitudinal analysis. Thus the principal subject matter of this and the following chapter is static longitudinal stability and control. A brief discussion of the static aspects of directional and rolling motions is contained in Secs. 3.9 and 3.10.

We shall be concerned with two aspects of the equilibrium state. Under the heading *stability* we shall consider the pitching moment that acts on the airplane when its angle of attack is changed from the equilibrium value, as by a vertical gust. We focus our attention on whether or not this moment acts in such a sense as to restore the airplane to its original angle of attack. Under the heading *control* we discuss the use of a longitudinal control (elevator) to change the equilibrium value of the angle of attack.

The restriction to angle-of-attack disturbances when dealing with stability must be noted, since the applicability of the results is thereby limited. When the

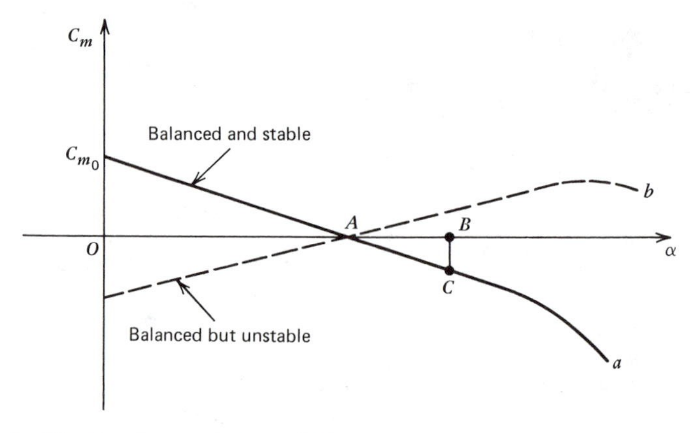

FIG. 2.1 Pitching moment of an airplane.

aerodynamic characteristics of an airplane change with speed, owing to compressibility effects, structural distortion, or the influence of the propulsive system, then the airplane may be unstable with respect to disturbances in speed. Such instability is not predicted by a consideration of angle-of-attack disturbances only. (See Fig. 1.1*d*, and identify speed with x, angle of attack with y.) A more general point of view than that adopted in this chapter is required to assess that aspect of airplane stability. Such a viewpoint is taken in Chap. 6.

Although the major portion of this and the following chapter treats a rigid airplane in gliding flight, an introduction to the effects of propulsive power and airframe distortion is contained in Chap. 3.

Balance, or Equilibrium

An airplane can continue in steady unaccelerated flight only when the resultant external force and the moment about the mass center both vanish. In particular, this requires that the pitching moment be zero. This is the condition of longitudinal *balance*. If the pitching moment were not zero, the machine would experience a rotational acceleration in the direction of the unbalanced moment. Figure 2.1 shows a typical graph of the pitching-moment coefficient versus the angle of attack for an airplane with a fixed elevator (curve *a*). The angle of attack is measured from the zero-lift line of the airplane. The graph is a straight line except near the stall. Since zero C_m is required for balance, the airplane can fly only at the angle of attack marked A, for the given elevator angle.

Stability

Suppose that the airplane of Fig. 2.1 is disturbed from its equilibrium attitude, the angle of attack being increased to that at B while its speed remains unaltered. It is now subject to a negative, or nose-down, moment, whose magnitude corresponds to BC. This moment tends to reduce the angle of attack to its equilibrium value, and hence is a *restoring* moment. In this case, the airplane is statically

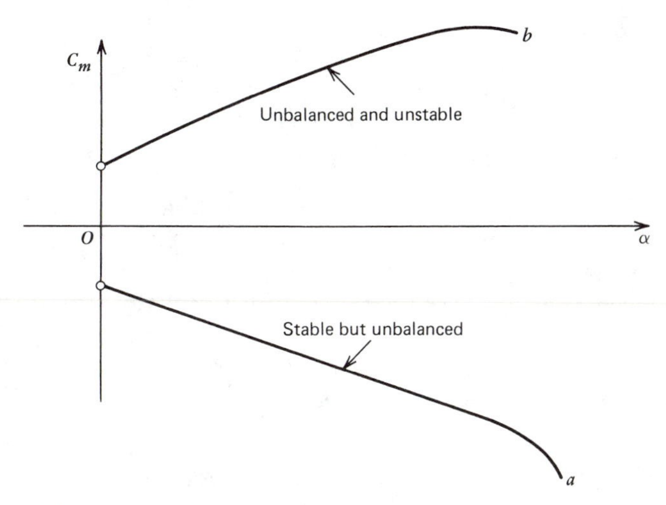

FIG. 2.2 Other possibilities.

stable. On the other hand, if C_m were given by the curve b, the moment acting when disturbed would be positive, or nose-up, and would tend to rotate the airplane still farther from its equilibrium attitude. We see that the static stability is determined by the sign and magnitude of the slope $\partial C_m/\partial \alpha$. *At the equilibrium* α, C_m *must be zero, and* $\partial C_m/\partial \alpha$ *must be negative.* It will be appreciated from Fig. 2.1 that an alternative statement is "C_{m_0} *must be positive, and* $\partial C_m/\partial \alpha$ *negative* if the airplane is to be in a condition of stable equilibrium." The various possibilities corresponding to the possible signs of C_{m_0} and $\partial C_m/\partial \alpha$ are shown in Figs. 2.1 and 2.2.

Possible Configurations

The possible solutions for a suitable configuration are readily discussed in terms of the requirements on C_{m_0} and $\partial C_m/\partial \alpha$. We state here without proof (this is given in Sec. 2.3) that $\partial C_m/\partial \alpha$ can be made negative for virtually any combination of lifting surfaces and bodies by placing the center of gravity far enough forward. Thus it is not the stability requirement, taken by itself, that restricts the possible configurations, but rather the requirement that the airplane must be *simultaneously* balanced and stable. Since a proper choice of the C.G. location can ensure a negative $\partial C_m/\partial \alpha$, *then any configuration with a positive* C_{m_0} *can satisfy the conditions for balanced and stable flight.*

Figure 2.3 shows the C_{m_0} of conventional airfoil sections. If an airplane were to consist of a straight wing alone (flying wing), then the wing camber would

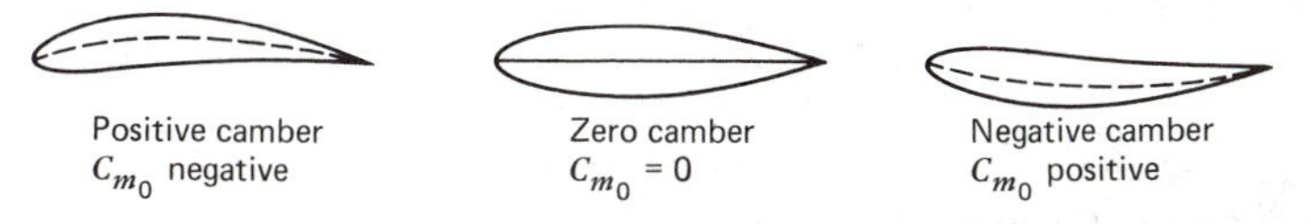

FIG. 2.3 C_{m_0} of airfoil sections.

determine the airplane characteristics as follows:

Negative camber—flight possible at $\alpha > 0$; i.e., $C_L > 0$ (Fig. 2.1*a*).
Zero camber—flight possible only at $\alpha = 0$, or $C_L = 0$.
Positive camber—flight not possible at any positive α or C_L.

For straight-winged tailless airplanes, only the negative camber satisfies the conditions for stable, balanced flight. Effectively the same result is attained if a flap, deflected upward, is incorporated at the trailing edge of a symmetrical airfoil. A conventional low-speed airplane, with essentially straight wings and positive camber, could fly upside down without a tail, provided the C.G. were far enough forward (ahead of the wing mean aerodynamic center). Flying wing airplanes based on a straight wing with negative camber are not in general use for three main reasons:

1. The dynamic characteristics tend to be unsatisfactory.
2. The permissible C.G. range is too small.
3. The drag and $C_{L_{max}}$ characteristics are not good.

Notwithstanding these objections, successful aircraft of this type have been built.

The positively cambered straight wing can be used only in conjunction with an auxiliary device that provides the positive C_{m_0}. The solution adopted by experimenters as far back as Samuel Henson (1842) and John Stringfellow (1848) was to add a tail behind the wing. The Wright brothers (1903) used a tail ahead of the wing (Canard configuration). Either of these alternatives can supply a positive C_{m_0}, as illustrated in Fig. 2.4. When the wing is at zero lift, the auxiliary surface must provide a nose-up moment. The conventional tail must therefore be at a negative angle of attack, and the Canard tail at a positive angle.

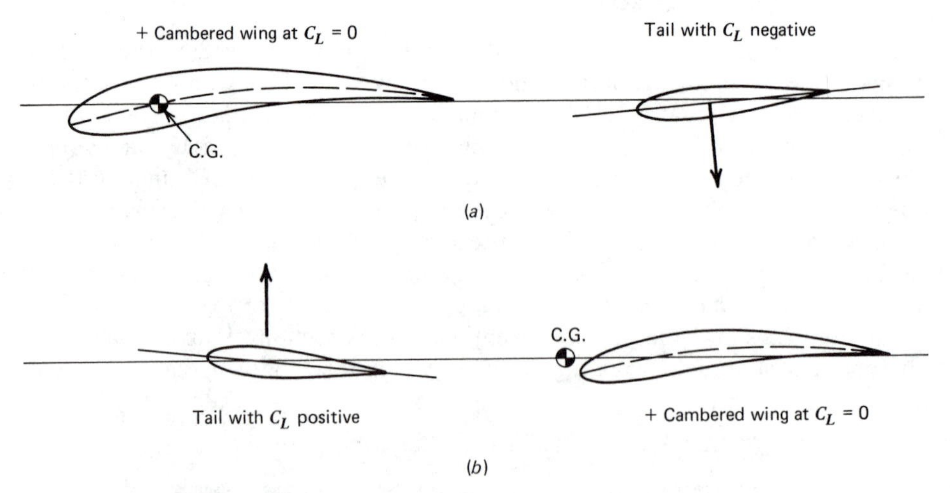

FIG. 2.4 Wing-tail arrangements with positive C_{m_0}. (*a*) Conventional arrangement. (*b*) Tail-first or Canard arrangement.

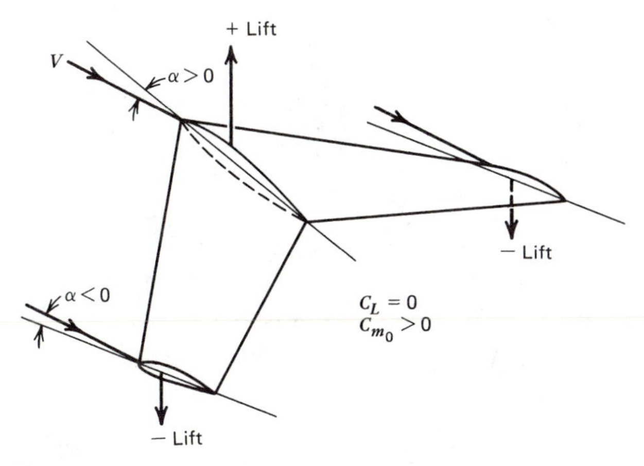

FIG. 2.5 Swept-back wing with twisted tips.

An alternative to the wing-tail combination is the swept-back wing with twisted tips (Fig. 2.5). When the net lift is zero, the forward part of the wing has positive lift, and the rear part negative. The result is a positive couple, as desired.

A variant of the swept-back wing is the delta wing. The positive C_{m_0} can be achieved with such planforms by twisting the tips, by employing negative camber, or by incorporating an upturned tailing edge flap.

2.2 LIFT AND PITCHING MOMENT IN GLIDING FLIGHT

The total lift and pitching moment of an airplane are, in general, functions of angle of attack, control-surface angle(s), Mach number, Reynolds number, thrust coefficient, and dynamic pressure.[1] (The last-named quantity enters because of aeroelastic effects. Changes in the dynamic pressure ($\frac{1}{2}\rho V^2$), when all the other parameters are constant, may induce enough distortion of the structure to alter C_m significantly.) An accurate determination of the lift and pitching moment is one of the major tasks in a static stability analysis. Extensive use is made of wind-tunnel tests, supplemented by aerodynamic and aeroelastic analyses.

For purposes of estimation, the total lift and pitching moment may be synthesized from the contributions of the various parts of the airplane, i.e., wing, body, nacelles, propulsive system, and tail, and their mutual interferences. Data for estimating the various aerodynamic parameters involved are contained in Appendix B, while the general formulation of the equations, in terms of these parameters, follows here. In this chapter propulsive system and aeroelastic effects are not included. Hence the analysis applies to a *rigid* airplane in *gliding flight*.

[1] When partial derivatives are taken in the following equations with respect to one of these variables, e.g., $\partial C_m/\partial\alpha$, it is to be understood that all the others are held constant.

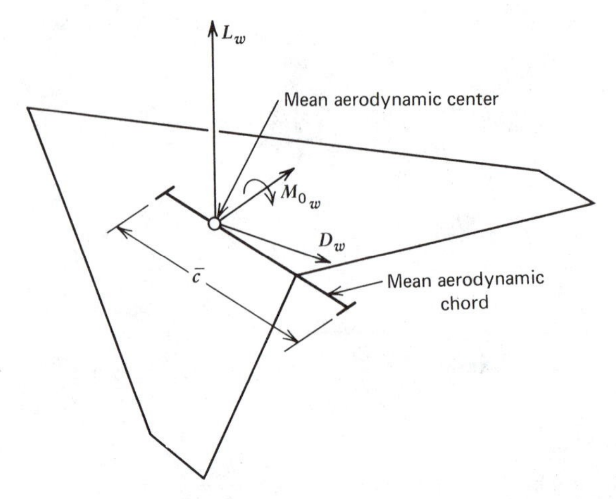

FIG. 2.6 Aerodynamic forces on the wing.

Lift and Pitching Moment of the Wing

The aerodynamic forces on any lifting surface can be represented as a lift and drag acting at the aerodynamic center, together with a pitching couple independent of the angle of attack (Fig. 2.6). The pitching moment of this force system about the center of gravity is given by (Fig. 2.7)

$$M_w = M_{0_w} + (L_w \cos \alpha_w + D_w \sin \alpha_w)(h - h_{n_w})\bar{c} + (L_w \sin \alpha_w - D_w \cos \alpha_w)z\bar{c} \tag{2.2,1}$$

It is assumed that the angle of attack is sufficiently small to justify the approximations

$$\cos \alpha_w = 1, \qquad \sin \alpha_w = \alpha_w$$

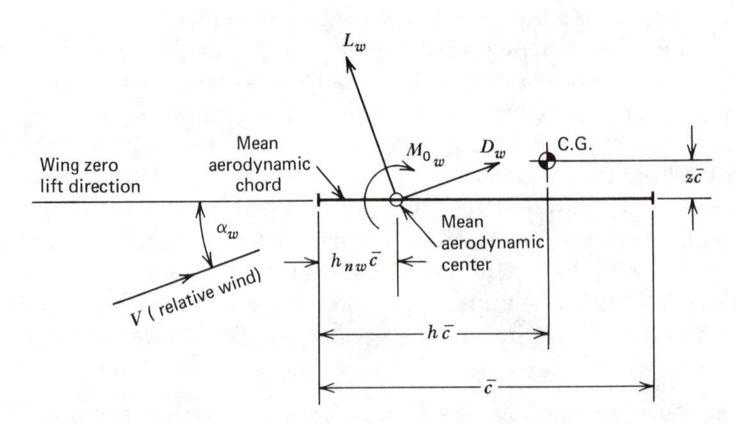

FIG. 2.7 Moment about the C.G. in the plane of symmetry.

and the equation is made nondimensional by dividing through by $\frac{1}{2}\rho V^2 S\bar{c}$. It then becomes

$$C_{m_w} = C_{m_{0_w}} + (C_{L_w} + C_{D_w}\alpha_w)(h - h_{n_w}) + (C_{L_w}\alpha_w - C_{D_w})z \qquad (2.2,2)$$

Although it may occasionally be necessary to retain all the terms in Eq. 2.2,2, experience has shown that the last term is frequently negligible, and that $C_{D_w}\alpha_w$ may be neglected in comparison with C_{L_w}. With these simplifications, we obtain

$$\begin{aligned} C_{m_w} &= C_{m_{0_w}} + C_{L_w}(h - h_{n_w}) \\ &= C_{m_{0_w}} + \alpha_w a_w(h - h_{n_w}) \end{aligned} \qquad (2.2,3)$$

Equation 2.2,3 will be used to represent the wing pitching moment in the discussions that follow.

Lift and Pitching Moment of the Body and Nacelles

The influences of the body and nacelles are complex. A body alone in an airstream is subjected to aerodynamic forces. These, like those on the wing, may be represented over moderate ranges of angle of attack by lift and drag forces at an aerodynamic center, and a pitching couple independent of α. Also as for a wing alone, the lift-incidence relation is approximately linear. When the wing and body are put together, however, a simple superposition of the aerodynamic forces that act upon them separately does not give a correct result. Strong interference effects are usually present, the flow field of the wing affecting the forces on the body, and vice versa. The result of adding a body and nacelles to a wing may usually be interpreted as a shift (forward) of the mean aerodynamic center, an increase in the lift-curve slope, and a negative increment in C_{m_0}. The equation that corresponds to Eq. 2.2,3 for a wing-body-nacelle combination is then of the same form as Eq. 2.2,3, but with different values of the parameters. The subscript *wb* is used to denote these values.

$$\begin{aligned} C_{m_{wb}} &= C_{m_{0_{wb}}} + C_{L_{wb}}(h - h_{n_{wb}}) \\ &= C_{m_{0_{wb}}} + a_{wb}\alpha_{wb}(h - h_{n_{wb}}) \end{aligned} \qquad (2.2,4)$$

Lift and Pitching Moment of the Tail

The forces on an isolated tail are represented just like those on an isolated wing. When the tail is mounted on an airplane, however, important interferences occur. The most significant of these, and one that is usually predictable by aerodynamic theory, is a downward deflection of the flow at the tail caused by the wing. This is characterized by the mean downwash angle ε. Blanking of part of the tail by the body is a second effect, and a reduction of the relative wind when the tail lies in the wing wake is the third.

Figure 2.8 depicts the forces acting on the tail. **V** is the relative wind vector of the airplane, and **V**′ is the average or effective relative wind at the tail. The tail lift and drag forces are, respectively, perpendicular and parallel to **V**′. The reader should note the tail angle i_t.

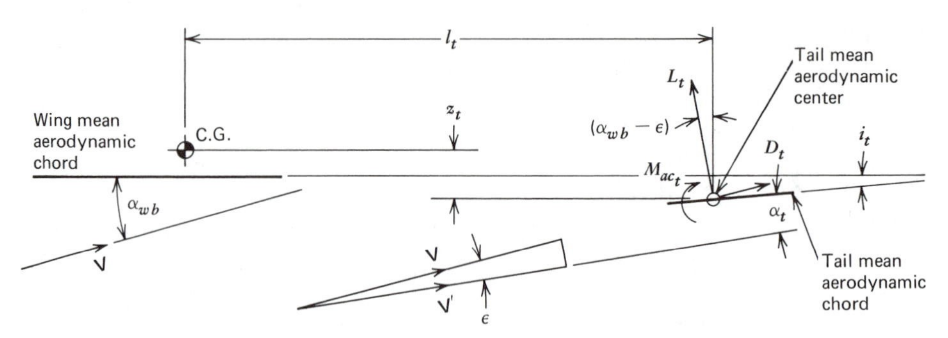

FIG. 2.8 Forces acting on the tail.

The contribution of the tail to the airplane lift, which by definition is perpendicular to **V**, is

$$L_t \cos \varepsilon - D_t \sin \varepsilon$$

ε is always a small angle, and $D_t\varepsilon$ may be neglected compared with L_t. The contribution of the tail to the airplane lift then becomes simply L_t. We introduce the symbol C_{L_t} to represent the lift coefficient of the tail, based on the airplane dynamic pressure $\frac{1}{2}\rho V^2$ and the tail area S_t.

$$C_{L_t} = \frac{L_t}{\frac{1}{2}\rho V^2 S_t} \tag{2.2,5}$$

The reader should note that the lift coefficient of the tail is often based on the local dynamic pressure at the tail, which differs from $\frac{1}{2}\rho V^2$ when the tail lies in the wing wake. This practice entails carrying the ratio V'/V in many subsequent equations. The definition employed here amounts to incorporating V'/V into the tail lift-curve slope a_t. This quantity is in any event different from that for the isolated tail, owing to the interference effects previously noted. This circumstance is handled in various ways in the literature. Sometimes a tail efficiency factor η_t is introduced, the isolated tail lift slope being multiplied by η_t. In other treatments, η_t is used to represent $(V'/V)^2$. In the convention adopted here, a_t is the lift-curve slope of the tail, as measured in situ on the airplane, and based on the dynamic pressure $\frac{1}{2}\rho V^2$. This is the quantity that is directly obtained in a wind-tunnel test.

From Fig. 2.8 we find the pitching moment of the tail about the C.G. to be

$$\begin{aligned} M_t = &-l_t[L_t \cos(\alpha_{wb} - \varepsilon) + D_t \sin(\alpha_{wb} - \varepsilon)] \\ &- z_t[D_t \cos(\alpha_{wb} - \varepsilon) - L_t \sin(\alpha_{wb} - \varepsilon)] + M_{ac_t} \end{aligned} \tag{2.2,6}$$

Experience has shown that in the majority of instances the dominant term in this equation is the first one, and that all others are negligible by comparison. Only this case will be dealt with here. The reader is left to extend the analysis to cases in which this approximation is not valid. With the above approximation, and that of small angles,

$$M_t = -l_t L_t = -l_t C_{L_t} \tfrac{1}{2}\rho V^2 S_t$$

Upon conversion to coefficient form, we obtain

$$C_{m_t} = \frac{M_t}{\frac{1}{2}\rho V^2 S\bar{c}} = -\frac{l_t}{\bar{c}}\frac{S_t}{S} C_{L_t} \tag{2.2,7}$$

The combination $l_t S_t / S\bar{c}$ is the ratio of two volumes characteristic of the airplane's geometry. It is commonly called the "horizontal-tail volume ratio," or more simply, the "tail volume." It is denoted here by V_H. Thus

$$C_{m_t} = -V_H C_{L_t} \tag{2.2,8}$$

From Fig. 2.8 it can be seen that

$$\alpha_t = \alpha_{wb} - \varepsilon - i_t \tag{2.2,9}$$

It follows that (for zero elevator angle)

$$C_{L_t} = a_t \alpha_t = a_t(\alpha_{wb} - \varepsilon - i_t) \tag{2.2,10}$$

The downwash angle ε may usually be approximated by a linear function of α_{wb}, and is expressed as

$$\varepsilon = \varepsilon_0 + \frac{\partial \varepsilon}{\partial \alpha} \alpha_{wb} \tag{2.2,11}$$

The zero lift downwash ε_0 will not normally be predictable. It must be found from a wind-tunnel test. Upon eliminating ε from Eq. 2.2,10, we obtain

$$C_{L_t} = a_t \alpha_{wb}\left(1 - \frac{\partial \varepsilon}{\partial \alpha}\right) - a_t(\varepsilon_0 + i_t) \tag{2.2,12}$$

and from Eq. 2.2,8 the expression for the pitching moment becomes

$$C_{m_t} = -a_t V_H \alpha_{wb}\left(1 - \frac{\partial \varepsilon}{\partial \alpha}\right) + a_t V_H(\varepsilon_0 + i_t) \tag{2.2,13}$$

Total Lift and Pitching Moment

The total lift is the sum of that on the tail and that on the wing-body–nacelle combination.

$$L = L_{wb} + L_t$$

Hence the total airplane lift coefficient is

$$C_L = \frac{L}{\frac{1}{2}\rho V^2 S} = C_{L_{wb}} + \frac{C_{L_t}\frac{1}{2}\rho V^2 S_t}{\frac{1}{2}\rho V^2 S}$$

$$= C_{L_{wb}} + \frac{S_t}{S} C_{L_t} \tag{2.2,14}$$

When expressed in terms of α_{wb} as the independent variable, by making use of Eq. 2.2,12, this becomes

$$C_L = a_{wb}\alpha_{wb} + a_t \frac{S_t}{S}\alpha_{wb}\left(1 - \frac{\partial\varepsilon}{\partial\alpha}\right) - a_t\frac{S_t}{S}(\varepsilon_0 + i_t)$$

$$= a_{wb}\alpha_{wb}\left[1 + \frac{a_t}{a_{wb}}\frac{S_t}{S}\left(1 - \frac{\partial\varepsilon}{\partial\alpha}\right)\right] - a_t\frac{S_t}{S}(\varepsilon_0 + i_t) \tag{2.2,15}$$

Using Eq. 2.2,15 we may write the lift curve as $C_L = a\alpha$, where the lift-curve slope of the airplane is

$$a = \frac{\partial C_L}{\partial\alpha} = a_{wb}\left[1 + \frac{a_t}{a_{wb}}\frac{S_t}{S}\left(1 - \frac{\partial\varepsilon}{\partial\alpha}\right)\right] \tag{2.2,16}$$

and the angle of attack of the zero lift line of the airplane is

$$\alpha = \alpha_{wb} - \frac{a_t}{a}\frac{S_t}{S}(\varepsilon_0 + i_t) \tag{2.2,17}$$

Equation 2.2,15 may also be written as

$$C_L = a\alpha_{wb} - a_t\frac{S_t}{S}(\varepsilon_0 + i_t) \tag{2.2,18}$$

The pitching moment is obtained in terms of the component lift coefficients by adding Eqs. 2.2,4 and 2.2,8:

$$C_m = C_{m_{0_{wb}}} + C_{L_{wb}}(h - h_{n_{wb}}) - V_H C_{L_t} \tag{2.2,19}$$

In terms of α_{wb}, this equation becomes

$$C_m = C_{m_{0_{wb}}} + a_{wb}\alpha_{wb}\left[(h - h_{n_{wb}}) - V_H\frac{a_t}{a_{wb}}\left(1 - \frac{\partial\varepsilon}{\partial\alpha}\right)\right]$$
$$+ a_t V_H(\varepsilon_0 + i_t) \tag{2.2,20}$$

The value of the pitching moment at zero airplane lift is found by setting $\alpha = 0$ in Eq. 2.2,17 and substituting the value of α_{wb} so obtained into Eq. 2.2,20. The result is

$$C_{m_0} = C_{m_{0_{wb}}} + a_t V_H(\varepsilon_0 + i_t)$$
$$+ \frac{a_{wb}a_t}{a}\frac{S_t}{S}(\varepsilon_0 + i_t)\left[(h - h_{n_{wb}}) - V_H\frac{a_t}{a_{wb}}\left(1 - \frac{\partial\varepsilon}{\partial\alpha}\right)\right]$$

When due allowance is made for the fact that V_H varies with C.G. position, it can be shown from this equation that $\partial C_{m_0}/\partial h = 0$. Since C_{m_0} is independent of h, we may obtain its value by setting $h = h_{n_{wb}}$; i.e.,

$$C_{m_0} = C_{m_{0_{wb}}} + a_t V_H'(\varepsilon_0 + i_t)\left[1 - \frac{a_t}{a}\frac{S_t}{S}\left(1 - \frac{\partial\varepsilon}{\partial\alpha}\right)\right] \tag{2.2,21}$$

where V'_H is the value of V_H corresponding to $h = h_{n_{wb}}$. From Eq. 2.2,20 we obtain also the slope of the C_m vs. α curve

$$C_{m\alpha} = \frac{\partial C_m}{\partial \alpha} = a_{wb}\left[(h - h_{n_{wb}}) - V_H \frac{a_t}{a_{wb}}\left(1 - \frac{\partial \varepsilon}{\partial \alpha}\right)\right] \tag{2.2,22}$$

Finally Eqs. 2.2,21 and 22 are combined to yield the C_m–α curve

$$C_m = C_{m_0} + C_{m_\alpha}\alpha \tag{2.2,23}$$

2.3 THE STICK-FIXED NEUTRAL POINT

The value of $\partial C_m/\partial\alpha$ as given by Eq. 2.2,22 is seen to be strongly dependent on the value of h (Fig. 2.9). The term containing h is positive, and all the others are negative. Hence the magnitude of h controls the sign of $\partial C_m/\partial\alpha$, and there is one particular value for which $\partial C_m/\partial\alpha = 0$. This is the condition when the pitching moment is independent of the angle of attack; i.e., when the static stability is neutral. This particular value of h is denoted by h_n, and the corresponding C.G. position is defined herein to be the *stick-fixed neutral point.*

From Eq. 2.2,22, h_n is found to be

$$h_n = h_{n_{wb}} + V_{H_n} \frac{a_t}{a_{wb}}\left(1 - \frac{\partial \varepsilon}{\partial \alpha}\right) \tag{2.3,1}$$

where V_{H_n} is the value of the tail volume ratio corresponding to $h = h_n$. When the variation of V_H with C.G. position is small enough to be neglected, and a mean constant value is used, then Eq. 2.3,1 gives an explicit value for h_n. However, for airplanes having a large C.G. range, it may be worthwhile to allow for the variation of V_H. This is done as follows. From Fig. 2.10 we find that

$$l_t = l'_t - (h - h_{n_{wb}})\bar{c}$$

and thus

$$V_H = V'_H - \frac{S_t}{S}(h - h_{n_{wb}}) \tag{2.3,1a}$$

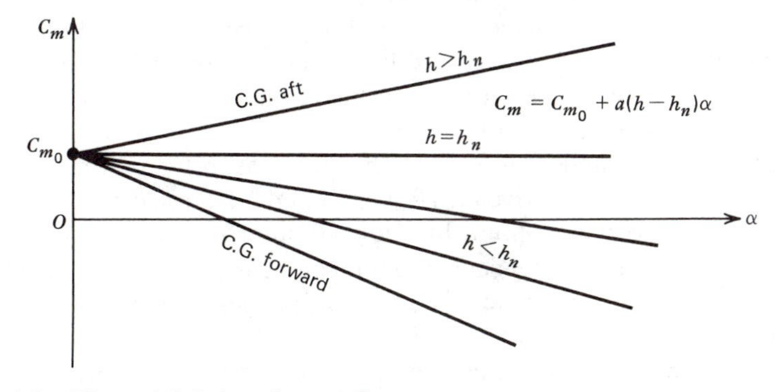

FIG. 2.9 Effect of C.G. location on C_m curve.

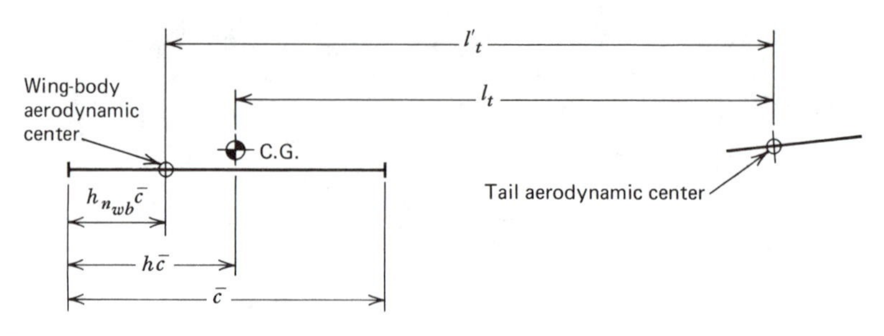

FIG. 2.10

and

$$V_{H_n} = V'_H - \frac{S_t}{S}(h_n - h_{n_{wb}})$$

When this value is substituted into Eq. 2.3,1 and Eq. 2.2,16 is used to eliminate a_{wb}, the result is

$$h_n - h_{n_{wb}} = V'_H \frac{a_t}{a}\left(1 - \frac{\partial \varepsilon}{\partial \alpha}\right) \tag{2.3,2}$$

The pitching moment equations take on a simpler form than given in Eqs. 2.2,21 and 2.2,22 when the neutral point is introduced. If the exact value of V_H is used, (Eq. 2.3,1a) then after some algebraic reduction, we find the results

$$\begin{aligned} C_m &= C_{m_0} + C_{m_\alpha}\alpha \\ C_{m_0} &= C_{m_{0_{wb}}} + a_t V_{H_n}(\varepsilon_0 + i_t) \\ C_{m_\alpha} &= a(h - h_n) \end{aligned} \tag{2.3,3}$$

The quantity $(h_n - h)$ is defined here as the *stick-fixed static margin.* It is a measure of the static stability of the airplane with respect to incidence disturbances.

Equation 2.3,3 provides the proof of the statement made in Sec. 2.1, namely, that $\partial C_m/\partial\alpha$ can always be made negative. This requires simply that h be less than h_n, or that the C.G. be forward of the neutral point.

The reason for the notations h_{n_w} and $h_{n_{wb}}$ to define the locations of the mean aerodynamic centers of the wing and wing-body–nacelle combinations may now be given. It is that as defined in this book the concepts *neutral point* and *aerodynamic center* are one and the same, both being determined by the criterion that the pitching moment is invariant with α.

Although it was implied in Sec. 2.1 that the reason for putting a tail on an airplane was to provide a positive C_{m_0}, Eqs. 2.3,1 and 2.3,2 reveal a second beneficial effect: control of the neutral-point location. By appropriate selection of the tail parameters, principally V'_H, the neutral point can be located at will. Whatever the neutral point location, the value of C_{m_0} is still at the command of the designer through the free choice of i_t (see Eq. 2.3,3).

The neutral point has sometimes been defined as the C.G. location at which the derivative $dC_m/dC_L = 0$. When this definition is applied to the gliding flight of a rigid airplane at low Mach number, the neutral point obtained is identical with that defined in this book. This is so because under these restricted conditions C_L is a unique function of α, and $dC_m/dC_L = (\partial C_m/\partial\alpha)/(\partial C_L/\partial\alpha)$. Then dC_m/dC_L and $\partial C_m/\partial\alpha$ are simultaneously zero. In general, however, C_m and C_L are both functions of several variables, as pointed out at the beginning of Sec. 2.2. For fixed values of δ_e and h, and neglecting Reynolds number effects (these are usually very small), we may write

$$C_L = f(\alpha, \mathbf{M}, T_c, \tfrac{1}{2}\rho V^2), \qquad C_m = g(\alpha, \mathbf{M}, T_c, \tfrac{1}{2}\rho V^2) \tag{2.3,4}$$

where T_c is the thrust coefficient, defined in Sec. 3.11.

Mathematically speaking, the derivative dC_m/dC_L does not exist unless $\mathbf{M}$, T_c, and $\frac{1}{2}\rho V^2$ are functions of C_L. When that is the case, then

$$\frac{dC_m}{dC_L} = \frac{\partial C_m}{\partial\alpha}\frac{\partial\alpha}{\partial C_L} + \frac{\partial C_m}{\partial \mathbf{M}}\frac{\partial \mathbf{M}}{\partial C_L} + \frac{\partial C_m}{\partial T_c}\frac{\partial T_c}{\partial C_L} + \frac{\partial C_m}{\partial(\frac{1}{2}\rho V^2)}\frac{\partial(\frac{1}{2}\rho V^2)}{\partial C_L} \tag{2.3,5}$$

Equation 2.3,5 has meaning only when a specific kind of flight is prescribed: e.g., horizontal unaccelerated flight, or rectilinear climbing flight at full throttle. When a condition of this kind is imposed, then $\mathbf{M}$, T_c, and the dynamic pressure are definite functions of C_L, dC_m/dC_L exists, and a neutral point may be calculated. The neutral point so found is not an index of stability with respect to incidence disturbances, and the question arises as to what it does relate to. It can be shown that it relates to the trim curves of the airplane. A plot of the elevator angle to trim versus speed will have a zero slope when dC_m/dC_L is zero, and a negative slope when the C.G. lies aft of the neutral point so defined. As shown in Sec. 2.5, this reversal of slope indicates a *tendency* toward instability with respect to speed, but only a dynamic analysis can show whether or not the airplane is stable in this condition. There are cases when the application of the "trim-slope" criterion can be definitely misleading as to stability. One such is level unaccelerated flight, during which the throttle must be adjusted every time the flight speed or C_L is altered. This is treated in more detail in Sec. 3.5.

It can be seen from the foregoing remarks that the "trim-slope" criterion for the neutral point does not lead to any definite and clear-cut conclusions, either about the stability with respect to incidence disturbances, or about the general static stability involving both speed and incidence disturbances. It is mainly for this reason that the neutral point has been defined herein on the basis of $\partial C_m/\partial\alpha$. A second cogent reason for this choice is the nature of the instabilities that can occur when the C.G. lies forward and aft of this neutral point. This aspect is discussed by means of an example in Chap. 6.

2.4 STATIC LONGITUDINAL CONTROL

It was pointed out in Sec. 2.1 that an airplane with fixed controls can fly only at the value of α or C_L for which $C_m = 0$. In order to fly at various speeds, some form of longitudinal control is necessary. The earliest method, used by Lilienthal

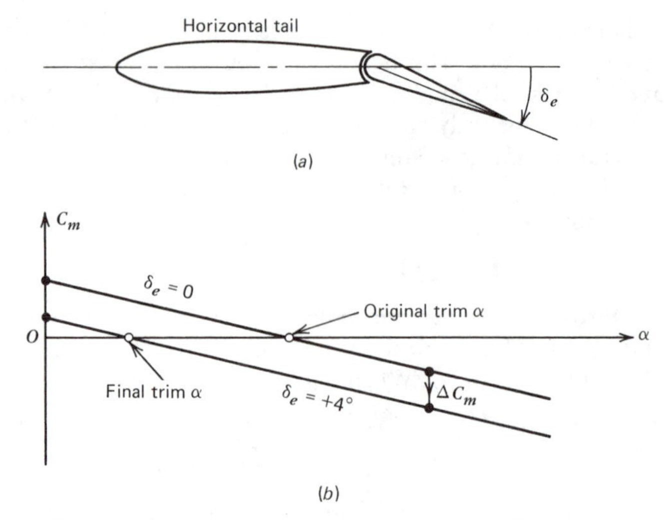

FIG. 2.11 Effect of elevator angle on C_m curve. (*a*) Elevator angle. (*b*) $C_m - \alpha$ curve.

in his gliding flights (1891–96) was to shift the C.G. by movement of the body. As indicated in Fig. 2.9, this will indeed change the trim.[2] Moving the C.G. forward reduces the trim α or C_L, resulting in an increase in the trim speed. However, this form of control also changes the slope $\partial C_m/\partial \alpha$; i.e., it changes the static stability. This is an undesirable effect. The practical difficulties involved in this form of control are also such as to inhibit its use.

The longitudinal control now generally used is aerodynamic. A variable pitching moment is provided by moving the elevator, which may be all or part of the tail, or a trailing-edge flap in a tailless design. Deflection of the elevator through an angle δ_e produces increments in both the C_m and C_L of the airplane. The ΔC_L caused by the elevator of aircraft with tails is small enough to be neglected for many purposes. This is not so for tailless aircraft, where the ΔC_L due to elevators is usually significant. We shall assume that the lift and moment increments for both kinds of airplane are linear in δ_e, which is a fair representation of the characteristics of typical controls at high Reynolds number. Therefore

$$\Delta C_L = C_{L_\delta}\delta_e \quad (a)$$
$$C_L = C_{L_\alpha}\alpha + C_{L_\delta}\delta_e \quad (b)$$
$$\Delta C_m = C_{m_\delta}\delta_e \quad (c) \qquad (2.4,1)$$

and

$$C_m = C_{m_0} + C_{m_\alpha}\alpha + C_{m_\delta}\delta_e \quad (d)$$

where $C_{L_\delta} = \partial C_L/\partial \delta_e$ and $C_{m_\delta} = \partial C_m/\partial \delta_e$. The usual convention is to take down elevator as positive (Fig. 2.11*a*). This leads to positive C_{L_δ} and negative C_{m_δ}. The

[2] The word *trim* as used here refers to the angle-of-attack or lift coefficient at zero C_m.

deflection of the elevator through a constant positive angle then shifts the C_m–α curve downward, without change of slope (Fig. 2.11*b*). At the same time the zero-lift angle of the airplane is slightly changed.

2.5 ELEVATOR ANGLE TO TRIM

Airplanes with Tails

Let the *elevator lift effectiveness* be given by

$$a_e = \frac{\partial C_{L_t}}{\partial \delta_e}$$

The tail-lift coefficient is then given by (cf. Eq. 2.2,10)

$$C_{L_t} = a_t \alpha_t + a_e \delta_e \tag{2.5,1}$$

From Eq. 2.2,14 we find the increment in airplane lift

$$\Delta C_L = \frac{S_t}{S} \Delta C_{L_t} = a_e \frac{S_t}{S} \delta_e \qquad (a)$$

and thus (2.5,2)

$$C_{L_\delta} = a_e \frac{S_t}{S} \qquad (b)$$

Likewise, the increment in airplane pitching moment, from Eq. 2.2,19, is

$$\Delta C_m = -V_H \Delta C_{L_t} = -a_e V_H \delta_e \qquad (a)$$

thus (2.5,3)

$$C_{m_\delta} = -a_e V_H \qquad (b)$$

The elevator angle to trim is obtained by setting $C_m = 0$ in Eq. 2.4,1*d*. This gives

$$\delta_{\text{trim}} = -\frac{C_{m_0} + C_{m_\alpha}\alpha}{C_{m_\delta}} \tag{2.5,4}$$

The corresponding C_L is

$$C_{L_{\text{trim}}} = C_{L_\alpha}\alpha + C_{L_\delta}\delta_{\text{trim}} \tag{2.5,5}$$

The parameter α can be eliminated from Eqs. 2.5,4 and 5 to obtain

$$\delta_{\text{trim}} = -\frac{C_{m_0}C_{L_\alpha} + C_{m_\alpha}C_{L_{\text{trim}}}}{C_{m_\delta}C_{L_\alpha} - C_{m_\alpha}C_{L_\delta}} \tag{2.5,6}$$

Tailless Airplanes

Equations 2.5,2 and 2.5,3 are special to the airplane with a tail. For tailless airplanes C_{L_δ} replaces a_e as the primary quantity to be estimated or determined from experiment. In addition, the change in C_{m_0} of the configuration caused by

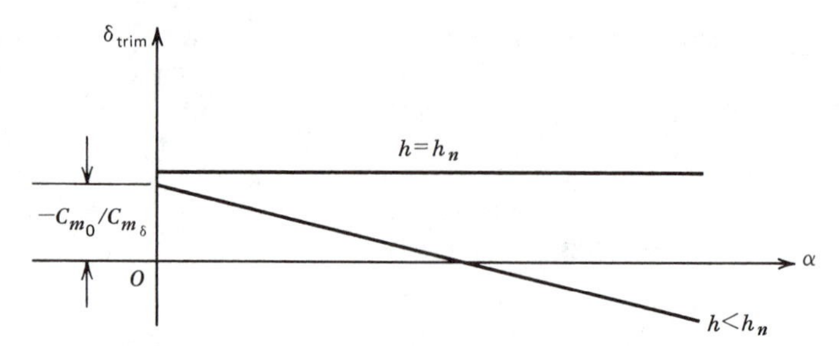

FIG. 2.12 Elevator angle to trim at various C.G. positions.

the elevator is usually large, and must be considered. Reference to Fig. 2.7 shows that, with the same assumptions as employed in Sec. 2.2,

$$\Delta C_m = \frac{\partial C_{m_0}}{\partial \delta} \delta_e + \Delta C_L(h - h_n)$$

(The identifying subscript *wb* is not needed when dealing with tailless machines.) Thus

$$C_{m_\delta} = \frac{\partial C_{m_0}}{\partial \delta} + C_{L_\delta}(h - h_n) \tag{2.5,7}$$

Equations 2.5,4, 5, and 6 for δ_{trim} and $C_{L_{trim}}$ apply to the tailless airplane as well.

For a fixed value of h, Eq. 2.5,4 leads to a linear relation between δ_{trim} and α for both tailed and tailless airplanes. This is illustrated in Fig. 2.12.

Variation with Speed

When, in the absence of compressibility, aeroelastic effects, and propulsive system effects, the aerodynamic coefficients of Eq. 2.5,6 are constant, the variation

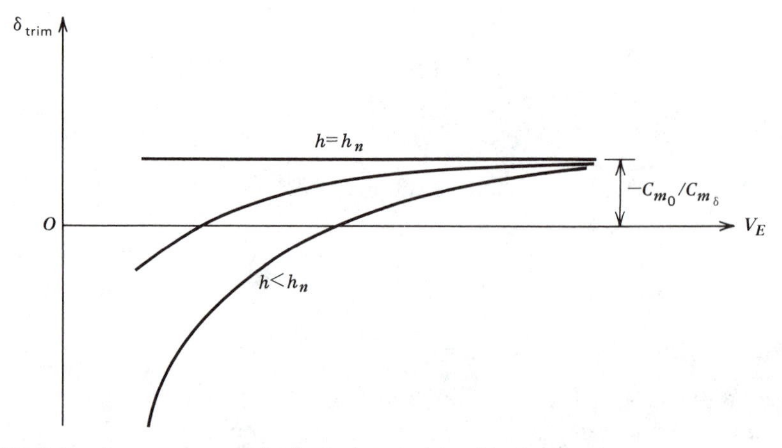

FIG. 2.13 δ_{trim} versus equivalent airspeed (no Mach effect).

of δ_{trim} with speed is simple. Then δ_{trim} is a unique function of $C_{L_{\text{trim}}}$ for each C.G. position. Since $C_{L_{\text{trim}}}$ is in turn fixed by the equivalent airspeed

$$C_{L_{\text{trim}}} = \frac{W}{\frac{1}{2}\rho_0 V_E{}^2 S} \tag{2.5,8}$$

then δ_{trim} becomes a unique function of V_E. The form of the curves is shown in Fig. 2.13.

When, owing to compressibility or the other effects mentioned above, the coefficients of Eq. 2.5,6 vary with speed (e.g., $C_{m_\alpha} = f(\mathbf{M})$), the shape of the curve may be quite different. Moreover, δ_{trim} is no longer a unique function of V_E, since $\mathbf{M}$ is a function of both V_E and altitude. The design of transonic and supersonic airplanes is too variable to permit displaying any typical δ_{trim} curves for such machines. One possibility in the transonic range is shown in Fig. 2.14, where the slope $d\delta_{\text{trim}}/dV$ reverses. The reason for this reversal is revealed by the following equation:

$$\frac{d\delta_{\text{trim}}}{dV} = \frac{\partial \delta_{\text{trim}}}{\partial C_{L_{\text{trim}}}} \frac{\partial C_{L_{\text{trim}}}}{\partial V} + \frac{\partial \delta_{\text{trim}}}{\partial \mathbf{M}} \frac{\partial \mathbf{M}}{\partial V} \tag{2.5,9}$$

From Eq. 2.5,6 we find that, as long as $C_{m_\alpha} < 0$, i.e., $h < h_n$, then $\partial\delta_{\text{trim}}/\partial C_{L_{\text{trim}}} < 0$. Also $\partial C_{L_{\text{trim}}}/\partial V < 0$. Hence the first term of Eq. 2.5,9 is positive as long as the airplane is statically stable with respect to incidence disturbance. This term ordinarily dominates the equation; however, at transonic speeds such quantities as C_{m_0} and C_{m_δ} may vary with $\mathbf{M}$ sufficiently rapidly to cause the second term to reverse the sign of $d\delta_{\text{trim}}/dV$. Notwithstanding the static stability referred to, Fig. 2.14 shows a potential instability with respect to speed. Let the airplane be in equilibrium flight at the point A, and be subsequently perturbed so that its speed increases to that of B with no change in α or δ. Now at B the elevator angle is too positive for trim; i.e., there is an unbalanced nose-down moment on the airplane. This *tends* to put the airplane into a dive and increase its speed still further. If not inhibited by other factors, the speed will increase until a new equilibrium is attained at C. The "other factors" referred to that may completely nullify the tendency

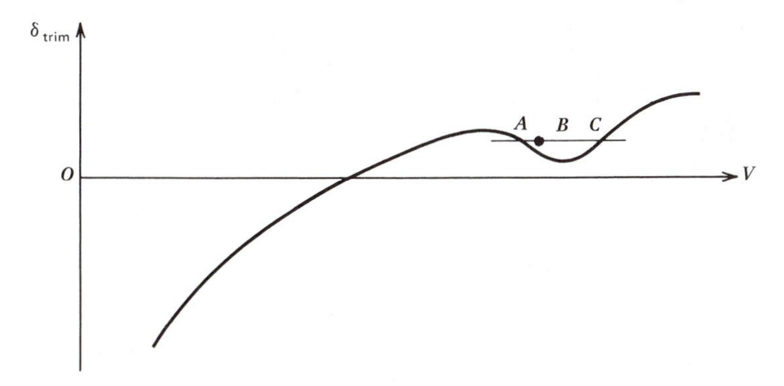

FIG. 2.14 Reversal of δ_{trim} slope at transonic speeds.

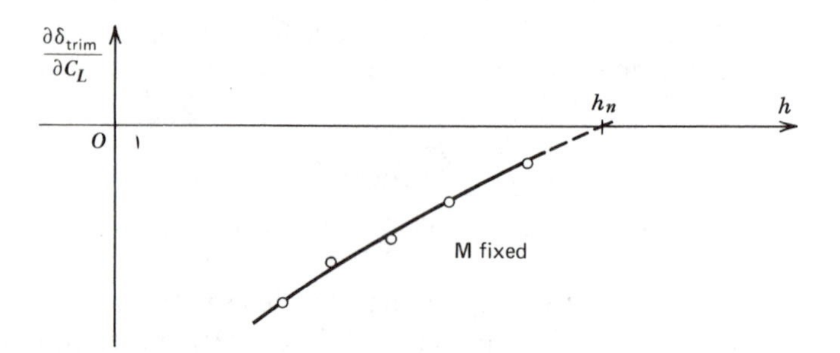

FIG. 2.15 Determination of stick-fixed neutral point from flight test.

noted are: the inherent stability with respect to speed associated with a positive slope of the C_D–V curve; the excess lift corresponding to the fixed α and increased speed, which tends to induce an upward curvature of the flight path.

The foregoing remarks are not intended to be an analysis of stability with respect to speed, but to exemplify a limitation of the concept of static stability based on incidence disturbances. As pointed out in Sec. 2.1, a more general point of view, adopted in Sec. 6.10 is required to assess fully the static stability of an airplane.

Flight Deterination of h_n

The equations for δ_{trim} point to a way of determining h_n from flight tests. From Eq. 2.5,6, we find

$$\frac{\partial \delta_{trim}}{\partial C_{L_{trim}}} = -\frac{C_{m\alpha}}{C_{L_\alpha} C_{m_\delta} - C_{m_\alpha} C_{L_\delta}}$$

Thus $\partial \delta_{trim}/\partial C_{L_{trim}}$ is zero when C_{m_α} is zero, or, from Eq. 2.3,2 when $h = h_n$. The technique is to make measurements of δ_{trim} and $C_{L_{trim}}$ in gliding flight, and obtain the slope $\partial \delta_{trim}/\partial C_{L_{trim}}$ for a range of C.G. positions. A plot such as that shown in Fig. 2.15 then permits finding h_n by extrapolating the curve to zero. The reader should note the use of the partial derivative employed here. It implies that $C_{L_{trim}}$ is altered without change of **M** or $\frac{1}{2}\rho V^2$. If for any reason, the coefficients occurring in Eq. 2.5,6 vary with speed, then a result obtained from flights at different speeds would be invalid

2.6 THE ELEVATOR HINGE MOMENT

The aerodynamic forces on any control surface produce a moment about the hinge. Figure 2.16 shows a typical tail surface incorporating an elevator with a tab. The tab usually exerts a negligible effect on the lift of the aerodynamic surface to which it is attached, although its influence on the hinge moment is large.

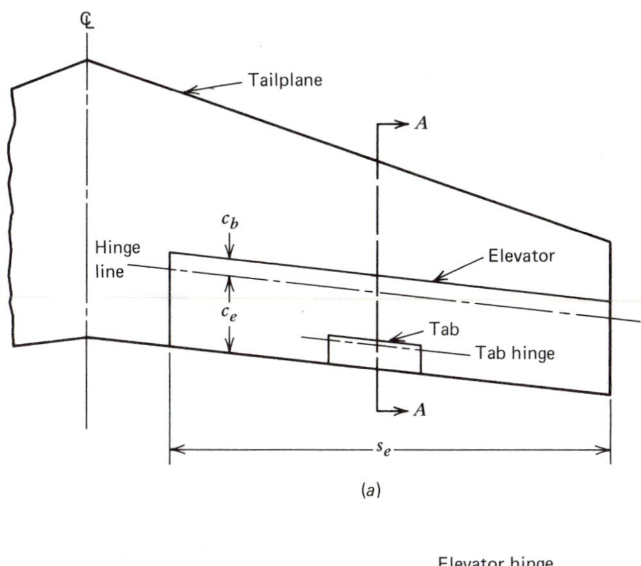

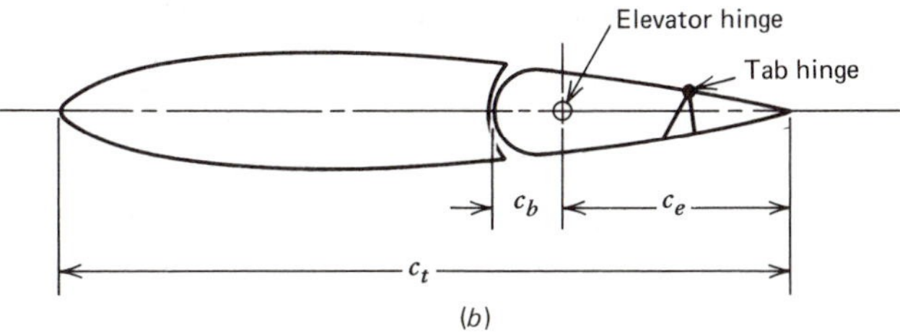

FIG. 2.16 Elevator and tab geometry. (*a*) Plan view. (*b*) Section *A–A*.

The coefficient of elevator hinge moment is defined by

$$C_{he} = \frac{H_e}{\frac{1}{2}\rho V^2 S_e \bar{c}_e}$$

Here H_e is the moment, about the elevator hinge line, of the aerodynamic forces on the elevator and tab, S_e is the area of that portion of the elevator and tab that lies *aft of the elevator hinge line*, and $\bar{c}_e$ is a mean chord of the same portion of the elevator and tab. In British practice, $\bar{c}_e$ is the geometric mean value, i.e., $\bar{c}_e = S_e/2s_e$, and, in the United States, $\bar{c}_e$ is the root-mean square of c_e. The taper of elevators is usually slight, and the difference between the two values is generally small. The reader is cautioned to note which definition is employed when using reports on experimental measurements of C_{he}.

Of all the aerodynamic parameters required in stability and control analysis, the hinge-moment coefficients are most difficult to determine with precision. A large number of geometrical parameters influence these coefficients, and the range of design configurations is wide. Scale effects tend to be larger than for many other parameters, owing to the sensitivity of the hinge moment to the state of the

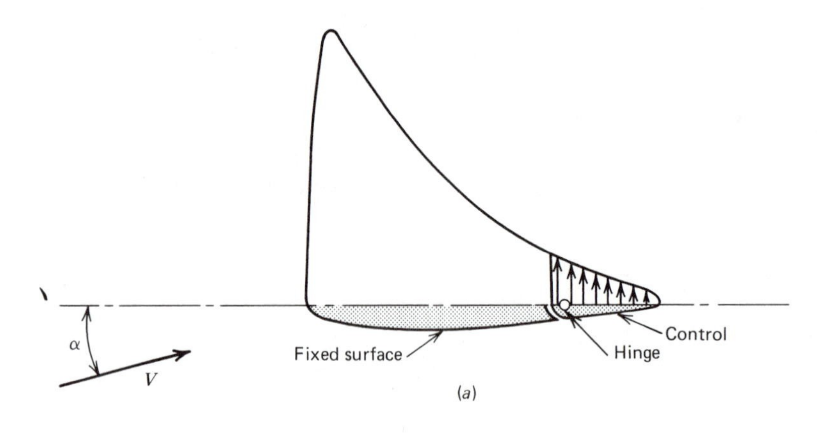

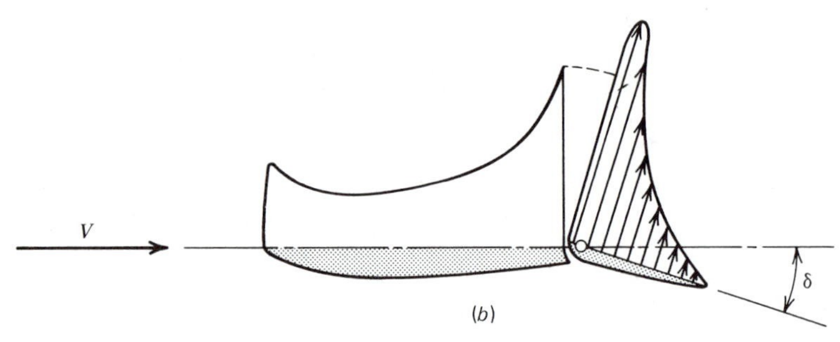

FIG. 2.17 Normal-force distribution over control surface at subsonic speed. (a) Force distribution over control associated with α at $\delta = 0$. (b) Force distribution over control associated with δ at zero α.

boundary layer at the trailing edge. Two-dimensional airfoil theory shows that the hinge moment of simple flap controls is linear with angle of attack and control angle in both subsonic and supersonic flow.

The normal-force distributions typical of subsonic flow associated with changes in α and δ are shown qualitatively in Fig. 2.17. The force acting on the movable flap has a moment about the hinge that is quite sensitive to its location. Ordinarily the hinge moments in both cases (a) and (b) shown are negative.

In the majority of practical cases it is a satisfactory engineering approximation to assume that for finite surfaces C_{he} is a linear function of α_t, δ_e, and δ_t. The reader should note however that there are important exceptions in which strong non-linearities are present. An example is the Frise aileron, shown with a typical C_h curve, in Fig. 2.18.

We assume that C_{he} is linear, as follows,

$$C_{he} = b_0 + b_1\alpha_t + b_2\delta_e + b_3\delta_t \tag{2.6,1}$$

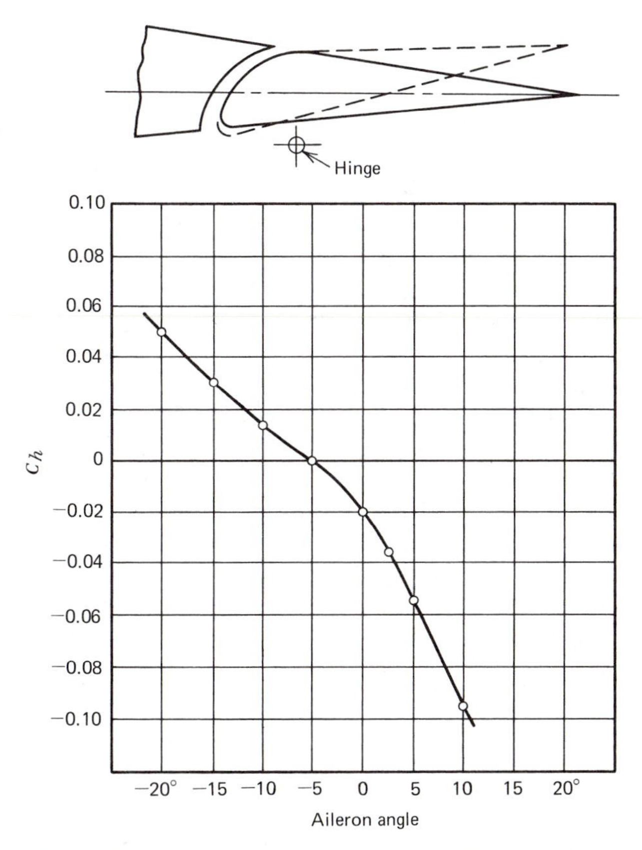

FIG. 2.18 Typical hinge moment of Frise aileron. Wing $\alpha = 2°$. R.N. $= 3.3 \times 10^6$.

where

$$b_1 = \frac{\partial C_{he}}{\partial \alpha_t} = C_{h_{\alpha_t}}$$

$$b_2 = \frac{\partial C_{he}}{\partial \delta_e} = C_{h_\delta}$$

$$b_3 = \frac{\partial C_{he}}{\partial \delta_t} = C_{h_{\delta_t}}$$

The determination of the hinge moment then resolves itself into the determination of b_0, b_1, b_2, and b_3. Some data for estimating these are contained in Appendix B. The geometrical variables that enter are elevator chord ratio c_e/c_t, balance ratio c_b/c_e, nose shape, hinge location, gap, trailing-edge angle, and planform. When a set-back hinge is used, some of the pressure acts ahead of the hinge, and the hinge moment is less than that of a simple flap with a hinge at its leading edge. The force that the control system must exert to hold the elevator at the desired angle is in direct proportion to the hinge moment. The hinge moment itself, as can be deduced from the definition of C_{he} is roughly proportional to the square of the speed, and the cube of the airplane size. The advent of large high-speed airplanes therefore brought serious control problems, since the forces required became too large for a human pilot to supply. Much development has gone into attempts to arrive at purely aerodynamic solutions to this difficulty. The devices employed include various forms of nose balance, and the use of geared and spring tabs. Closely balanced controls have experienced difficulties because of the sensitivity of the hinge moment to such factors as nose shape and gap, which are inevitably subject to variations in manufacture.

An alternative solution is to relieve the pilot of some or all of the aerodynamic load through the use of power controls. These may be designed so that the pilot supplies a fixed proportion of the control force, the power system supplying the remainder. A system of this kind is illustrated in Fig. 4.4. With such "ratio"-type controls, the *feel* has the same character as when power is absent, i.e., the stick forces vary with speed, and, in maneuvers, in the same way. Alternatively, the power controls may be irreversible, in that none of the aerodynamic load is carried directly to the pilot. Such systems are fitted with devices that produce a synthetic feel at the stick. The stick-force characteristics can then be made virtually whatever the designer wishes. In the remaining sections of this chapter, the controls are assumed to be fully manual, or of the ratio type.

2.7 INFLUENCE OF A FREE ELEVATOR

In Secs. 2.2 and 2.3 we have dealt with the static stability of an airplane the controls of which are fixed in position. Even with a completely rigid structure, which never exists, a manually operated control cannot be regarded as fixed. A human pilot is incapable of supplying an ideal rigid constraint. When irreversible power controls are fitted, however, the stick-fixed condition is closely approximated. A characteristic of interest from the point of view of flying qualities is the stability of the airplane when the elevator is completely free to rotate about its hinge under the influence of the aerodynamic pressures that act upon it. Normally, the stability in the control-free condition is less than with fixed controls. It is desirable that this difference should be small. Since friction is always present in the control system, the free control is never realized in practice either. However, the two ideal conditions, free control and fixed control, represent the possible extremes.

Airplanes with Tails

We shall assume that the tail airfoil is symmetrical and the tab angle is zero. Then b_0 and δ_t (Eq. 2.6,1) are zero. When the elevator is free, the hinge moment is also zero, and hence

$$b_1\alpha_t + b_2\delta_e = 0 \tag{2.7,1}$$

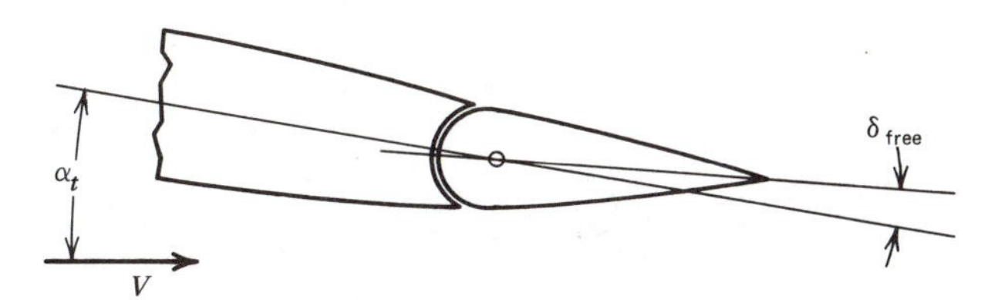

FIG. 2.19 Elevator floating angle.

The free-floating angle of the elevator is then

$$\delta_{\text{free}} = -\frac{b_1}{b_2}\alpha_t \tag{2.7,2}$$

This is illustrated in Fig. 2.19. $b_1(C_{h\alpha})$ and $b_2(C_{h\delta})$ are usually negative, although b_1 may be positive. When both are negative, then a positive α_t causes the elevator to float up, as shown. The opposite occurs when b_1 is positive. In the following equations, a prime is used to denote the elevator-free condition.

Using Eq. 2.7,2, the tail lift with elevator free is, from Eq. 2.5,1,

$$\begin{aligned} C'_{L_t} &= a_t\alpha_t - a_e\frac{b_1}{b_2}\alpha_t \\ &= a_t\alpha_t\left(1 - \frac{a_e}{a_t}\frac{b_1}{b_2}\right) \end{aligned} \tag{2.7,3}$$

The factor in parentheses is the "free-elevator factor" F, (2.7,4)

$$F = \left(1 - \frac{a_e}{a_t}\frac{b_1}{b_2}\right)$$

so that

$$C'_{L_t} = Fa_t\alpha_t$$

and

$$\frac{\partial C'_{L_t}}{\partial \alpha_t} = Fa_t$$

The free-elevator factor F represents a reduction in the tail effectiveness when it is less than unity, and an increase when it is greater. Since a_e and a_t are always positive, and b_2 always negative, it is the sign of b_1 that determines whether F is greater or less than unity. As noted above, b_1 is usually negative, in which case F is less than 1. When the control surface is very closely balanced with a set-back hinge, or when a so-called "horn" balance is used (Fig. 2.20), then b_1 may be positive, and the tail effectiveness increased when the control is freed.

The lift and pitching-moment equations of the airplane with elevator free are deduced from those for elevator fixed by replacing a_t with Fa_t. The following

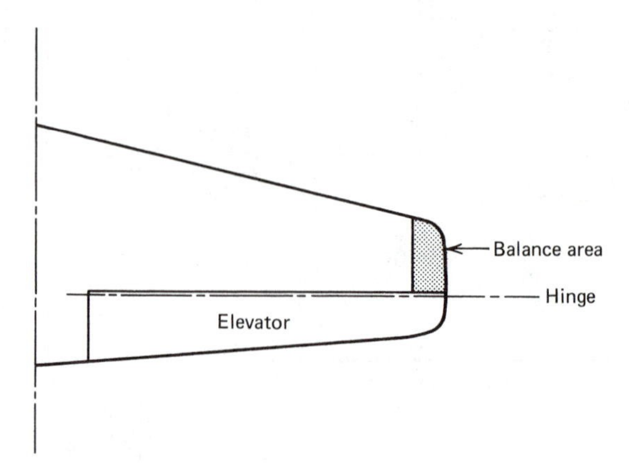

FIG. 2.20 Horn balance.

results are obtained, a prime being used to represent the elevator-free values:

$$C'_L = a'\alpha'$$

$$a' = a_{wb}\left[1 + F\frac{a_t}{a_{wb}}\frac{S_t}{S}\left(1 - \frac{\partial\varepsilon}{\partial\alpha}\right)\right] \tag{2.7,5}$$

$$\alpha' = \alpha_{wb} - F\frac{a_t}{a'}\frac{S_t}{S}(\varepsilon_0 + i_t)$$

$$C'_m = C'_{m_0} + C'_{m_\alpha}\alpha' \tag{2.7,6}$$

$$C'_{m_0} = C_{m_{0_{wb}}} + Fa_tV'_{H_n}(\varepsilon_0 + i_t) \tag{2.7,7}$$

$$C'_{m_\alpha} = a'(h - h'_n) \tag{2.7,8}$$

$$h'_n = h_{n_{wb}} + FV'_H\frac{a_t}{a'}\left(1 - \frac{\partial\varepsilon}{\partial\alpha}\right)$$

$$= h_{n_{wb}} + FV'_{H_n}\frac{a_t}{a_{wb}}\left(1 - \frac{\partial\varepsilon}{\partial\alpha}\right) \tag{2.7,9}$$

h'_n is the *stick-free neutral point*, $(h'_n - h)$ is the *stick-free static margin*, and V'_{H_n} is the value of V_H when $h = h'_n$. If we neglect the variation of V_H with h, a comparison of Eqs. 2.3,1 and 2.7,9 gives the neutral-point shift caused by freeing the elevator as approximately

$$h_n - h'_n \doteqdot (1 - F)\,V_{H_n}\frac{a_t}{a_{wb}}\left(1 - \frac{\partial\varepsilon}{\partial\alpha}\right) \tag{2.7,10}$$

When the exact value of V_H is used, (Eq. 2.3,1a) the exact relation between the two neutral points is found to be given by either of

$$h_n - h'_n = \frac{a_t}{a}V'_{H_n}(1 - F)\left(1 - \frac{\partial\varepsilon}{\partial\alpha}\right)$$

or

$$h_n - h_n' = \frac{a_t}{a'} V_{H_n}(1 - F)\left(1 - \frac{\partial \varepsilon}{\partial \alpha}\right) \tag{2.7,11}$$

Representative numerical values applicable to Eq. 2.7,10 are $F = 0.7$, $V'_{H_n} = 0.5$, $a_t/a_{wb} = 0.8$, $(1 - \partial\varepsilon/\partial\alpha) = 0.7$, giving $h_n - h_n' = 0.084$. This represents a decrease in the static margin of 0.08.

Tailless Airplanes

On a tailless airplane, the airfoil section may be cambered, and the constant b_0 different from zero. The tab angle is assumed zero. When the elevator is free, it will take up a position determined by the condition of zero hinge moment; i.e.,

$$b_0 + b_1\alpha_{wb} + b_2\delta_e = 0$$

It should be noted that, since the elevator is part of the wing, the appropriate angle of attack to use is that of the wing. Since there can be no confusion, the identifying subscript *wb* is dropped. Then we have

$$\delta_{\text{free}} = -\frac{b_0}{b_2} - \frac{b_1}{b_2}\alpha \tag{2.7,12}$$

The lift of the airplane with elevator free is

$$\begin{aligned} C_L' &= a\alpha + C_{L_\delta} \cdot \delta_{\text{free}} \\ &= a\alpha - C_{L_\delta}\left(\frac{b_0}{b_2} + \frac{b_1}{b_2}\alpha\right) \\ &= Fa\alpha - \frac{b_0}{b_2} C_{L_\delta} \end{aligned} \tag{2.7,13}$$

where

$$F = 1 - \frac{C_{L_\delta}}{a}\frac{b_1}{b_2} \tag{2.7,14}$$

F is the free-elevator factor for the tailless airplane. The lift curve is given by

$$C_L' = a'\alpha' \tag{2.7,15}$$

where

$$a' = Fa \tag{2.7,16}$$

and

$$\alpha' = \alpha - \frac{b_0 G_{L_\delta}}{Fab_2} \tag{2.7,17}$$

The pitching moment with the elevator free becomes

$$\begin{aligned} C_m' &= C_{m_0} + C_{m_\alpha}\alpha + C_{m_\delta} \cdot \delta_{\text{free}} \\ &= C_{m_0} + \alpha\left(C_{m_\alpha} - \frac{b_1}{b_2} C_{m_\delta}\right) - \frac{b_0}{b_2} C_{m_\delta} \end{aligned} \tag{2.7,18}$$

We substitute for C_{m_α} from Eq. 2.3,2, and for C_{m_δ} from Eq. 2.5,7, to obtain, after some manipulation

$$C'_m = C'_{m_0} + C'_{m_\alpha}\alpha' \tag{2.7,19}$$

where

$$C'_{m_0} = C_{m_0} - \frac{1}{F}\frac{b_0}{b_2}\frac{\partial C_{m_0}}{\partial \delta} \tag{2.7,20}$$

and

$$C'_{m_\alpha} = Fa(h - h_n) - \frac{b_1}{b_2}\frac{\partial C_{m_0}}{\partial \delta} \tag{2.7,21}$$

From Eqs. 2.7,21 and 2.7,14 it follows that the stick-free neutral point is

$$h'_n = h_n + \frac{1 - F}{FC_{L_\delta}}\frac{\partial C_{m_0}}{\partial \delta} \tag{2.7,22}$$

2.8 THE USE OF TABS

Trim Tabs

In order to fly at a given speed, or C_L, it has been shown in Sec. 2.5 that a certain elevator angle δ_{trim} is required. When this differs from the free-floating angle δ_{free}, a force is required to hold the elevator. When flying for long periods at a constant speed, it is very fatiguing for the pilot to maintain such a force. The trim tabs are used to relieve the pilot of this load by causing δ_{trim} and δ_{free} to coincide. The trim-tab angle required is calculated below for aircraft with and without tails.

Aircraft with Tails. From Eq. 2.6,1, we find that, for symmetrical tail surfaces with deflected tab, the floating angle of the elevator is

$$\delta_{\text{free}} = -\frac{b_1}{b_2}\alpha_t - \frac{b_3}{b_2}\delta_t \tag{2.8,1}$$

From Eq. 2.5,4 we have

$$\delta_{\text{trim}} = -\frac{C_{m_0} + C_{m_\alpha}\alpha}{C_{m_\delta}} \tag{2.5,4}$$

Equating δ_{free} and δ_{trim}, we obtain the required tab angle as

$$\delta_t = \frac{b_2}{b_3}\left(\frac{C_{m_0}}{C_{m_\delta}} + \frac{C_{m_\alpha}}{C_{m_\delta}}\alpha - \frac{b_1}{b_2}\alpha_t\right) \tag{2.8,2}$$

Equation 2.8,2 can be thrown into a more convenient form by the use of Eqs. 2.2,9,11, and 17 to eliminate α_t. The result is

$$\alpha_t = \alpha\left(1 - \frac{\partial\varepsilon}{\partial\alpha}\right) - (\varepsilon_0 + i_t)\left[1 - \frac{a_t}{a}\frac{S_t}{S}\left(1 - \frac{\partial\varepsilon}{\partial\alpha}\right)\right] \tag{2.8,3}$$

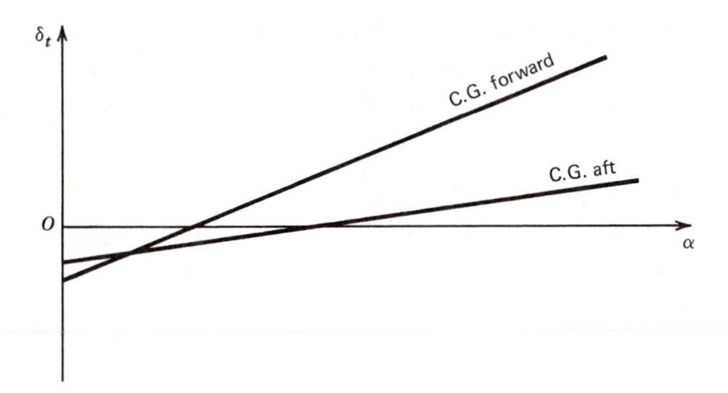

FIG. 2.21 Tab angle required for trim.

and

$$\delta_t = \frac{b_2}{b_3}\left\{\left[\frac{C_{m_\alpha}}{C_{m_\delta}} - \frac{b_1}{b_2}\left(1 - \frac{\partial\varepsilon}{\partial\alpha}\right)\right]\alpha + \frac{C_{m_0}}{C_{m_\delta}} + (\varepsilon_0 + i_t)\frac{b_1}{b_2}\left[1 - \frac{a_t}{a}\frac{S_t}{S}\left(1 - \frac{\partial\varepsilon}{\partial\alpha}\right)\right]\right\} \tag{2.8,4}$$

δ_t is seen to be linear in α. It will appear typically as in Fig. 2.21. The gradient of the curve is found to be related to the stick-free static margin as follows. From Eq. 2.8,4

$$\frac{\partial\delta_t}{\partial\alpha} = \frac{b_2}{b_3 C_{m_\delta}}\left[C_{m_\alpha} - \frac{b_1}{b_2}C_{m_\delta}\left(1 - \frac{\partial\varepsilon}{\partial\alpha}\right)\right] \tag{2.8,5}$$

Upon substituting for C_{m_α} from Eq. 2.3,3, for C_{m_δ} from Eq. 2.5,3, and using the definition of F, Eq. 2.7,3, we get

$$\frac{\partial\delta_t}{\partial\alpha} = \frac{b_2 a}{b_3 C_{m_\delta}}\left[h - h_n + (1 - F)V_H\frac{a_t}{a}\left(1 - \frac{\partial\varepsilon}{\partial\alpha}\right)\right] \tag{2.8,6}$$

If we neglect the difference between V_H and V'_{Hn}, then from Eq. 2.7,11 we get

$$\frac{\partial\delta_t}{\partial\alpha} \doteqdot \frac{b_2 a}{b_3 C_{m_\delta}}(h - h'_n) \tag{2.8,7}$$

This relation is exact in the limit $h = h'_n$, for then $V_H = V'_{Hn}$.

Tailless Aircraft. The elevator floating angle in this case is

$$\delta_{\text{free}} = -\frac{b_0}{b_2} - \frac{b_1}{b_2}\alpha - \frac{b_3}{b_2}\alpha_t \tag{2.8,8}$$

Upon equating it to δ_{trim} (Eq. 2.5,4), we get

$$\delta_t = \frac{b_2}{b_3}\left[\left(\frac{C_{m_\alpha}}{C_{m_\delta}} - \frac{b_1}{b_2}\right)\alpha + \frac{C_{m_0}}{C_{m_\delta}} - \frac{b_0}{b_2}\right] \tag{2.8,9}$$

Equation 2.8,9 is of the same form as 2.8,4, and Fig. 2.21 applies here as well. A similar dependence of the gradient on the stick-free static margin is obtained, with the difference that the relation is exact, not approximate. From Eq. 2.8,9,

$$\frac{\partial \delta_t}{\partial \alpha} = \frac{b_2}{b_3 C_{m_\delta}} \left(C_{m_\alpha} - \frac{b_1}{b_2} C_{m_\delta} \right)$$

From Eq. 2.2,22 $C_{m_\alpha} = a(h - h_n)$ for a tailless airplane. After substituting for C_{m_α}, for C_{m_δ} from Eq. 2.5,7, and using the definition of F for a tailless airplane, Eq. 2.7,14, we get

$$\frac{\partial \delta_t}{\partial \alpha} = \frac{b_2}{b_3 C_{m_\delta}} \left[Fa(h - h_n) - \frac{b_1}{b_2} \frac{\partial C_{m_0}}{\partial \delta} \right] \tag{2.8,10}$$

Equation 2.7,22 is now used to eliminate h_n, leading to the result

$$\frac{\partial \delta_t}{\partial \alpha} = \frac{b_2}{b_3} \frac{Fa}{C_{m_\delta}} (h - h_n') \tag{2.8,11}$$

Flight Determination of h_n'

Equations 2.8,7 and 2.8,11 show that the trim-tab slope bears the same relation to the stick-free neutral point as the elevator-angle slope does to the stick-fixed neutral point. That is, for aircraft with or without tails, $\partial \delta_t / \partial \alpha = \partial \delta_t / \partial C_{L_{\text{trim}}} = 0$ when $h = h_n'$. Hence a possible method for finding h_n' in flight is to measure the trim-tab slopes $\partial \delta_t / \partial C_{L_{\text{trim}}}$, and plot versus h in a manner similar to Fig. 2.15. The intercept in this case is h_n'. The same proviso holds here as in the determination of h_n; namely that the result obtained is invalid if any of the coefficients of Eqs. 2.8,7 and 11 vary significantly with speed.

Servo Tabs (Geared Tabs)

The coefficient C_{h_δ} dominates the hinge moment of a control, and hence the control force. It gives the rate at which the hinge moment increases with control angle. The need for reduction of C_{h_δ} by aerodynamic means was referred to in Sec. 2.6. One such means, which is very effective, is the geared or servo tab. The geometry of such a tab is illustrated in Fig. 2.22. The angle of the tab relative to the control surface is determined by the rigid link AB. When arranged as shown,

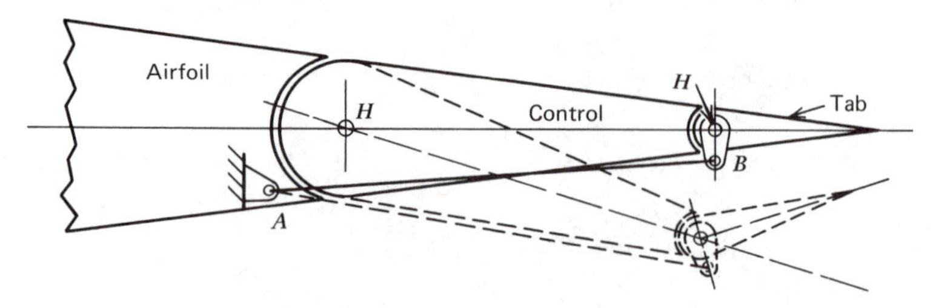

FIG. 2.22 Geometry of geared tab.

downward movement of the control is accompanied by an automatic upward movement of the tab. The hinge moment caused by the tab is then of the sense that assists the control movement. If B were moved to the upper surface of the tab, so that AB crossed HH, then the opposite effect would be obtained. This arrangement, known as an antiservo, or antibalance tab can be used when a control is otherwise overbalanced, or too closely balanced. It provides a means of achieving a zero or positive C_{h_α} without any detrimental effect on C_{h_δ}, as follows. The balance, c_b (Fig. 2.16), is chosen large enough so that C_{h_α} becomes zero or positive. The control will then have C_{h_δ} either too small or even positive. This is then corrected by introducing an antiservo geared tab.

Suppose that, when the elevator moves through an angle δ_e, the tab displacement is $-\gamma\delta_e$. γ, called the "tab gearing," is positive for a servo tab and negative for an antiservo tab. The hinge-moment coefficient will then be

$$\begin{aligned} C_{he} &= b_0 + b_1\alpha_t + b_2\delta_e + b_3\delta_t \\ &= b_0 + b_1\alpha_t + b_2\left(1 - \frac{b_3}{b_2}\gamma\right)\delta_e \end{aligned} \tag{2.8,12}$$

The servo tab thus in effect reduces the value of $b_2 = C_{h_\delta}$ by the factor

$$\left(1 - \frac{b_3}{b_2}\gamma\right).$$

Spring Tabs

The effect of the "speed-squared law" on control forces at high speeds has led to the development of the "spring tab." The effect of this device is to mitigate the influence of speed. Figure 2.23 shows the application to a rudder control. The

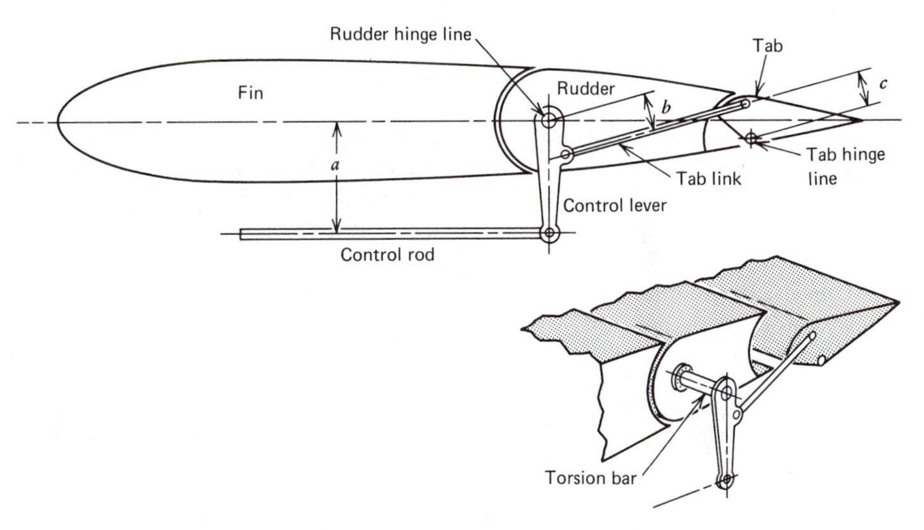

FIG. 2.23 Spring tab applied to a rudder.

system functions as follows. When a force is applied through the control rod to the control lever, the latter rotates through some angle θ. The rudder would rotate through the same angle, and the tab not move at all, if the control lever were *rigidly* connected to the rudder. However, this is not so, and the torsion bar twists through some angle ϕ. The rudder displacement is then $\delta_r = \theta - \phi$. The movement of the control lever *relative* to the rudder (angle ϕ) causes the tab link to move and deflect the tab to the left, just as though it were a geared tab. Now with all other factors equal, an increase in speed will require an increase in the control-rod load to hold the same rudder angle. But an increase in this force introduces extra twist into the torsion bar, and hence increases the tab deflection. Thus, as the speed increases, an increasing *proportion* of the rudder hinge moment is balanced by the tab, and a decreasing proportion by the pilot or control system. In effect, the system behaves like a geared servo tab, the gearing of which increases with speed.

2.9 STICK FORCE TO TRIM

One of the important handling characteristics of an airplane is the force required of the pilot to hold the elevator at the angle required for trim, and the manner in which this force varies with speed. If friction in the control system is neglected, the stick force is simply related to the elevator hinge moment. Figure 2.24 is a schematic representation of a control system. We assume that the elements of the linkage and the structure to which it is attached are rigid. Then the principle of virtual displacements[3] states that, when it is in equilibrium, the net work done by the external forces P and H_e during a virtual displacement is zero. Hence, if Δs and $\Delta\delta_e$ are the displacements of the control force and the elevator,

$$P\,\Delta s + H_e\,\Delta\delta_e = 0$$

or

$$P = GH_e \tag{2.9,1}$$

where $G = -\Delta\delta_e/\Delta s$ is the *elevator gearing*. It has the unit radians per foot and is positive. (A backward movement of the control column produces upward movement of the elevator.) G is usually constant, or nearly so, over the whole range of movement. However, the possibility should be noted of designing special linkages with variable gearing.

Introduction of the hinge-moment coefficient enables the expression for P to be written

$$P = GC_{he}S_e\bar{c}_e\tfrac{1}{2}\rho V^2 \tag{2.9,2}$$

In view of the different character of C_{he} for airplanes with and without tails, we shall deal with these two cases separately.

[3] See, for example, Chap. VIII of S. Timoshenko and D. H. Young, *Engineering Mechanics, Statics*, McGraw-Hill Book Co., 1937.

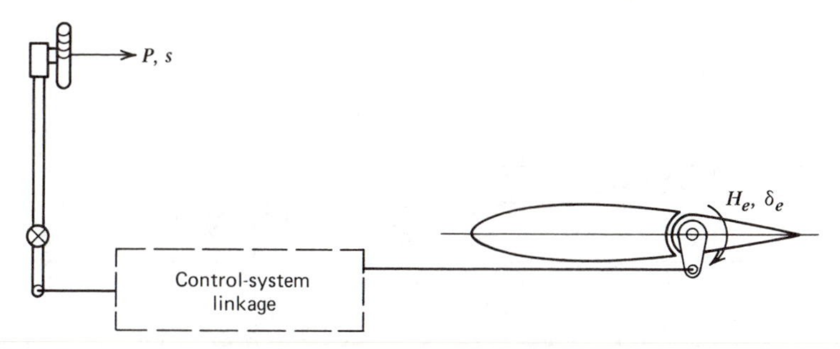

FIG. 2.24 Schematic diagram of an elevator control system.

Airplanes with Tails

Assuming as before that $b_0 = 0$, we have for C_{he}

$$C_{he} = b_1\alpha_t + b_2\delta_{\text{trim}} + b_3\delta_t \tag{2.9,3}$$

If we use the exact expression for δ_{trim}, the equations for C_{he} and P become very complicated. In order to bring out clearly the principal characteristics of the variation of P with speed, we shall, in the following, neglect the variation of V_H with h, and the contribution of the elevator to the lift.

We have then from Eq. 2.8,3

$$\alpha_t = \alpha\left(1 - \frac{\partial\varepsilon}{\partial\alpha}\right) - (\varepsilon_0 + i_t)\left[1 - \frac{a_t}{a}\frac{S_t}{S}\left(1 - \frac{\partial\varepsilon}{\partial\alpha}\right)\right]$$

and, from Eq. 2.5,6 with $C_{L_\delta} = 0$,

$$\delta_{\text{trim}} = -\frac{C_{m_0}}{C_{m_\delta}} - \frac{C_{m_\alpha}}{C_{m_\delta}C_{L_\alpha}}C_{L_{\text{trim}}} \tag{2.9,4}$$

When these are substituted into Eq. 2.9,3, α is replaced by $C_{L_{\text{trim}}}/a$, and the approximate Eq. 2.7,10 is used, we obtain, after some reduction,

$$C_{he} = -\frac{b_2}{C_{m_\delta}}(h - h_n')C_{L_{\text{trim}}} + \left\{b_3\delta_t - b_2\frac{C_{m_0}}{C_{m_\delta}} - b_1(\varepsilon_0 + i_t)\left[1 - \frac{a_t}{a}\frac{S_t}{S}\left(1 - \frac{\partial\varepsilon}{\partial\alpha}\right)\right]\right\} \tag{2.9,5}$$

We further assume that the lift equals the weight, so that

$$C_{L_{\text{trim}}} = \frac{w}{\frac{1}{2}\rho V^2} \tag{2.9,6}$$

where w is the wing loading. We now substitute Eqs. 2.9,5 and 2.9,6 into Eq. 2.9,2, with the following result

$$P = A + B\tfrac{1}{2}\rho V^2 \tag{2.9,7}$$

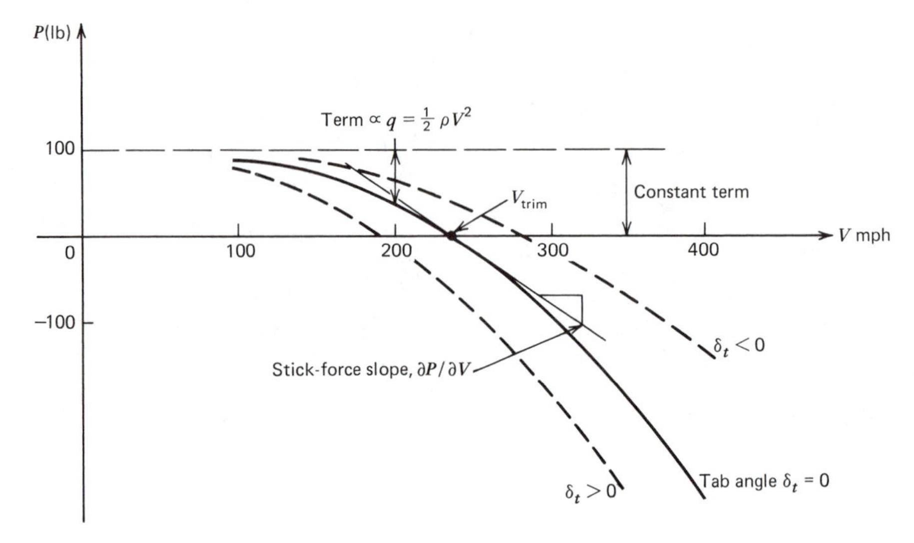

FIG. 2.25 Typical low-speed stick force.

where

$$A = -GS_e\bar{c}_e \frac{b_2}{C_{m_\delta}} w(h - h_n')$$

$$B = GS_e\bar{c}_e b_2 \left\{ \frac{b_3}{b_2} \delta_t - \frac{C_{m0}}{C_{m_\delta}} - \frac{b_1}{b_2}(\varepsilon_0 + i_t)\left[1 - \frac{a_t}{a}\frac{S_t}{S}\left(1 - \frac{\partial \varepsilon}{\partial \alpha}\right)\right]\right\}$$

The typical parabolic variation of P with V (in the absence of **M** effects) is shown in Fig. 2.25. The following conclusions may be drawn.

1. Other things remaining equal, $P \propto S_e\bar{c}_e$, i.e., to the cube of the airplane size. This indicates a very rapid increase in stick forces with size.
2. P is directly proportional to the gearing G.
3. The C.G. position only affects the constant term (apart from a second-order influence on C_{m_δ}). A forward movement of the C.G. produces an upward translation of the curve.
4. The weight of the airplane enters only through the wing loading, a quantity that tends to be constant for airplanes serving a given function, regardless of weight. An increase in wing loading has the same effect as a forward shift of the C.G.
5. The part of P that varies with $\frac{1}{2}\rho V^2$ decreases with height, and increases as the speed squared.
6. Of the terms contained in B, none can be said in general to be negligible. All of them are "built-in" constants except for δ_t.

7. The effect of the trim tab is to change the coefficient of $\frac{1}{2}\rho V^2$, and hence the curvature of the parabola in Fig. 2.24. Thus it controls the intercept of the curve with the V axis. This intercept is denoted V_{trim}; it is the speed for zero stick force.

Tailless Airplanes

If we allow for wing camber, $b_0 \neq 0$, and the hinge moment is given by

$$C_{he} = b_0 + b_1\alpha + b_2\delta_{\text{trim}} + b_3\delta_t \tag{2.9,8}$$

To obtain C_{he} as a function of C_L, we set

$$\alpha = \frac{C_L - C_{L_\delta}\delta_{\text{trim}}}{C_{L_\alpha}} \tag{2.9,9}$$

and use Eq. 2.5,6 for δ_{trim}. We then introduce the value $C_{m_\alpha} = a(h - h_n)$, and use Eq. 2.5,7 for C_{m_δ}. After some algebraic reduction, the following result is obtained

$$C_{he} = b_0 + b_3\delta_t - \frac{Fb_2C_{m_0}}{\partial C_{m_0}/\partial\delta_e} + C_{L_{\text{trim}}}\frac{Fb_2(h_n' - h)}{\partial C_{m_0}/\partial\delta_e} \tag{2.9,10}$$

When this value of C_{he} is put into Eq. 2.9,2, making use of Eq. 2.9,6, we get an equation similar to Eq. 2.9,7. (Note, however, that no approximations were needed in this case.)

$$P = A' + B'\tfrac{1}{2}\rho V^2 \tag{2.9,11}$$

where

$$A' = GS_e\bar{c}_e b_2 \frac{F}{\partial C_{m_0}/\partial\delta_e} w(h_n' - h)$$

$$B' = GS_e\bar{c}_e\left(b_0 + b_3\delta_t - b_2\frac{F}{\partial C_{m_0}/\partial\delta_e}C_{m_0}\right)$$

All the conclusions drawn from Eq. 2.9,7 apply as well to Eq. 2.9,11.

2.10 STICK-FORCE GRADIENT

Section 2.9 demonstrated how the trim tabs can be used to reduce the stick force to zero. A significant handling characteristic is the gradient of P with V at $P = 0$. The manner in which this changes as the C.G. is moved aft is illustrated in Fig. 2.26. The trim tab is assumed to be set so as to keep V_{trim} the same. The gradient $\partial P/\partial V$ is seen to decrease in magnitude as the C.G. moves backward. When it is at the stick-free neutral point, $A = 0$ for aircraft with or without tails, and, under the stated conditions, the P–V graph becomes a straight line lying on the V axis. This is an important characteristic of the stick-free neutral point; i.e., when the C.G. is at that point, no force is required to change the trim speed.

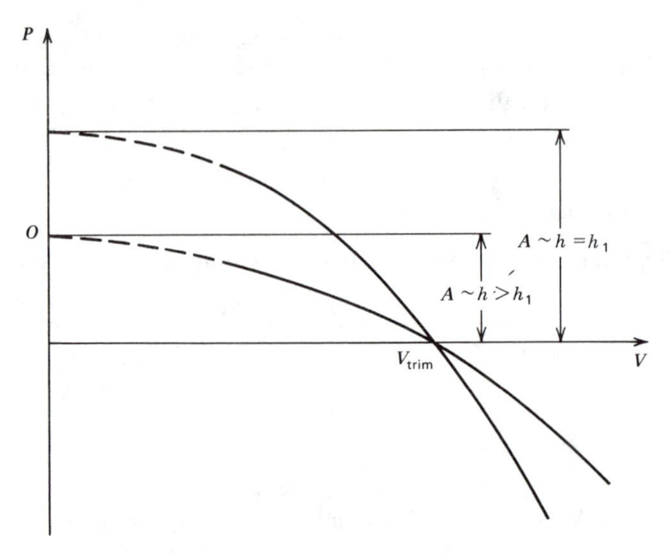

FIG. 2.26 Effect of C.G. location on stick-force gradient.

A quantitative analysis of the stick-force gradient follows.

The stick force is given by Eq. 2.9,7. From it we obtain the derivative

$$\frac{\partial P}{\partial V} = B\rho V$$

At the speed V_{trim}, $P = 0$, and $B = -A/\frac{1}{2}\rho V_{\text{trim}}^2$, thus

$$\frac{\partial P}{\partial V} = -\frac{2A}{V_{\text{trim}}} \tag{2.10,1}$$

Airplanes with Tails

A is given following Eq. 2.9,7. Substituting its value into Eq. 2.10,1, we get

$$\frac{\partial P}{\partial V} = 2GS_e\bar{c}_e \frac{b_2}{C_{m_\delta}} \frac{w}{V_{\text{trim}}}(h - h_n') \tag{2.10,2}$$

Tailless Airplanes

A is given after Eq. 2.9,11, from which we obtain by a procedure similar to that above

$$\frac{\partial P}{\partial V} = 2GS_e\bar{c}_e \frac{Fb_2}{\partial C_{m_0}/\partial \delta_e} \frac{w}{V_{\text{trim}}}(h - h_n') \tag{2.10,3}$$

From Eqs. 2.10,2 and 3 we deduce the following for both kinds of aircraft:

1. The stick-force gradient is proportional to $S_e\bar{c}_e$; i.e., to the cube of airplane size.

2. It is inversely proportional to the trim speed; i.e., it increases with decreasing speed. This effect is also evident in Fig. 2.24.
3. It is directly proportional to wing loading.
4. It is *independent* of height for a given true speed, but *decreases* with height for a fixed V_E.
5. It is directly proportional to the stick-free static margin.

Thus, in the absence of compressibility, the elevator control will be "heaviest" at sea-level, low-speed, forward C.G. and maximum weight.

2.11 ADDITIONAL SYMBOLS INTRODUCED IN CHAPTER 2

$a = C_{L_\alpha}$	airplane lift-curve slope, $\partial C_L/\partial\alpha$
a_w	wing lift-curve slope, $\partial C_{L_w}/\partial\alpha$
a_{wb}	wing-body lift-curve slope $\partial C_{L_{wb}}/\partial\alpha$
a_t	tail lift-curve slope, $\partial C_{L_t}/\partial\alpha_t$
$a_e = C_{L_{t_\delta}}$	$\partial C_{L_t}/\partial\delta_e$
$b_0 = C_{h_0}$	see Eq. 2.6,1
$b_1 = C_{h_{\alpha_t}}$	$\partial C_{h_e}/\partial\alpha_t$
$b_2 = C_{h_\delta}$	$\partial C_{h_e}/\partial\delta_e$
$b_3 = C_{h_{\delta_t}}$	$\partial C_{h_e}/\partial\delta_t$
$\bar{c}$	length of mean aerodynamic chord
$\bar{c}_e$	mean elevator chord (see Sec. 2.6)
C_D	$D/\frac{1}{2}\rho V^2 S$
C_L	$L/\frac{1}{2}\rho V^2 S$
$C_{L_{wb}}$	$L_{wb}/\frac{1}{2}\rho V^2 S$
C_{L_t}	tail lift coefficient, $L_t/\frac{1}{2}\rho V^2 S_t$
C_{L_δ}	$\partial C_L/\partial\delta_e$
C_m	$M/\frac{1}{2}\rho V^2 S\bar{c}$
C_{m_0}	airplane pitching-moment coefficient at zero airplane lift
$C_{m_{0_{wb}}}$	wing-body pitching-moment coefficient at zero wing-body lift
C_{m_t}	$M_t/\frac{1}{2}\rho V^2 S\bar{c}$
C_{m_α}	$\partial C_m/\partial\alpha$
C_{m_δ}	$\partial C_m/\partial\delta_e$
C_{he}	elevator hinge-moment coefficient, $H_e/\frac{1}{2}\rho V^2 S_e\bar{c}_e$
D_w	wing drag
D_t	drag of the tail

F	free-elevator factor $(1 - a_e b_1/a_t b_2)$ for aircraft with tails and $(1 - C_{L_\delta} b_1/ab_2)$ for tailless aircraft
G	elevator gearing
H_e	elevator hinge moment
h	C.G. position, fraction of mean chord (see Fig. 2.7)
h_n	neutral point of airplane, fraction of mean chord
h_{n_w}	neutral point of wing, fraction of mean chord
$h_{n_{wb}}$	neutral point of the wing-body combination
i_t	tail-setting angle (see Fig. 2.8)
L	airplane lift
L_w	wing lift
L_{wb}	lift of wing-body combination
L_t	lift of the tail
l_t	distance between C.G. and tail aerodynamic center
l'_t	see Fig. 2.10
$\mathbf{M}$	Mach number
M	pitching moment about the C.G.
M_w	pitching moment of the wing about the C.G.
M_{0_w}	pitching moment of the wing at zero wing lift
M_{wb}	pitching moment, about the C.G., of the wing-body combination
M_t	moment of tail about C.G.
P	stick force, positive to the rear
S	wing area
S_e	area of elevator aft of hinge line
S_t	area of tail
V	true airspeed,
V_E	equivalent airspeed (EAS), $V\sqrt{\rho/\rho_0}$
V_H	horizontal tail volume, $S_t l_t/S\bar{c}$
V'_H	$S_t l'_t/S\bar{c}$
W	aircraft weight
w	wing loading (W/S)
α	angle of attack of the zero lift line of the airplane, (elevator angle zero)
α_w	angle of attack of the zero lift line of the wing
α_{wb}	angle of attack of the zero lift line of the wing-body combination
α_t	angle of attack of the tail

γ	tab gearing ratio
δ_e	elevator angle
δ_t	tab angle
ε	downwash angle
ε_0	downwash when $L_{wb} = 0$
ρ	air density
ρ_0	standard sea-level value of ρ, 0.002378 slugs/ft^3

2.12 BIBLIOGRAPHY

2.1 R. R. Gilruth and M. D. White. Analysis and Prediction of Longitudinal Stability of Airplanes. *NACA Rept. 711*, 1941.

2.2 H. C. Garner and A. S. Batson. Measurement of Lift, Pitching Moment and Hinge Moment on a Two-Dimensional Cambered Airfoil to Assist the Estimation of Camber Derivatives. *ARC R&M 2946*, 1955.

2.3 S. B. Gates. Notes on the Transonic Movement of Wing Aerodynamic Centre. *ARC R&M 2785*, 1956.

2.4 J. Weil, G. S. Campbell, and M. S. Diederich. An Analysis of Estimated and Experimental Transonic Downwash Characteristics as Affected by Plan Form and Thickness for Wing and Wing-Fuselage Configurations. *NACA TN 3628*, 1956.

2.5 J. L. Decker. Prediction of Downwash at Various Angles of Attack for Arbitrary Tail Locations. *Aero. Eng. Rev.*, Aug. 1956.

2.6 S. B. Gates and H. M. Lyon. A Continuation of Longitudinal Stability and Control Analysis. Part I, General Theory. *ARC R&M 2027*, 1944.

2.7 A. Silverstein and S. Katzoff. Design Charts for Predicting Downwash Angles and Wake Characteristics behind Plain and Flapped Wings. *NACA TR 648*, 1940.

2.8 H. Multhopp. Aerodynamics of the Fuselage. *NACA TM 1036*, 1942.

2.9 S. B. Gates. An Analysis of Static Longitudinal Stability in Relation to Trim and Control Forces; Part I, Gliding. *RAE Rept. BA 1531*, 1939.

2.10 R. I. Sears. Wind-Tunnel Data on the Aerodynamic Characteristics of Airplane Control Surfaces. *NACA Wartime Rept. L-663*, 1943.

2.11 H. B. Glauert. Theoretical Relationships for an Aerofoil with Hinged Flap. *ARC R&M 1095*, 1927.

2.12 H. J. Goett and J. P. Reeder. Effect of Elevator Nose Shape, Gap, Balance, and Tabs on the Aerodynamic Characteristics of a Horizontal Tail Surface. *NACA TR 675*, 1939.

2.13 W. Jacobs. Downwash Behind Wings at Supersonic Speeds. *Aero. Res. Inst. Sweden Rept. 61*, 1955.

2.14 K. L. Goin. Equations and Charts for the Rapid Estimation of Hinge Moment and Effectiveness Parameters for Trailing Edge Controls Having Leading and Trailing Edges Swept Ahead of the Mach lines. *NACA Rept. 1041*, 1951.

STATIC STABILITY AND CONTROL—PART 2

CHAPTER 3

3.1 MANEUVERABILITY—ELEVATOR ANGLE PER g

In this and the following sections we investigate the elevator angle and stick force required to hold the airplane in a steady pull-up with load factor[1] n (Fig. 3.1). The concepts discussed here were introduced by S. B. Gates, ref. 3.16. The flight-path tangent is horizontal at the point under analysis, and hence the net normal force is $L - W = (n - 1)W$ vertically upward. The normal acceleration is therefore $(n - 1)g$.

When the airplane is in straight horizontal flight at the same speed and altitude, the elevator angle and stick force to trim are δ_e and P, respectively. When in the pull-up, these are changed to $\delta_e + \Delta\delta_e$ and $P + \Delta P$. The ratios $\Delta\delta_e/(n - 1)$ and $\Delta P/(n - 1)$ are known, respectively, as the *elevator angle per g*, and the *stick force per g*. These two quantities provide a measure of the maneuverability of the airplane; the smaller they are, the more maneuverable it is.

The angular velocity of the airplane is fixed by the speed and normal acceleration (Fig. 3.1).

$$q = \frac{(n - 1)g}{V} \tag{3.1,1}$$

As a consequence of this angular velocity, the field of the relative air flow past the airplane is curved. It is as though the machine were attached to the end of a whirling arm pivoted at O (Fig. 3.1). This curvature of the flow field alters the pressure distribution and the aero dynamic forces from their values in translational flight. The change is large enough that it must be taken into account in the equations describing the motion.

[1] The load factor is the ratio of lift to weight, $n = L/W$. It is unity in straight horizontal flight.

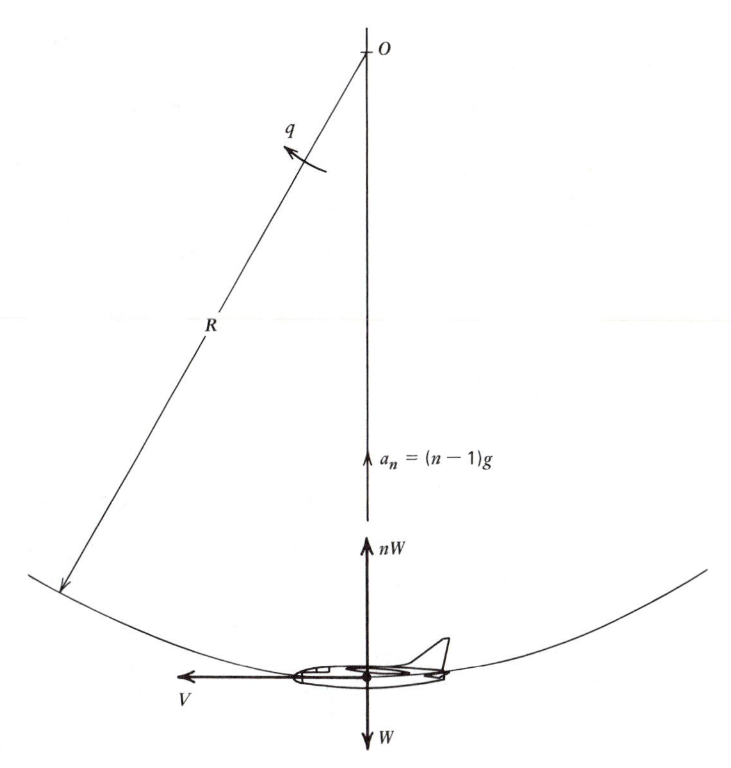

FIG. 3.1 Airplane in a pull-up.

In the analysis to follow, it is assumed that the $C_{L-\alpha}$ relation of the airplane is not affected by the curvature, but that the pitching moment and elevator hinge moment are. This assumption entails some error, but not sufficient to prevent the essential features of the problem from emerging.

Elevator Angle per *g*

In the absence of the curvature effect and for a given height and speed, the airplane pitching moment is described by the functional relation

$$C_m = f(\alpha, \delta_e)$$

With curvature added, this relation is generalized to

$$C_m = f(\alpha, \delta_e, q) \tag{3.1,2}$$

If we assume that the dependence on q is linear, the change in C_m from straight to rotational flight may be expressed as

$$\Delta C_m = \frac{\partial C_m}{\partial \alpha} \Delta\alpha + \frac{\partial C_m}{\partial \delta_e} \Delta\delta_e + \frac{\partial C_m}{\partial q} q \tag{3.1,3}$$

Now C_m is zero in straight flight, as discussed in Chap. 2. In the maneuver described here, although the angular velocity is not zero, the angular acceleration is. Hence C_m must again be zero, and therefore ΔC_m as well. The elevator-angle increment is therefore obtained from Eq. 3.1,3 as

$$\Delta\delta_e = -\frac{\dfrac{\partial C_m}{\partial\alpha}\Delta\alpha + \dfrac{\partial C_m}{\partial q}q}{\partial C_m/\partial\delta_e} \tag{3.1,4}$$

The change in α is related to that in C_L by

$$\Delta\alpha = \frac{1}{a}(\Delta C_L - C_{L_\delta}\Delta\delta_e) \tag{3.1,5}$$

and

$$\Delta C_L = \frac{\Delta L}{\frac{1}{2}\rho V^2 S} = \frac{nW - W}{\frac{1}{2}\rho V^2 S} = (n-1)C_L \tag{3.1,6}$$

Here $C_L = W/\frac{1}{2}\rho V^2 S$ is the lift coefficient corresponding to unity load factor at the given speed and height. With the use of Eqs. 3.1,1, 5, and 6, the elevator angle per g is found from Eq. 3.1,4 to be

$$\frac{\Delta\delta_e}{n-1} = -\frac{C_{m_\alpha}C_L + \dfrac{g}{V}\dfrac{\partial C_m}{\partial q}C_{L_\alpha}}{C_{L_\alpha}C_{m_\delta} - C_{m_\alpha}C_{L_\delta}} \tag{3.1,7}$$

The Derivative $\partial C_m/\partial q$

This derivative represents the rate of change of airplane pitching moment with angular velocity in pitch, α remaining constant. It is known as *damping in pitch*, owing to the fact that it is usually negative and hence represents a resistance to rotation in pitch. The value of $\partial C_m/\partial q$ is simply related to the rotary stability derivative C_{m_q} (American notation), or m_q (British notation). This stability derivative is discussed fully in Chap. 5, but the definition is repeated here for convenience.

$$C_{m_q} = \frac{\partial C_m}{\partial(q\bar{c}/2V)} \tag{3.1,8}$$

thus

$$\frac{\partial C_m}{\partial q} = \frac{\bar{c}}{2V}C_{m_q} \tag{3.1,9}$$

The expression for elevator angle per g can conveniently be rewritten by replacing C_{m_α} in the numerator by $a(h - h_n)$, and $\partial C_m/\partial q$ by Eq. 3.1,9. When this is done the result is

$$\frac{\Delta\delta_e}{n-1} = -\frac{aC_L}{C_{L_\alpha}C_{m_\delta} - C_{L_\delta}C_{m_\alpha}}\left[(h - h_n) + \frac{1}{4}\frac{\rho S\bar{c}}{m}C_{m_q}\right] \tag{3.1,10}$$

The quantity $2m/\rho S\bar{c}$ is called the *relative mass parameter*, and is denoted μ in this book. (The significance of this parameter is discussed in Sec. 4.15.) With the introduction of these symbols, Eq. 3.1,10 becomes

$$\frac{\Delta\delta_e}{n-1} = -\frac{aC_L}{C_{L_\alpha}C_{m_\delta} - C_{L_\delta}C_{m_\alpha}}\left(h - h_n + \frac{1}{2\mu}C_{m_q}\right) \tag{3.1,11}$$

Stick-Fixed Maneuver Point

Equation 3.1,11 shows that the elevator angle per g is a nearly linear function of h (not exactly linear because of the dependence of C_{m_α} in the denominator on h). The C.G. position for which the elevator angle per g vanishes is called the *stick-fixed maneuver point*, and is denoted h_m. By setting Eq. 3.1,11 equal to zero, we find

$$h_m = h_n - \frac{1}{2\mu}C_{m_q} \tag{3.1,12}$$

When h_m is introduced into Eqs. 3.1,10 and 11, they both reduce to the same form: viz.

$$\frac{\Delta\delta_e}{n-1} = -\frac{aC_L}{C_{L_\alpha}C_{m_\delta} - C_{L_\delta}C_{m_\alpha}}(h - h_m) \tag{3.1,13}$$

$(h_m - h)$ is called the *stick-fixed maneuver margin.*

Tailless Airplanes. For tailless airplanes, $\bar{c}$ is used for the representative length instead of l_t and C_{m_δ} is obtained from Eq. 2.5,7. Then

$$h_m - h_n = -\frac{1}{2\mu}C_{m_q} \tag{3.1,14}$$

and

$$\frac{\Delta\delta_e}{n-1} = -\frac{C_L}{\partial C_{m_0}/\partial\delta_e}(h - h_m) \tag{3.1,15}$$

No generally applicable formula can be given for the damping in pitch of wing-body combinations. It depends very much on the wing plan form and whether the speed is subsonic or supersonic.

Airplanes with Tails. The main contribution to the C_{m_q} of airplanes with tails comes from the tail. It is shown in Chap. 5 that this is given by $-2a_t(l_t/\bar{c})V_H$. In order to allow for the increase provided by the wing and body, the tail contribution is multiplied by a factor K which is of the order 1.1 for straight-winged airplanes. Thus

$$C_{m_q} = -2Ka_t\frac{l_t}{\bar{c}}V_H \tag{3.1,16}$$

With this value of C_{m_q}, Eq. 3.1,12 takes on the form

$$h_m = h_n + \frac{K}{\mu}a_t\frac{l_t}{\bar{c}}V_H \tag{3.1,17}$$

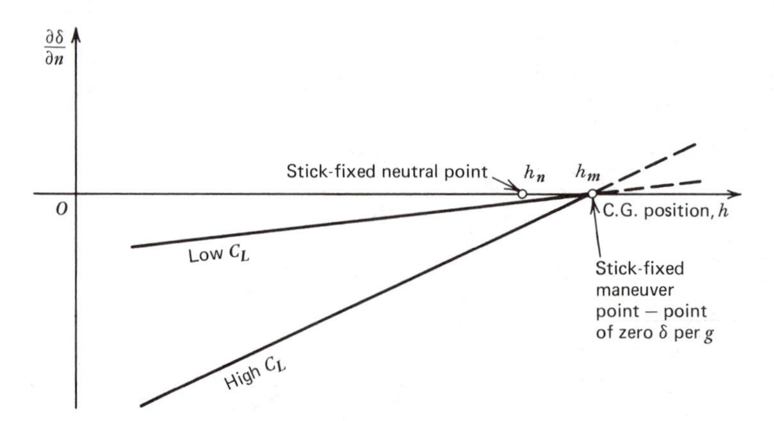

FIG. 3.2 Elevator angle per g.

Significance of the Maneuver Point

On airplanes with tails, the maneuver point stick-fixed is typically about 6% of $\bar{c}$ aft of the stick-fixed neutral point; on tailless machines, the difference may be much less. Aside from the purely geometrical factors that determine this difference, there is the relative mass parameter μ. μ increases with altitude, and hence the maneuver point approaches the neutral point at high altitude. It also contains the ratio $m/S = (W/S)/g$. For constant wing loading, this means that μ decreases with aircraft size, because of the factor $\bar{c}$ in the denominator. The distance between the stick-fixed neutral and maneuver points will therefore be greatest for large airplanes with tails, having a low wing loading and flying at sea level.

Equation 3.1,13 shows the elevator angle per g to be almost linear in h, and negative for $h < h_m$. This is illustrated in Fig. 3.2. It has been shown in Sec. 2.5 that the gradient of the elevator angle to trim versus α reverses when the C.G. is aft of the stick-fixed neutral point (Figs. 2.12 and 2.13). We find now that the elevator angle required to hold a steady pull-up does not reverse until the C.G. moves aft of the maneuver point h_m.

3.2 STICK FORCE PER g

Since the stick force is given by $P = GH$, and the hinge moment is $H = C_{he}\frac{1}{2}\rho V^2 S_e \bar{c}_e$, then the incremental force is

$$\Delta P = G\tfrac{1}{2}\rho V^2 S_e \bar{c}_e \, \Delta C_{he} \tag{3.2,1}$$

The details of the analysis differ in the two cases of aircraft with and without tails. They are therefore dealt with separately below.

Tailless Aircraft

In the absence of airplane rotation, the elevator hinge moment of a tailless aircraft is

$$C_{he} = b_0 + b_1\alpha + b_2\delta_e$$

When an angular velocity q is superimposed, we may allow for its effect by adding another term to C_{he}: viz.

$$C_{he} = b_0 + b_1\alpha + b_2\delta_e + \frac{\partial C_{he}}{\partial q} q \tag{3.2,2}$$

Just as we described the pitching-moment derivative $\partial C_m/\partial q$ in terms of the stability derivative C_{m_q}, so we relate $\partial C_{he}/\partial q$ to the nondimensional coefficient C_{h_q} defined as

$$C_{h_q} = \frac{\partial C_{he}}{\partial(q\bar{c}/2V)} \tag{3.2,3}$$

Then

$$\frac{\partial C_{he}}{\partial q} = \frac{\bar{c}}{2V} C_{h_q} \tag{3.2,4}$$

No explicit formula can be given for C_{h_q} since it is strongly dependent on the wing geometry and the Mach number. The curved relative stream has the same effect on the wing pressure distribution as a positive circular arc camber plus a symmetric twist dependent on the sweepback (see Sec. 5.4). For some wing plan-forms and Mach numbers the effect of this curvature on the elevator hinge moment can be calculated theoretically.

When Eq. 3.2,4 is introduced into Eq. 3.2,2, we may write the increment in C_{he}, which occurs in the pull-up as

$$\Delta C_{he} = b_1\,\Delta\alpha + b_2\,\Delta\delta_e + \frac{\bar{c}}{2V} C_{h_q} q \tag{3.2,5}$$

We eliminate $\Delta\alpha$ by the use of Eqs. 3.1,5 and 3.1,6, and q by Eq. 3.1,1, to obtain

$$\frac{\Delta C_{he}}{n-1} = \frac{b_1}{a} C_L + F b_2 \frac{\Delta\delta_e}{n-1} + \frac{\bar{c}}{2} C_{h_q} \frac{g}{V^2}$$

where $F = 1 - (C_{L_\delta}/a)(b_1/b_2)$. g/V^2 can be replaced by $\rho S C_L/2m$, so that

$$\frac{\Delta C_{he}}{n-1} = C_L\left[\frac{b_1}{a} + F b_2 \frac{\Delta\delta_e}{(n-1)C_L} + \frac{1}{2\mu} C_{h_q}\right] \tag{3.2,6}$$

where $\mu = 2m/\rho S\bar{c}$. We now eliminate $\Delta\delta_e$ by Eq. 3.1,15, and obtain

$$\frac{\Delta C_{he}}{n-1} = C_L\left[\frac{b_1}{a} + \frac{1}{2\mu} C_{h_q} - \frac{F b_2}{\partial C_{m_0}/\partial\delta_e}(h - h_m)\right]$$

Now we introduce the value of h_m given by Eq. 3.1,14, and h_n from Eq. 2.7,22. After some reduction, the result obtained is

$$\frac{\Delta C_{he}}{n-1} = -C_L \frac{Fb_2}{\partial C_{m_0}/\partial \delta_e}\left[h - h_n' + \frac{1}{2\mu}\left(C_{m_q} - \frac{\partial C_{m_0}/\partial \delta_e}{Fb_2} C_{h_q}\right)\right] \tag{3.2,7}$$

Applying the above result in Eq. 3.2,1 gives the expression for the stick force per g:

$$\frac{\partial P}{\partial n} = \frac{\Delta P}{n-1} = -G\tfrac{1}{2}\rho V^2 C_L S_e \bar{c}_e \frac{Fb_2}{\partial C_{m_0}/\partial \delta_e}$$

$$\times\left[h - h_n' + \frac{1}{2\mu}\left(C_{m_q} - \frac{\partial C_{m_0}/\partial \delta_e}{Fb_2} C_{h_q}\right)\right] \tag{3.2,8}$$

The *stick-free maneuver point* is that value of h for which $Q = 0$; i.e.,

$$h_m' = h_n' - \frac{1}{2\mu}\left(C_{m_q} - \frac{\partial C_{m_0}/\partial \delta_e}{Fb_2} C_{h_q}\right) \tag{3.2,9}$$

Upon noting that $\frac{1}{2}\rho V^2 C_L$ is equal to the wing loading w, the stick force per g is expressed more simply as

$$\frac{\partial P}{\partial n} = -GwS_e\bar{c}_e \frac{Fb_2}{\partial C_{m_0}/\partial \delta_e}(h - h_m') \tag{3.2,10}$$

The quantity $(h_m' - h)$ is called the *stick-free maneuver margin.*

Airplanes with Tails

For this case, the effect of airplane rotation on the elevator hinge moment is taken into account with sufficient accuracy by including the change in α_t that is produced by q (Fig. 3.3). The change in C_{he} is then

$$\Delta C_{he} = b_1\,\Delta\alpha_t + b_2\,\Delta\delta_e \tag{3.2,11}$$

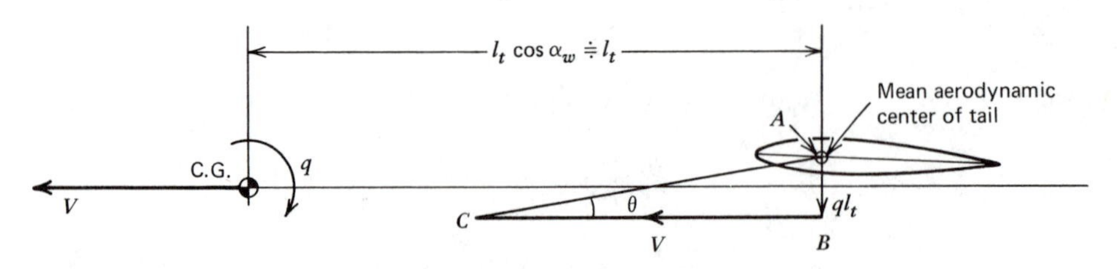

FIG. 3.3 Effect of rotation on tail angle of attack.

$\overrightarrow{AB}$ = velocity vector of tail due to rotation q.

$\overrightarrow{BC}$ = velocity vector of tail due to translation V

θ = change in α_t due to q

$= \tan^{-1}\dfrac{ql_t}{V} \doteqdot \dfrac{ql_t}{V}$ radians

The change in α_t comes partly from the rotation and partly from the increase in airplane angle of attack. Hence

$$\Delta\alpha_t = \Delta\alpha_w\left(1 - \frac{\partial\varepsilon}{\partial\alpha}\right) + \frac{ql_t}{V}$$

We may relate $\Delta\alpha_w$ to ΔC_L approximately by neglecting the change in lift associated with $\Delta\delta_e$:

$$\Delta\alpha_w \doteqdot \frac{\Delta C_L}{a}$$

We use the previously determined values of ΔC_L and q (Eqs. 3.1,1 and 3.1,6) to get

$$\Delta\alpha_t = (n-1)\left[\frac{C_L}{a}\left(1 - \frac{\partial\varepsilon}{\partial\alpha}\right) + \frac{gl_t}{V^2}\right] \tag{3.2,12}$$

The value of $\Delta\delta_e$ is that given by Eq. 3.1,13. However, since the elevator lift has been neglected in Eq. 3.2,12, it should for consistency be neglected in Eq. 3.1,13 as well. This entails setting $C_{L_\delta} = 0$ in the denominator of that equation. When this is done, we get

$$\frac{\Delta C_{he}}{n-1} = \frac{b_1}{a}C_L\left(1 - \frac{\partial\varepsilon}{\partial\alpha}\right) + \frac{b_1 g l_t}{V^2} - b_2\frac{C_L}{C_{m_\delta}}(h - h_m) \tag{3.2,13}$$

We now introduce the fact that $C_L = mg/\frac{1}{2}\rho V^2 S$, to obtain

$$\frac{\Delta C_{he}}{n-1} = C_L\frac{b_2}{a_e V_H}\left[\frac{a_e}{a}\frac{b_1}{b_2}V_H\left(1 - \frac{\partial\varepsilon}{\partial\alpha}\right) + \frac{b_1}{b_2}\frac{a_e}{\mu}\frac{l_t}{\bar{c}}V_H + h - h_m\right] \tag{3.2,14}$$

From Eq. 2.7,11 we identify the first term in the bracket as approximately $h_n - h_n'$. Furthermore, upon using Eq. 3.1,17 for the difference $h_m - h_n$, we get an expression for ΔC_{he} in terms of the stick-free neutral point

$$\frac{\Delta C_{he}}{n-1} = C_L\frac{b_2}{a_e V_H}\left[h - h_n' - \frac{a_t V_H}{\mu}\frac{l_t}{\bar{c}}(K + F - 1)\right] \tag{3.2,15}$$

Finally, the stick force per g is obtained by setting Eq. 3.2,15 into Eq. 3.2,1,

$$\frac{\partial P}{\partial n} = GwS_e\bar{c}_e\frac{b_2}{a_e V_H}(h - h_m') \tag{3.2,16}$$

where the stick-free maneuver point is

$$h_m' = h_n' + (K + F - 1)\frac{a_t V_H}{\mu}\frac{l_t}{\bar{c}} \tag{3.2,17}$$

Discussion

Examination of the equations that describe the stick force per g for both tailed and tailless airplanes shows the following significant points.

1. The stick force per g increases linearly from zero as the C.G. is moved forward from the stick-free maneuver point, and reverses sign for $h > h_m'$.

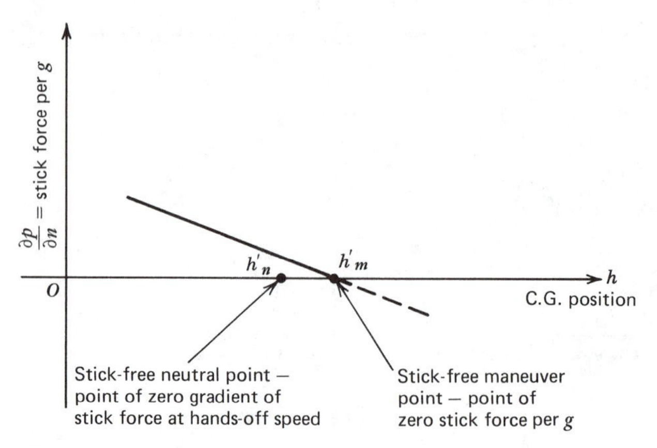

FIG. 3.4 Stick force per *g*.

2. It is directly proportional to the wing loading. High wing loading produces "heavier" controls.
3. For similar aircraft of different size but equal wing loading, $\partial P/\partial n \propto S_e\bar{c}_e$; i.e., to the cube of the linear size.
4. Neither C_L nor V enters the expression for $\partial P/\partial n$ explicitly. Thus, apart from **M** and Reynolds number effects, $\partial P/\partial n$ is independent of speed.
5. The factor μ, which appears in Eqs. 3.2,9 and 3.2,17, causes the separation of the stick-free neutral and maneuver points to vary with altitude, size, and wing loading, in the same manner as the interval $(h_m - h_n)$.

Figure 3.4 shows a typical variation of $\partial P/\partial n$ with C.G. position. The statement made above that the stick force per g is "reversed" when $h > h'_m$ must be interpreted correctly. In the first place this does not necessarily mean a reversal of stick travel per g, for this is governed by the elevator angle per g. If $h'_m < h < h_m$, then there would be reversal of force without reversal of stick travel. In the second place, the analysis given applies only to the *steady state* at load factor n, and throws no light whatsoever on the transition between unaccelerated flight and the pull-up condition. No matter what the value of h, the *initial* stick force and movement required to start the maneuver will be in the normal direction (backward for a pull-up), although one or both of them may have to be reversed before the final steady state is reached.

3.3 BOB WEIGHTS AND SPRINGS

The stick-force characteristics of manual-control systems can be modified by the introduction of weights and springs, as illustrated schematically in Fig. 3.5. When a spring, or bungee, is used as in Fig. 3.5*b*, it is usually so designed that it exerts a nearly constant force on the control column. Thus both weight and spring

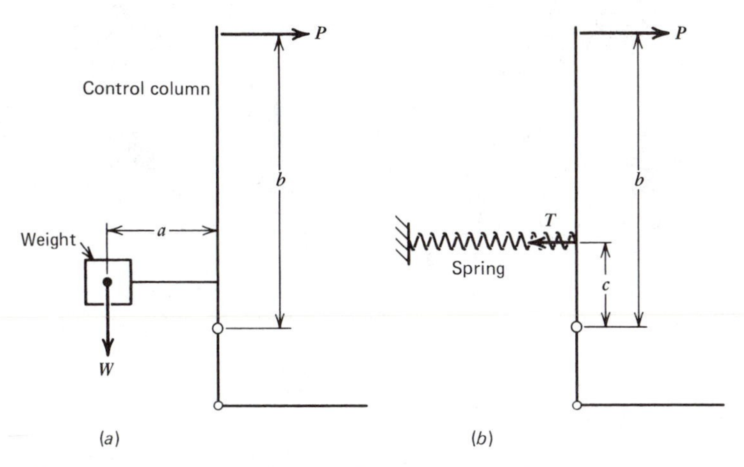

FIG. 3.5 Bob weight and spring. (*a*) Bob weight. (*b*) Spring.

require an additive stick force ΔP to maintain equilibrium. These forces are

$$\Delta P = W\frac{a}{b} \quad \text{for the weight}$$

$$\Delta P = T\frac{c}{b} \quad \text{for the spring}$$

Effect upon Stick Force to Trim and h_n'

The added constant term in the stick force will produce a change in the characteristic as shown in Fig. 3.6. The figure illustrates the case where the trim tab is set to produce the same trim speed as when the ΔP is absent. The parabolic

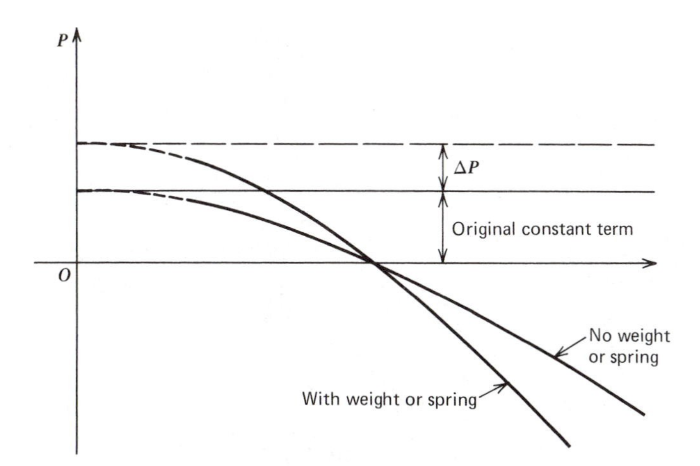

FIG. 3.6 Effect of bob weight and spring on the stick-force characteristic. The trim tab is set to trim at the same speed in both cases.

part of the variation is different for the two cases (see Eqs. 2.9,7 and 2.9,11) because of the altered trim-tab setting. It is clear from the figure that the net result of adding the ΔP and moving the tab is to produce a steeper stick-force gradient at the given trim speed. Now the stick-force gradient has been shown in Sec. 2.10 to depend on the stick-free static margin $(h_n' - h)$. Thus *the increased gradient corresponds to an apparent backward shift of the stick-free neutral point*. The same conclusion is reached by consideration of the constant terms of Eqs. 2.9,7 and 2.9,11, both of which are proportional to $(h_n' - h)$. The apparent shift of the neutral point may be calculated directly from these constant terms. The result obtained for an airplane with tail is

$$\Delta h_n' = \frac{C_{m_\delta}}{b_2} \frac{\Delta P}{S_e \bar{c}_e G w} \tag{3.3,1}$$

and for a tailless airplane

$$\Delta h_n' = \frac{\partial C_{m_0}/\partial \delta_e}{F b_2} \frac{\Delta P}{S_e \bar{c}_e G w} \tag{3.3,2}$$

The term "apparent shift" of the neutral point has been used above since in the real sense of the term (i.e., as defined by $\partial C_m'/\partial \alpha$) it is not displaced at all by the force ΔP. This statement is proved below.

Let the elevator stick force be zero, so that the elevator system is acted upon only by ΔP and the hinge moment. Then, for an aircraft with a tail

$$\Delta P = GH = G\tfrac{1}{2}\rho V^2 S_e \bar{c}_e (b_1 \alpha_t + b_2 \delta_e)$$

whence the floating angle of the elevator is (cf. Eq. 2.7,2)

$$\delta_{\text{free}} = \frac{1}{b_2} \frac{\Delta P}{G\frac{1}{2}\rho V^2 S_e \bar{c}_e} - \frac{b_1}{b_2} \alpha_t$$

The tail lift coefficient is then (cf. Eq. 2.7,3)

$$C_{L_t}' = F a_t \alpha_t + \frac{a_e}{b_2} \frac{\Delta P}{G\frac{1}{2}\rho V^2 S_e \bar{c}_e}$$

and the tail moment coefficient is

$$C_{m_t}' = -V_H C_{L_t}' = -V_H F a_t \alpha_t - V_H \frac{a_e}{b_2} \frac{\Delta P}{G\frac{1}{2}\rho V^2 S_e \bar{c}_e}$$

Since the term containing ΔP is not a function of α, then it contributes zero to $\partial C_m'/\partial \alpha$, and therefore has no influence on the location of the stick-free neutral point. A similar argument holds for the tailless airplane.

Effect upon Stick Force per g and h_m'

When ΔP is provided by a spring, then it is not dependent in any way on acceleration of the airplane. Hence the addition of a spring does not alter the stick force per g or the maneuver point. The bob weight, on the other hand, is affected by airplane acceleration. At load factor n, the effective weight of the bob is increased from W to nW, and hence induces an additional stick force of $(n-1)\,\Delta P$. The stick force per g is thereby increased by the amount

$$\frac{(n-1)\,\Delta P}{n-1} = \Delta P$$

Since stick force per g is proportional to $h_m' - h$, this increase moves the maneuver point aft. Consideration of Eqs. 3.2,10 and 3.2,16 shows this shift to be

$$\Delta h_m' = \frac{\Delta P}{GwS_e\bar{c}_eFb_2}\frac{\partial C_{m_0}}{\partial \delta_e} \quad \text{for tailless aircraft} \tag{3.3,3}$$

and

$$\Delta h_m' = -\frac{\Delta P}{GwS_e\bar{c}_e}\frac{a_eV_H}{b_2} \quad \text{for tailed aircraft} \tag{3.3,4}$$

These are seen to be identical with the apparent neutral-point shift of Eqs. 3.3,1 and 3.3,2. The movement of the maneuver point is real, however, and not simply apparent. This is because the maneuver point is *defined* in terms of the stick force per g.

3.4 INFLUENCE OF WING FLAPS ON TRIM AND STABILITY

When partial-span wing flaps are operated on conventional aircraft, changes take place in both the trim and stability. These changes are the result of three main aerodynamic effects of the flaps, which are illustrated in Fig. 3.7.

1. The lowering of the flaps has the same effect on $C_{m_{0_{wb}}}$ as an increase in wing camber, producing a negative increment, $\Delta C_{m_{0_{wb}}}$.
2. The angle of zero lift changes, becoming more negative. Since the tail setting i_t is measured relative to the wing-body zero lift line, then it is subject to a positive increment Δi_t.
3. The distortion of the spanwise lift distribution on the wing produces changes in the trailing vortex sheet, and in the downwash at the tail. The values of ε_0 and $\partial\varepsilon/\partial\alpha$ may both increase.

The consequent changes in C_{m_0}, h_n, and C_{m_α} may be calculated from Eqs. 2.2,21, 2.3,2, and 2.3,3. Whether the airplane will trim at a higher or lower α after the flaps are lowered will depend upon the relative magnitudes of the effects noted. In the example of Fig. 3.7, the final trim is at a higher α and hence higher C_L. Down elevator would have to be applied to return the trim speed to the value for flaps retracted. Because of the rotation of the zero lift line, the airplane attitude will be altered in a nose-down direction when the original trim speed has been regained.

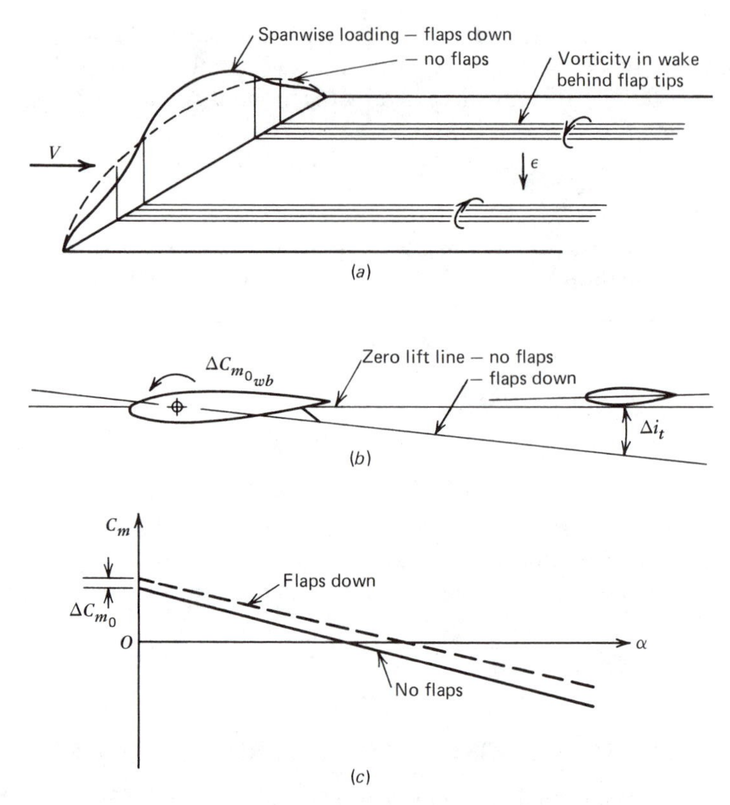

FIG. 3.7 Effect of flaps on pitching moment. (*a*) Increased down wash due to flaps. (*b*) Δi_t due to flaps. (*c*) $C_m - \alpha$ curves for $\delta = 0$.

3.5 INFLUENCE OF THE PROPULSIVE SYSTEM

The influence of the airplane propulsive system upon trim and stability may be both important and complex. The range of conditions to be considered in this connection is extremely wide. In the first place, there are several types of propulsive units in common use—reciprocating-engine-driven propellers, gas-turbine-driven propellers, turbojets, propeller jets, and rockets. In the second place, the operating condition may be anything from hovering to flight at high supersonic Mach number. Finally, the variations in engine-plus-airplane geometry are very great. The analyst may have to deal with such widely divergent cases as a high-aspect-ratio, straight-winged airplane with wing-mounted propellers or a low-aspect-ratio delta with buried jet engines. Owing to its complexity, the subject has not yet reached the state where a definite and comprehensive treatment of propulsive system influences on stability is possible. There does not exist sufficient theoretical or empirical information to enable reliable predictions to be made under all the above-mentioned conditions. However, certain of the major effects of propellers and propulsive jets are sufficiently well understood to make it worthwhile to discuss them.

Before embarking upon the details, attention is drawn to the necessity for distinguishing between the changes of trim produced by the application of power and the changes in static stability. For this purpose we introduce the thrust coefficient

$$T_c = \frac{T}{\rho V^2 d^2} \tag{3.5,1}$$

where T is the thrust of one engine, and d is the diameter of the propeller or jet. T_c is a suitable index for the effects of the propulsive system. They will be largest when T_c is largest, and vice versa.

When calculating the trim curves (i.e., elevator angle, tab angle, and stick force to trim), the thrust used must be that which maintains equilibrium at the speed and angle of climb being investigated. Thus, in level flight $T = D$, and in climbing flight at angle of climb θ, $T = D + W \sin\theta$. Consider the trim curves for level flight as an example. Then

$$T_c = \frac{D}{n\rho V^2 d^2} = C_D \frac{S}{2nd^2} \tag{3.5,2}$$

where n is the number of propulsion units. As the flight speed varies, C_L and hence C_D varies. Thus a significant variation of T_c with C_L will exist. It is given by

$$\frac{dT_c}{dC_L} = \frac{S}{2nd^2}\frac{dC_D}{dC_L} = \frac{S}{nd^2}\frac{C_L}{\pi Ae} \tag{3.5,3}$$

where A is the aspect ratio, and e the span efficiency factor. Since the change in C_m produced by the thrust increases with T_c, then the variation of T_c with C_L will produce certain changes in dC_m/dC_L as well. When the C_m curve is calculated under the conditions described, a very large positive increment in dC_m/dC_L is often found over the value for power-off flight. This has sometimes been interpreted as a decrease in static stability. Actually it does not necessarily represent a true stability change, since in order to change from one speed to another in level flight, the settings of the propulsion controls must be altered (e.g., throttle or rpm). The true inherent stability can only be assessed by considering flight at fixed settings of the engine and propeller controls.

Furthermore in accordance with the definition adopted in Sec. 2.1, our consideration of static stability is restricted to angle-of-attack disturbances at constant speed. Under this condition we must look for changes in $\partial C_m/\partial\alpha$, with T_c constant. It has been shown (ref. 3.1) that this is a valid measure of the behavior of the airplane in flight. The power-on neutral point will be that C.G. location for which this derivative is zero.

The Influence of Running Propellers

Direct Thrust Moment. As illustrated in Fig. 3.8, the thrust T introduces a pitching moment of amount $\Delta M = T \cdot Z_p$. The corresponding increment in C_m is,

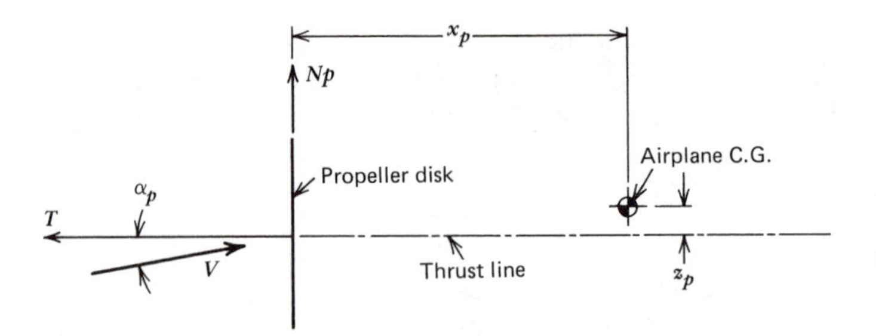

FIG. 3.8 Forces on a propeller.

for n propellers,

$$\Delta C_m = 2nT_c \frac{z_p}{\bar{c}} \frac{d^2}{S} \tag{3.5,4}$$

Since this increment is constant for constant T_c, then $\Delta(\partial C_m/\partial\alpha) = 0$, and the effect is on trim only, the neutral-point location remaining unaltered.

Moment from Propeller Normal Force. When the propeller axis is inclined to the flow, it experiences a normal force N_p (Fig. 3.8). The moment produced by this force is $\Delta M = N_p x_p$. The corresponding increment in C_m is, for n propellers,

$$\Delta C_m = nC_{N_p} \frac{x_p}{\bar{c}} \frac{S'}{S} \tag{3.5,5}$$

where C_{N_p} is the coefficient of N_p, $C_{N_p} = N_p/\frac{1}{2}\rho V^2 S'$, and S' is the propeller disk area, $S' = \pi d^2/4$. Theory (ref. 3.4) shows that, for small angles, C_{N_p} is proportional to α_p. Hence this force produces changes in both trim and stability. The stability change is given by

$$\Delta \frac{\partial C_m}{\partial \alpha} = n \frac{\chi_p}{\bar{c}} \frac{S'}{S} \frac{\partial C_{N_p}}{\partial \alpha_p} \frac{\partial \alpha_p}{\partial \alpha} \tag{3.5,6}$$

If the propeller were situated far from the induced flow field of the wing, then $\partial\alpha_p/\partial\alpha$ would be unity. However, for the common case of wing-mounted tractor propellers, there is a strong upwash ε_p at the propeller. Thus

$$\alpha_p = \alpha + \varepsilon_p + \text{const} \quad \text{and} \quad \partial\alpha_p/\partial\alpha = 1 + \partial\varepsilon_p/\partial\alpha$$

Thus

$$\Delta C_{m_\alpha} = n \frac{\chi_p}{\bar{c}} \frac{S'}{S} \left(1 + \frac{\partial \varepsilon_p}{\partial \alpha}\right) \frac{\partial C_{N_p}}{\partial \alpha_p} \tag{3.5,7}$$

Methods of estimating $(\partial C_{N_p}/\partial\alpha_p)$ and $(\partial\varepsilon_p/\partial\alpha)$ are contained in Appendix B.

Increase of Wing Lift. When a propeller is located ahead of a wing, the high-velocity slipstream causes an increase in the lift of the wing. A semiempirical theory of this phenomenon has been given by Smelt and Davies (ref. 3.5), which shows that the lift increment is essentially linear in α for constant T_c. Thus it has the effect of increasing the wing-body lift-curve slope a_{wb}. From Eq. 2.3,1 for the location of the neutral point, it is seen that this produces a decrease in stability: i.e., a forward shift of h_n. A summary of the method of estimating the wing lift increment is given in Appendix B.

Effects on the Tail. The slipstream from the propellers may influence the forces on the tail in two principal ways:

1. Increase the effective values of a_t and a_e owing to the augmented velocity in the slipstream.
2. Increase the downwash $\partial\varepsilon/\partial\alpha$.

Methods of estimating these effects are uncertain, and the testing of powered models is required to establish them with engineering precision for most new configurations. However, empirical methods (ref. 3.6, 3.7, 3.8) available in the literature are suitable for some cases.

The Influence of Jet Engines

Direct Thrust Moment. The relevant inlet and exit geometry of a jet-engine system is shown in Fig. 3.9. The argument relating to the moment of the thrust T is identical with that given above for propellers; i.e.,

$$\Delta C_m = 2nT_c \frac{z_j}{\bar{c}} \frac{d^2}{S} \tag{3.5,8}$$

Jet Normal Force. The air that passes through a propulsive duct experiences, in general, changes in the direction and magnitude of its velocity. The change in magnitude is the principal source of the thrust, and the direction change entails a force normal to the thrust line. The magnitude and line of action of this force can be found from momentum considerations. Let the mass flow through the duct be m'

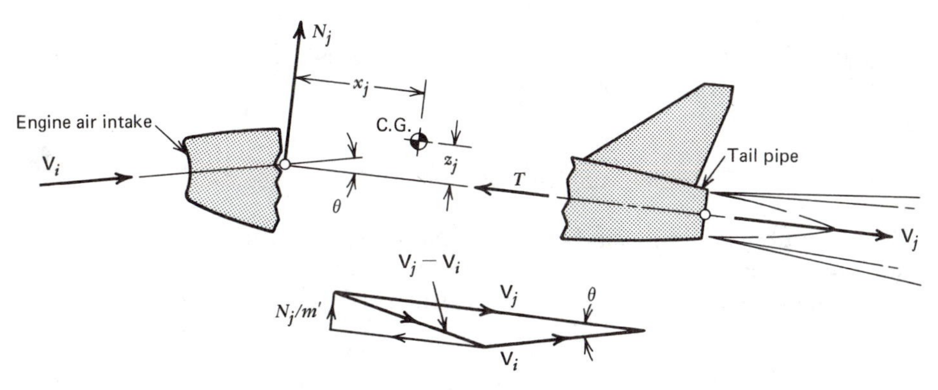

FIG. 3.9 Momentum change of engine air.

slugs per second, and the velocity vectors at the inlet and outlet be $\mathbf{V}_i$ and $\mathbf{V}_j$. Application of the momentum principle then shows that the reaction on the airplane of the air flowing through the duct is

$$\mathbf{F} = -m'(\mathbf{V}_j - \mathbf{V}_i) + \mathbf{F}'$$

where $\mathbf{F}'$ is the resultant of the pressure forces acting across the inlet and outlet areas. For the present purpose, $\mathbf{F}'$ may be neglected, since it is approximately in the direction of the thrust T. The component of $\mathbf{F}$ normal to the thrust line is then found as in Fig. 3.9. It acts through the intersection of $\mathbf{V}_i$ and $\mathbf{V}_j$. The magnitude is given by

$$N_j = m'V_i \sin\theta$$

or, for small angles,

$$N_j = m'V_i\theta \tag{3.5,9}$$

In order to use this relation, both V_i and θ are required. It is assumed that V_i has that direction which the flow would take in the absence of the engine; i.e., θ equals the angle of attack of the thrust line α_j plus the upwash angle due to wing induction ε_j.

$$\theta = \alpha_j + \varepsilon_j \tag{3.5,10}$$

It is further assumed that the magnitude V_i is determined by the mass flow and inlet area; thus

$$V_i = \frac{m'}{A_i\rho_i} \tag{3.5,11}$$

where A_i is the inlet area, and ρ_i the density in the inlet. We then get for N_j the expression

$$N_j = \frac{m'^2}{A_i\rho_i}(\alpha_j + \varepsilon_j)$$

The corresponding pitching-moment coefficient is

$$\Delta C_m = \frac{m'^2}{A_i\rho_i}\frac{x_j}{\frac{1}{2}\rho V^2 S\bar{c}}(\alpha_j + \varepsilon_j) \tag{3.5,12}$$

Since the pitching moment given by Eq. 3.5,12 varies with α at constant thrust, then there is a change in stability given by

$$\Delta C_{m_\alpha} = \frac{m'^2}{A_i\rho_i}\frac{1}{\frac{1}{2}\rho V^2 S\bar{c}}\left[x_j\left(1 + \frac{\partial\varepsilon_j}{\partial\alpha}\right) + \theta\frac{\partial x_j}{\partial\alpha}\right] \tag{3.5,13}$$

The quantities m' and ρ_i can be determined from the engine performance data, and for subsonic flow, $\partial\varepsilon_j/\partial\alpha$ is the same as the value $\partial\varepsilon_p/\partial\alpha$ used for propellors, and given in Appendix B. $\partial x_j/\partial\alpha$ can be calculated from the geometry.

Jet-Induced Inflow. A spreading jet entrains the air that surrounds it, as illustrated in Fig. 3.10, thereby inducing a flow toward the jet axis. If a tailplane

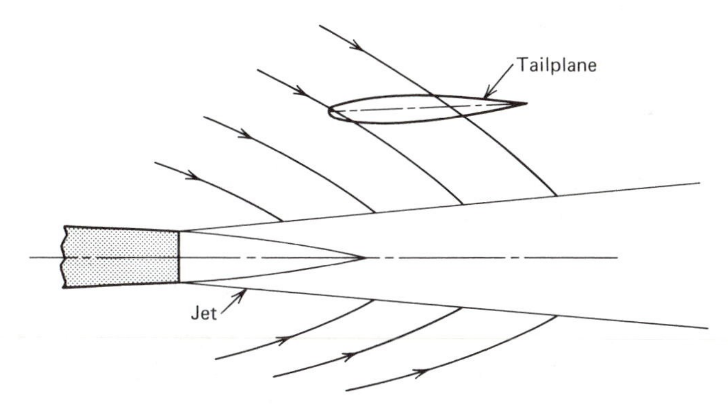

FIG. 3.10 Jet-induced inflow.

is placed in the induced flow field, the angle of attack will be modified by this inflow. A theory of this phenomenon that allows for the curvature of the jet due to angle of attack has been formulated by Ribner (ref. 3.2). The results required for estimating the induced angles are contained in Appendix B. This inflow at the tail may vary with α sufficiently to reduce the stability by a significant amount.

3.6 EFFECT OF AIRPLANE FLEXIBILITY

Modern airplanes flying at very high speeds are frequently subject to important aeroelastic phenomena. Broadly speaking, we may define these as the effects upon the aerodynamic forces of changes in the shape of the airframe—changes caused by these same aerodynamic forces. Up to this point, we have treated the airplane as a rigid body, whose shape is invariant under all flight conditions. Now no real structure is ideally rigid, and airplanes are no exception. Indeed the structure of an airplane is very flexible when compared with bridges, buildings, and earthbound machines. This flexibility is an inevitable characteristic of structures designed to be as light as possible. The aeroelastic phenomena that result may be subdivided under the headings static and dynamic. The static cases are those in which we have steady-state distortions associated with steady loads. Examples are aileron reversal, wing divergence, and the reduction of longitudinal stability. Dynamic cases include buffeting and flutter. In these the time dependence is an essential element. From the practical design point of view, the elastic behavior of the airplane affects all three of its basic characteristics: performance, stability, and structural integrity. This subject has developed very rapidly, until it now occupies a well-established position as a separate branch of aeronautical engineering. For further information the reader is referred to one of the books devoted to it (refs. 1.18 and 1.19).

In this section we take up an example of an aeroelastic effect; namely the influence of fuselage flexibility on static longitudinal stability and control. Assume that the tail load L_t bends the fuselage so that the tail rotates through the angle

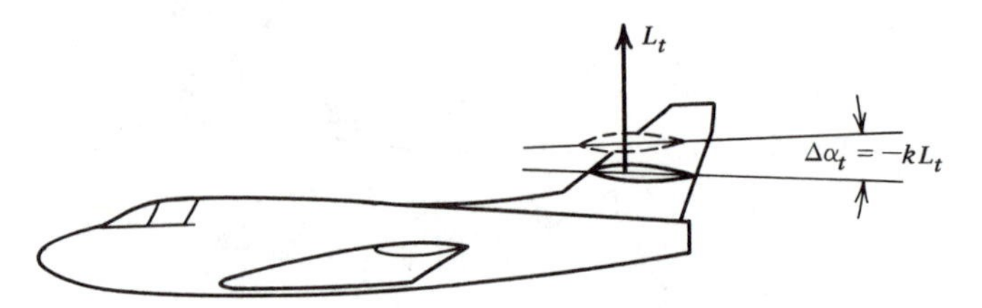

FIG. 3.11 Tail rotation due to fuselage bending.

$\Delta\alpha_t = -kL_t$ (Fig. 3.11). The net angle of attack of the tail will then be

$$\alpha_t = \alpha_{wb} - \varepsilon - i_t - kL_t$$

and the tail lift coefficient at $\delta_e = 0$ will be

$$C_{L_t} = a_t\alpha_t = a_t(\alpha_{wb} - \varepsilon - i_t - kL_t)$$

But $L_t = C_{L_t}\frac{1}{2}\rho V^2 S_t$, from which

$$C_{L_t} = a_t(\alpha_{wb} - \varepsilon - i_t - kC_{L_t}\tfrac{1}{2}\rho V^2 S_t) \tag{3.6,1}$$

Solving for C_{L_t}, we get

$$C_{L_t} = \frac{a_t}{1 + ka_t S_t \frac{1}{2}\rho V^2}(\alpha_{wb} - \varepsilon - i_t) \tag{3.6,2}$$

Comparison of Eq. 3.6,2 with 2.2,10 shows that the tail effectiveness has been reduced by the factor $1/[1 + ka_t(\rho/2)V^2 S_t]$. The main variable in this expression is V, and it is seen that the reduction is greatest at high speeds. From Eq. 2.3,1, we find that the reduction in tail effectiveness causes the neutral point to move forward. The shift is given by (neglecting the dependence of V_H on h)

$$\Delta h_n = \frac{\Delta a_t}{a_{wb}} V_H\left(1 - \frac{\partial\varepsilon}{\partial\alpha}\right) \tag{3.6,3}$$

where

$$\Delta a_t = a_t\left(\frac{1}{1 + ka_t\frac{1}{2}\rho V^2 S_t} - 1\right) \tag{3.6,4}$$

The elevator effectiveness is also reduced by the bending of the fuselage. For, if we consider the case when δ_e is different from zero, then Eq. 3.6,1 becomes

$$C_{L_t} = a_t(\alpha_{wb} - \varepsilon - i_t - kC_{L_t}\tfrac{1}{2}\rho V^2 S_t) + a_e\delta_e$$

and Eq. 3.6,2 becomes

$$C_{L_t} = \frac{a_t(\alpha_{wb} - \varepsilon - i_t) + a_e\delta_e}{1 + ka_t\frac{1}{2}\rho V^2 S_t}$$

Thus the same factor $1/(1 + ka_t\frac{1}{2}\rho V^2 S_t)$ that operates on the tail lift slope a_t also multiplies the elevator effectiveness a_e.

3.7 GROUND EFFECT

At landing and takeoff airplanes fly for very brief (but none the less extremely important) time intervals close to the ground. The presence of the ground modifies the flow past the airplane significantly, so that large changes can take place in the trim and stability. For conventional airplanes, the takeoff and landing cases provide some of the governing design criteria.

The presence of the ground imposes a boundary condition that inhibits the downward flow of air normally associated with the lifting action of the wing and tail. The reduced downwash has three main effects, being in the usual order of importance:

1. A reduction in ε, the downwash angle at the tail.
2. An increase in the wing-body lift slope a_{wb}.
3. An increase in the tail lift slope a_t.

The problem of calculating the stability and control near the ground then resolves itself into estimating these three effects. Data for estimating them are contained in Appendix B. When appropriate values of $\partial\varepsilon/\partial\alpha$, a_{wb}, and a_t have been found, their use in the equations of the foregoing sections will readily yield the required information. The most important items to be determined are the elevator angle and stick force required to maintain $C_{L_{\max}}$ in level flight close to the ground. It will usually be found that the ratio a_t/a_{wb} is decreased by the presence of the ground. Equation 2.3,1 shows that this would tend to move the neutral point forward. However the reduction in $\partial\varepsilon/\partial\alpha$ is usually so great that the net effect is a large rearward shift of the neutral point. Since the value of C_{m_0} is only slightly affected, it turns out that the elevator angle required to trim at $C_{L_{\max}}$ is much larger than in flight remote from the ground. It commonly happens that this is the critical design condition on the elevator, and it will govern the ratio S_e/S_t, or the forward C.G. limit (see Sec. 3.8).

3.8 C.G. LIMITS

One of the dominant parameters of longitudinal stability and control has been shown in the foregoing sections to be the fore-and-aft location of the C.G. (see Figs. 2.9, 2.12, 2.13, 2.15, 2.21, 2.26, 3.2, and 3.4). The question now arises as to what range of C.G. position is consistent with satisfactory flying qualities. This is a critical design problem, and one of the most important aims of stability and control analysis is to provide the answer to it. Since aircraft always carry some disposable load (e.g., fuel, armaments), and since they are not always loaded identically to begin with (variations in passenger and cargo load), it is always necessary to cater for a variation in the C.G. position. The range to be provided for is kept to a minimum by proper location of the items of variable load, but still it often becomes a difficult matter to keep the flying qualities acceptable over the whole C.G. range. Sometimes the problem is not solved, and the airplane is subjected to restrictions on the fore-and-aft distribution of its variable load when operating at part load.

The Aft Limit

The permissible aft C.G. limit is determined by stability considerations. It is based on the location of the neutral-point stick-free h_n' when normal manual or ratio-type controls are employed, and on the stick-fixed neutral point h_n if the elevator control is irreversible. Conservative practice is to keep the aft limit a small distance forward of the relevant neutral point, computed with due allowance for the effects of wing flaps, the propulsive system, and aeroelastic deformation. It should be pointed out here that by the use of automatic devices the stability may be increased so that the neutral point is artificially moved rearward. If the C.G. is located aft of the "natural" neutral point on the strength of such synthetic stability, then failure of the automatic stabilizer will result in an unstable airplane. This design practice may be acceptable if failure of the automatic stabilizer is no more probable than failure of the primary control system.

The Forward Limit

As the C.G. moves forward, the stability of the airplane increases, and larger control movements and forces are required to maneuver or change the trim. The forward C.G. limit is therefore based on control considerations and may be determined by any one of the following requirements:

1. The stick force per g shall not exceed a specified value.
2. The stick-force gradient at trim, dP/dV, shall not exceed a specified value.
3. The stick force required to land, from trim at the approach speed, shall not exceed a specified value.
4. The elevator angle required to land shall not exceed maximum up elevator.

3.9 DIRECTIONAL STABILITY AND CONTROL

The concept of static stability introduced in Sec. 2.1 with respect to the symmetric or longitudinal motion of an airplane may be applied as well to the lateral motions. The dividends to be gained from such an approach are, however, smaller than in the longitudinal case. A full appreciation of the lateral characteristics can be obtained only from dynamic analysis. The primary reason for this is the close coupling that exists between the yawing and rolling motions. When the airplane is yawed relative to its flight path, both yawing and rolling moments are developed, and, when the airplane rolls about its longitudinal axis, there appear once more both yawing and rolling moments. Furthermore, in many designs the controls are coupled as well. Operation of the ailerons sometimes introduces some yawing moment as well as a rolling moment. Also, significant rolling moments may be produced by deflection of the rudder.

Weathercock Stability

Application of the static stability principle to rotation about the z axis shows that a stable airplane must have "weathercock" stability. That is, when the airplane is at an angle of sideslip β relative to its flight path (Fig. 3.12), the yawing moment produced must be such as to tend to restore it to symmetric flight. The yawing moment N is positive as shown. Hence the requirement for static stability is that $\partial N/\partial\beta$ must be positive. The nondimensional coefficient of N is

$$C_n = \frac{N}{\frac{1}{2}\rho V^2 Sb}$$

and hence for stability $\partial C_n/\partial\beta$ must be positive. The usual notation for this derivative is

$$\frac{\partial C_n}{\partial\beta} = C_{n_\beta} \tag{3.9,1}$$

This quantity is analogous in some respects to the longitudinal stability parameter C_{m_α}. It is estimated in a similar way by synthesis of the contributions of the various components of the airplane. The principal contributions are those of the body and the vertical-tail surface. By contrast with C_{m_α}, the wing has little influence in most cases, and the C.G. location is a weak parameter.

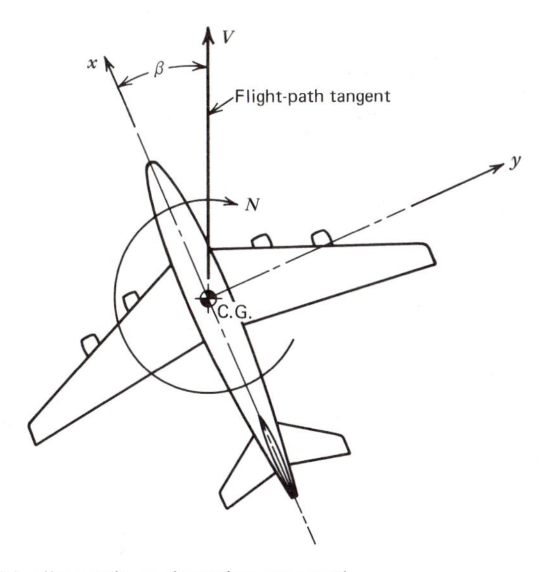

FIG. 3.12 Sideslip angle and yawing moment.

Data for estimating the contribution of the body to C_{n_β} is contained in Appendix B. There are also given data suitable for estimation of the lift-curve slope of the vertical-tail surface. This may be used to calculate the tail contribution as shown below.

In Fig. 3.13 are shown the relevant geometry and the lift force L_F acting on the vertical tail surface. If the surface were alone in an airstream, the velocity vector $\mathbf{V}_F$ would be that of the free stream, so that (cf. Fig. 3.12) α_F would be equal to $-\beta$. When installed on an airplane, however, changes in both magnitude and direction of the local flow at the tail take place. These changes may be caused by the propellor slipstream, and by the wing and fuselage when the airplane is yawed. The angular deflection is allowed for by introducing the sidewash angle σ, analogous to the downwash angle ε. σ is positive when it corresponds to a flow in the y direction; i.e., when it tends to increase α_F. Thus the angle of attack is

$$\alpha_F = -\beta + \sigma \tag{3.9,2}$$

and the lift coefficient of the vertical-tail surface is

$$C_{L_F} = a_F(-\beta + \sigma) + a_r\delta_r \tag{3.9,3}$$

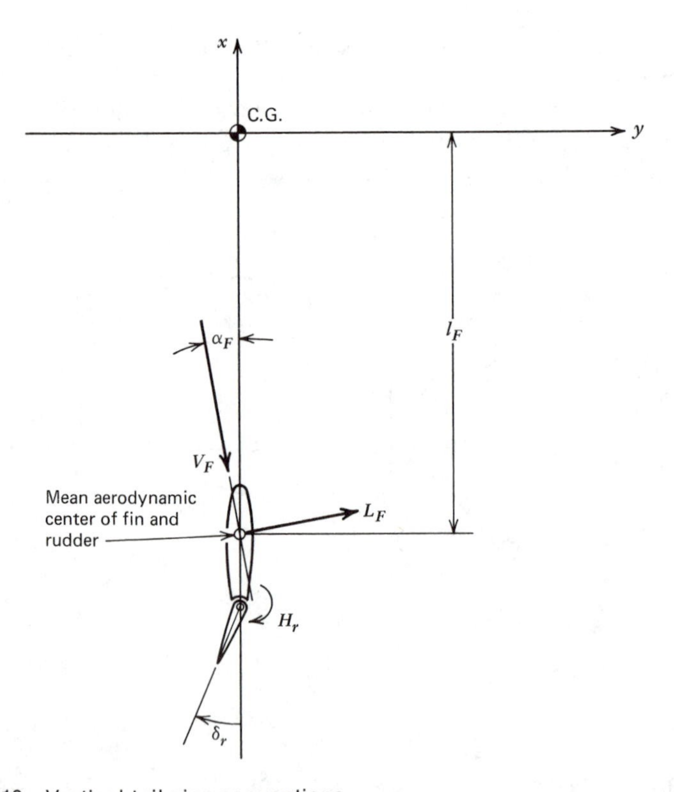

FIG. 3.13 Vertical-tail sign conventions.

The lift is then

$$L_F = C_{L_F} \frac{\rho}{2} V_F{}^2 S_F \tag{3.9,4}$$

and the yawing moment is

$$N_F = -C_{L_F} \frac{\rho}{2} V_F{}^2 S_F l_F$$

thus

$$C_{n_F} = -C_{L_F} \frac{S_F l_F}{Sb} \left(\frac{V_F}{V}\right)^2 \tag{3.9,5}$$

The ratio $S_F l_F / Sb$ is analogous to the horizontal-tail volume ratio, and is therefore called the *vertical-tail volume ratio*, denoted here by V_V. Equation 3.9,5 then reads

$$C_{n_F} = -V_V C_{L_F} \left(\frac{V_F}{V}\right)^2 \tag{3.9,6}$$

and the corresponding contribution to the weathercock stability is

$$\frac{\partial C_{N_F}}{\partial \beta} = -V_V \left(\frac{V_F}{V}\right)^2 \frac{\partial C_{L_F}}{\partial \beta} = V_V a_F \left(\frac{V_F}{V}\right)^2 \left(1 - \frac{\partial \sigma}{\partial \beta}\right) \tag{3.9,7}$$

The Sidewash Factor $\partial\sigma/\partial\beta$. Generally speaking, the sidewash is difficult to estimate with engineering precision. Suitable wind-tunnel tests are required for this purpose. The contribution from the fuselage arises through its behavior as a lifting body when yawed. Associated with the side force that develops is a vortex wake which induces a lateral-flow field at the tail. The sidewash from the propeller is associated with the side force which acts upon it when yawed, and may be estimated by the method of ref. 3.3, previously cited in Sec. 3.5. The contribution from the wing is associated with the asymmetric structure of the flow which develops when the airplane is yawed. This phenomenon is especially pronounced with low-aspect-ratio swept wings. It is illustrated in Fig. 3.14.

The Velocity Ratio V_F/V. When the vertical tail is not in a propellor slipstream, V_F/V is unity. When it is in a slipstream, the effective velocity increment may be dealt with as for a horizontal tail (see Sec. 3.5).

Contribution of Propeller Normal Force. The yawing moment produced by the normal force that acts on the yawed propeller is calculated in the same way as the pitching-moment increment dealt with in Sec. 3.5. The result is similar to Eq. 3.5,6:

$$\Delta \frac{\partial C_n}{\partial \beta} = -n \frac{x_p}{b} \frac{S'}{S} \frac{\partial C_{N_p}}{\partial \alpha_p}. \tag{3.9,8}$$

This is known as the propeller fin effect and is negative (i.e., destabilizing) when the propeller is forward of the C.G., but is usually positive for pusher propellers.

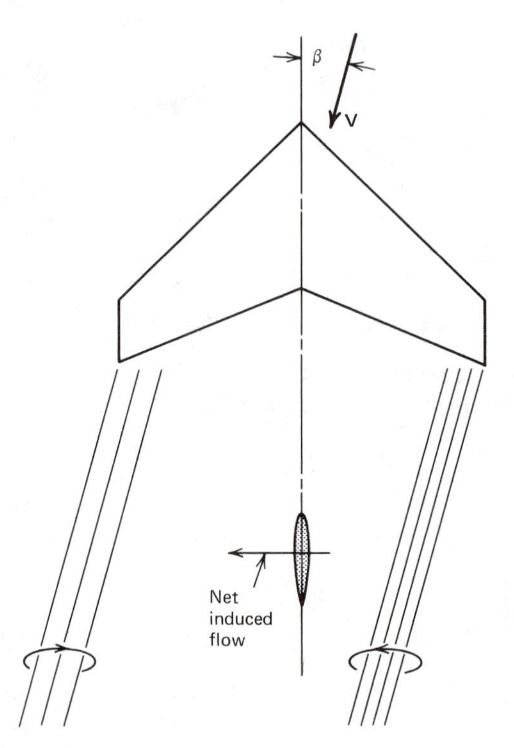

FIG. 3.14 Vortex wake of yawed wing.

Directional Control

In most flight conditions it is desired to maintain the sideslip angle at zero. If the airplane has positive weathercock stability, and is truly symmetrical, then it will tend to fly in this condition. However, yawing moments may act upon the airplane as a result of unsymmetrical thrust (e.g., one engine inoperative), slipstream rotation, or the unsymmetrical flow field associated with turning flight. Under these circumstances, β can be kept zero only by the application of a control moment. The control that provides this is the rudder. Another condition requiring the use of the rudder is the steady sideslip, a maneuver sometimes used, particularly with light aircraft, to increase the drag and hence the glide path angle. A major point of difference between the rudder and the elevator is that for the former trimming the airplane is a secondary and not a primary function. Apart from this difference, the treatment of the two controls is similar. From Eqs. 3.9,3 and 3.9,6, the rate of change of yawing moment with rudder deflection is given by

$$C_{n_\delta} = \frac{\partial C_n}{\partial \delta_r} = -V_V \left(\frac{V_F}{V}\right)^2 \frac{\partial C_{L_F}}{\partial \delta_r} = -a_r V_V \left(\frac{V_F}{V}\right)^2 \tag{3.9,9}$$

This derivative is sometimes called the "rudder power." It must be large enough to make it possible to maintain zero sideslip under the most extreme conditions of asymmetric thrust and turning flight.

A second useful index of the rudder control is the steady sideslip angle that can be maintained by a given rudder angle. The total yawing moment may be written

$$C_n = C_{n_\beta}\beta + C_{n_\delta}\delta_r \tag{3.9,10}$$

For steady motion, $C_n = 0$, and hence the desired ratio is

$$\frac{\beta}{\delta_r} = -\frac{C_{n_\delta}}{C_{n_\beta}} \tag{3.9,11}$$

The rudder hinge moment and control force are also treated in a manner similar to that employed for the elevator. Let the rudder hinge-moment coefficient be given by

$$C_{hr} = b_1\alpha_F + b_2\delta_r \tag{3.9,12}$$

The rudder pedal force will then be given by

$$\begin{aligned} P &= G\frac{\rho}{2}V_F{}^2S_rc_r(b_1\alpha_F + b_2\delta_r) \\ &= G\frac{\rho}{2}V_F{}^2S_rc_r[b_1(-\beta+\sigma) + b_2\delta_r] \end{aligned} \tag{3.9,13}$$

where G is the rudder system gearing.

The effect of a free rudder on the directional stability is found by setting $C_{hr} = 0$ in Eq. 3.9,12. Then the rudder floating angle is

$$\delta_{\text{free}} = -\frac{b_1}{b_2}\alpha_F \tag{3.9,14}$$

The vertical-tail lift coefficient with rudder free is found from Eq. 3.9,3 to be

$$\begin{aligned} C'_{L_F} &= a_F\alpha_F - a_r\frac{b_1}{b_2}\alpha_F \\ &= a_F\alpha_F\left(1 - \frac{a_r}{a_F}\frac{b_1}{b_2}\right) \end{aligned} \tag{3.9,15}$$

The free control factor for the rudder is thus seen to be of the same form as that for the elevator (see Sec. 2.7) and to have a similar effect.

3.10 ROLLING STABILITY AND CONTROL

It has been shown in the foregoing that airplanes can be and are provided with static stability about the pitching and yawing axes. No purely aerodynamic means has yet been devised for achieving this result for the rolling axis. When an airplane is rotated about its velocity vector (Fig. 3.15a), no aerodynamic restoring

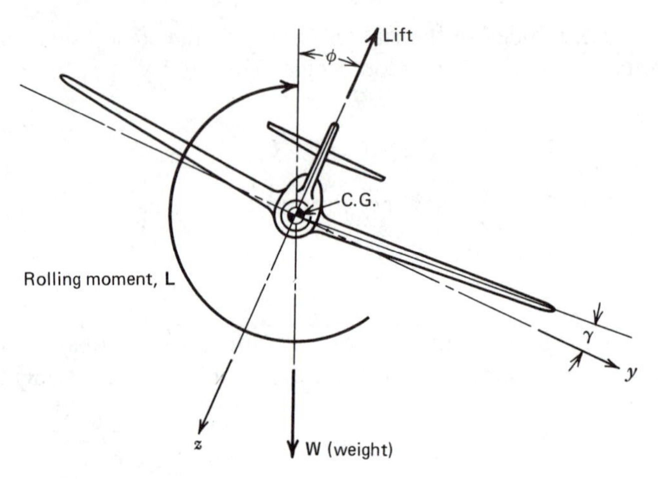

FIG. 3.15*a* Rolled airplane.

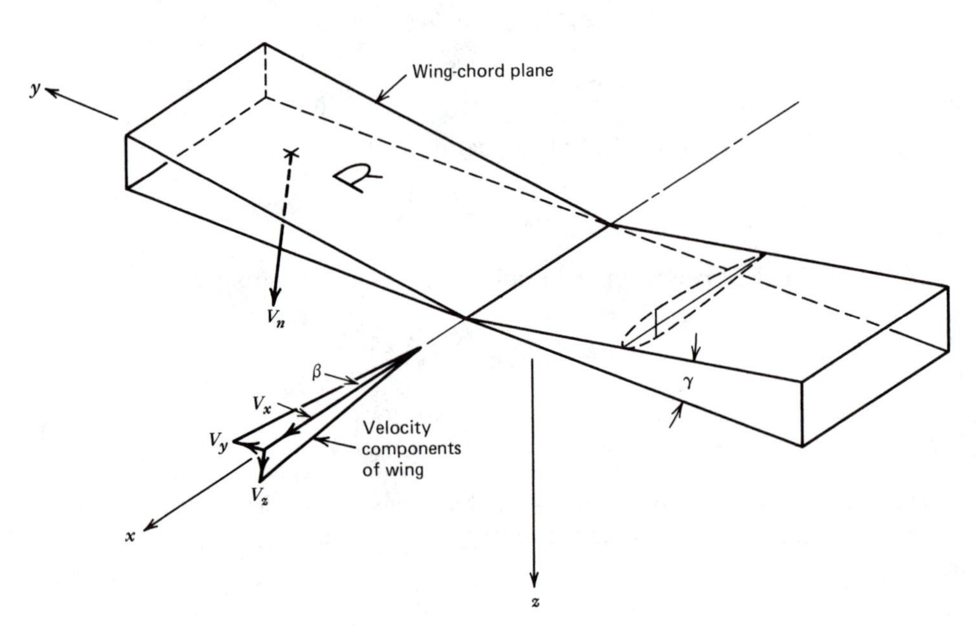

FIG. 3.15*b* Dihedral effect.

$$V_n = \text{normal velocity of panel } R$$
$$= V_x \cos \gamma + V_y \sin \gamma \doteqdot V_x + V_y \gamma$$

$$\therefore \Delta\alpha \text{ of } R \text{ due to dihedral} \doteqdot \frac{V_y \gamma}{V} = \frac{V\beta\gamma}{V} = \beta\gamma$$

moment arises. The airflow remains symmetrical with respect to the plane of symmetry, and the lift vector remains in that plane.

A secondary result of rolling is that a gravity component $W \sin \phi$ acts along the y axis. This induces sideslipping of the airplane. As has been shown in Sec. 3.9, slipping to the right will produce a yawing moment, which tends to swing the nose to the right. However, yet another aerodynamic moment comes into play when the airplane sideslips. This is a righting moment about the x axis owing to the wing dihedral γ. The dihedral has the effect of increasing the angle of attack on the right wing by $\sin \beta \cdot \sin \gamma \doteqdot \beta\gamma$, and decreasing it by the same amount on the left wing (Fig. 3.15*b*). The difference in lift that then occurs provides a negative rolling moment, which tends to bring the wings level. The tendency of an airplane to fly on an even keel is obviously related in some complicated fashion to the derivative $\partial C_l/\partial \beta = C_{l_\beta}$, which is known as the *dihedral effect*. Actually, as pointed out in Sec. 3.9, the rolling and yawing motions are inextricably coupled, and no significant conclusions can be drawn as to the lateral behavior of an airplane except by dynamic analysis. The dihedral effect is a very important factor in such dynamic analysis, however, and hence is discussed in further detail below. Data for estimation of C_{l_β} is contained in Appendix B.

Influence of Fuselage on C_{l_β}

The flow field of the body interacts with the wing in such a way as to modify its dihedral effect. To illustrate this, consider a long cylindrical body, of circular cross section, yawed with respect to the main stream. Consider only the cross-flow component of the stream, of magnitude $V\beta$, and the flow pattern which it produces about the body. This is illustrated in Fig. 3.16. It is clearly seen that the body induces vertical velocities that, when combined with the mainstream velocity, alter the local angle of attack of the wing. When the wing is at the top of the body (high-wing), then the angle-of-attack distribution is such as to produce a negative rolling moment: i.e., the dihedral effect is enhanced. Conversely, when the airplane has a low wing, the dihedral effect is diminished by the fuselage interference. The magnitude of the effect is dependent on the fuselage length ahead of the wing, its cross-section shape, and the planform and location of the wing. Generally, this explains why high-wing airplanes often have little or no dihedral, whereas low-wing airplanes may have dihedral angles of as much as 10°.

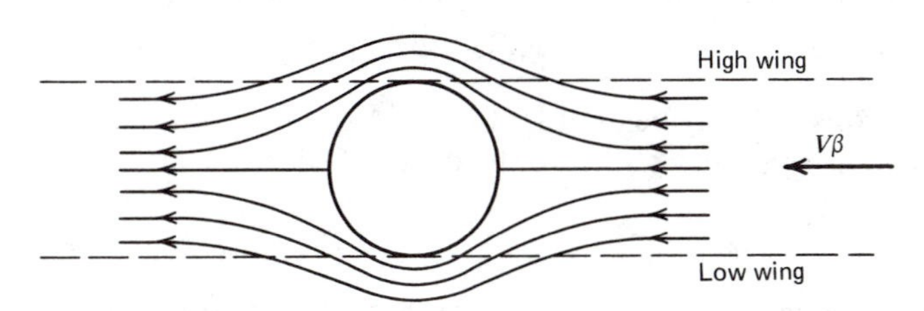

FIG. 3.16 Influence of body on C_{l_β}.

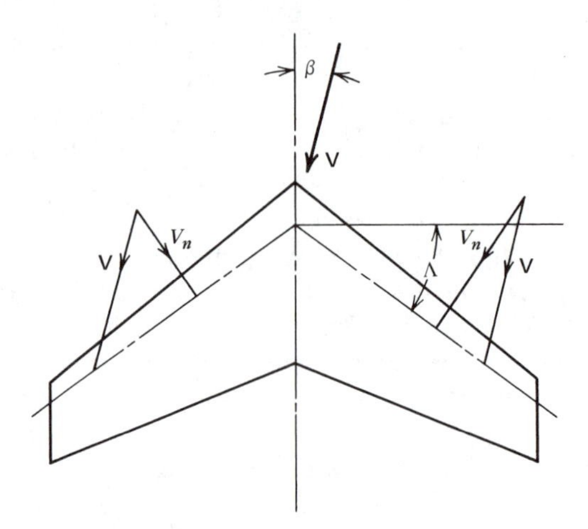

FIG. 3.17 Dihedral effect of a swept wing.

Influence of Sweep on C_{l_β}

The sweepback of the wing is just as important as γ in determining the wing dihedral effect. It is most easily demonstrated by considering the lift on an infinite yawed wing. Aerodynamic theory[2] shows that the lift of such a wing is determined by the component of the stream velocity normal to the wing, i.e., $L = C_L \frac{1}{2}\rho V_n^2 S$ where V_n is the normal velocity. Now, if we have a swept yawed wing as in Fig. 3.17, with the sweepback of the $\frac{1}{4}$ chord line equal to Λ, then the normal velocities on the two panels are $V \cos(\Lambda - \beta)$ and $V \cos(\Lambda + \beta)$. If each panel were to act as an infinite yawed wing, at lift coefficient C_L, then the difference in lift on the two would be

$$\Delta L = C_L \tfrac{1}{2}\rho \frac{S}{2} V^2[\cos^2(\Lambda - \beta) - \cos^2(\Lambda + \beta)]$$

Assuming that β is a small angle, and expanding the trigonometric functions, we may reduce this to

$$\Delta L = \beta C_L \tfrac{1}{2}\rho V^2 S \sin 2\Lambda$$

The rolling moment produced by the wing will be one-half the difference ΔL multiplied by the distance between the centers of pressure of the two panels. Finally the rolling-moment coefficient will be of the form

$$C_l = -\text{const}\,(\beta C_L \sin 2\Lambda) \tag{3.10,1}$$

in which the constant may be of order 0.2. C_l is seen to be linear in β, and negative in sign. Thus the sweepback adds to the dihedral effect the amount

$$\Delta C_{l_\beta} = -\text{const}\,(C_L \sin 2\Lambda) \tag{3.10,2}$$

[2] See A. M. Kuethe and J. D. Schetzer, *Foundations of Aerodynamics*, Sec. 11.31, p. 208, John Wiley & Sons, New York, 1950.

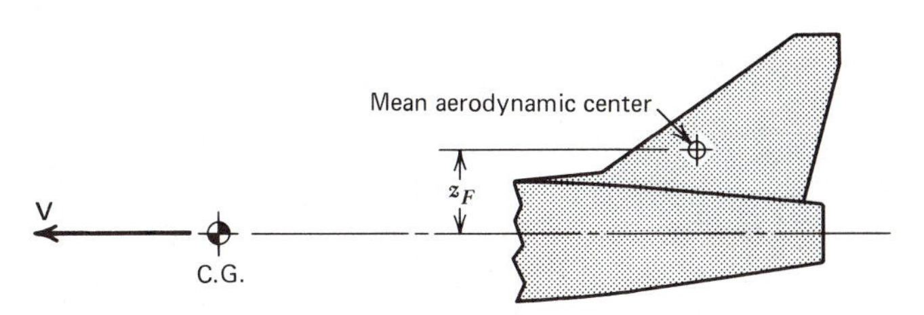

FIG. 3.18 Dihedral effect of the vertical tail.

The most important deductions to be drawn from Eq. 3.10,2 are the following:

1. For angles of sweep of the order of 45°, ΔC_{l_β} may be of the order $-\frac{1}{5}C_L$. For large values of C_L, this is a very large contribution, equivalent to some 10° of dihedral angle.
2. The effect of the sweepback varies with C_L, becoming vanishingly small at high speeds (low C_L) and quite large at low speeds. This poses a significant design problem for high-speed airplanes with swept wings.

Influence of Fin on C_{l_β}

The sideslipping airplane gives rise to a side force on the vertical tail as explained in Sec. 3.9. When the aerodynamic center of the vertical surface is appreciably offset from the rolling axis (see Fig. 3.18) then this force may produce a significant rolling moment. From Eqs. 3.9,3 and 4, with $\delta_r = 0$ this moment is found to be

$$a_F(-\beta + \sigma)\frac{\rho}{2} V_F{}^2 S_F z_F$$

thus

$$\Delta C_l = a_F(-\beta + \sigma)\frac{S_F z_F}{Sb}\left(\frac{V_F}{V}\right)^2$$

and

$$\Delta C_{l_\beta} = -a_F\left(1 - \frac{\partial\sigma}{\partial\beta}\right)\frac{S_F z_F}{Sb}\left(\frac{V_F}{V}\right)^2 \tag{3.10,3}$$

Rolling Control

The angle of bank of the airplane is controlled by the ailerons. The primary function of these controls is to produce a rolling moment, although they frequently introduce a yawing moment as well. The effectiveness of the ailerons in producing rolling and yawing moments is described by the two control derivatives $\partial C_l/\partial\delta_a$ and $\partial C_n/\partial\delta_a$. The aileron angle δ_a is defined as the mean value of the angular displacements of the two ailerons. It is positive when the right aileron movement is downward (see Fig. 3.19). The derivative $\partial C_l/\partial\delta_a$ is normally negative, right aileron down producing a roll to the left.

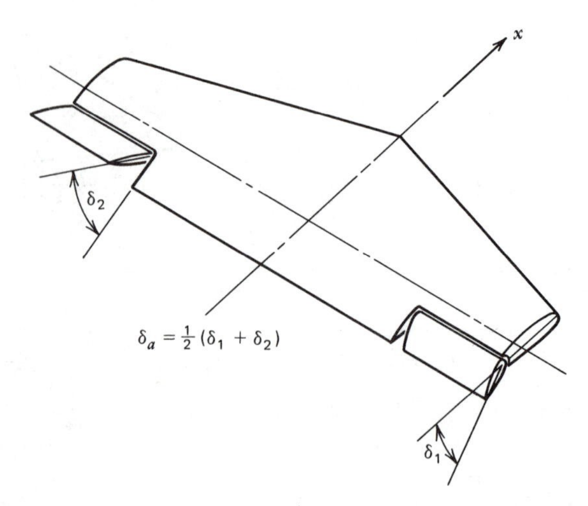

FIG. 3.19 Aileron angle.

For simple flap-type ailerons, the increase in lift on the right side and the decrease on the left side produce a drag differential that gives a positive (nose-right) yawing moment. Since the normal reason for moving the right aileron down is to initiate a turn to the left, then the yawing moment is seen to be in just the wrong direction. It is therefore called *aileron adverse yaw.* On high-aspect-ratio airplanes this tendency may introduce decided difficulties in lateral control. Means for avoiding this particular difficulty include the use of spoilers and Frise ailerons. Spoilers are illustrated in Fig. 3.20. They achieve the desired result by reducing the lift and increasing the drag on the side where the spoiler is raised. Thus the rolling and yawing moments developed are mutually complementary with respect to turning. Frise ailerons (Fig. 2.18) diminish the adverse yaw or eliminate it

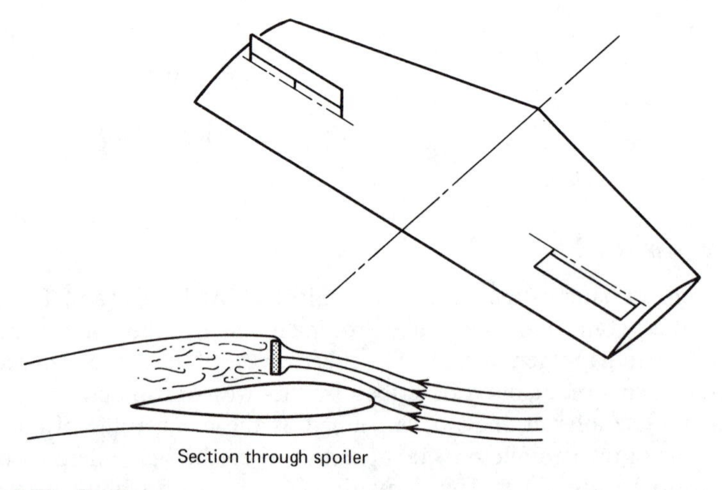

FIG. 3.20 Spoilers.

entirely by increasing the drag on the side of the upgoing aileron. This is achieved by the shaping of the aileron nose and the choice of hinge location. When deflected upward, the gap between the control surface and the wing is increased, and the relatively sharp nose protrudes into the stream. Both these geometrical factors produce a drag increase.

The deflection of the ailerons leads to still additional yawing moments once the airplane starts to roll. These are caused by the altered flow about the wing and tail. These effects are discussed in Sec. 5.9 (C_{n_p}), and are illustrated in Figs. 5.10 and 5.12.

A final remark about aileron controls is in order. They are functionally distinct from the other two controls in that they are *rate* controls. If the airplane is restricted only to rotation about the x axis, then the application of a constant aileron angle results in a steady *rate* of roll. The elevator and rudder, on the other hand, are *displacement* controls. When the airplane is constrained to the relevant singleaxis degree of freedom, a constant deflection of these controls produces a constant angular *displacement* of the airplane. It appears that both rate and displacement controls are acceptable to pilots.

3.11 ADDITIONAL SYMBOLS INTRODUCED IN CHAPTER 3

a_F $\partial C_{L_F}/\partial \alpha_F$

a_r $\partial C_{L_F}/\partial \delta_r$

b span of airplane

C_l rolling-moment coefficient, $L/\frac{1}{2}\rho V^2 Sb$

C_{l_β} $\partial C_l/\partial \beta$

C_{L_F} vertical-tail lift coefficient

C_{m_q} damping in pitch (see Eq. 3.1,8)

C_n yawing-moment coefficient, $N/\frac{1}{2}\rho V^2 Sb$

C_{n_β} $\partial C_n/\partial \beta$

C_{n_δ} $\partial C_n/\partial \delta_r$

d diameter of propeller or jet

g acceleration due to gravity

h_m maneuver point, stick fixed (see Eq. 3.1,12)

h'_m maneuver point, stick free (see Eq. 3.2,9)

L rolling moment

L_F vertical-tail lift

l_F see Fig. 3.13

m airplane mass

M_q $\partial M/\partial q$

n load factor, L/W
N yawing moment
q angular velocity in pitch, radians/sec
S_F vertical-tail area
T thrust of one propulsion unit
T_c thrust coefficient, $T/\rho V^2 d^2$
$\mathbf{V}_F$ effective velocity vector at the fin
V_V vertical-tail volume ratio, $S_F l_F/Sb$
w wing loading, W/S
z_F see Fig. 3.18
α_F effective angle of attack of the fin, radians (see Fig. 3.13)
β sideslip angle, radians
γ dihedral angle
δ_r rudder angle (see Fig. 3.13)
δ_a aileron angle (see Fig. 3.19)
Λ sweepback angle
σ sidewash angle, radians
μ $2m/\rho S\bar{c}$

3.12 BIBLIOGRAPHY

3.1 R. O. Schade. Free-Flight Tunnel Investigation of Dynamic Longitudinal Stability as Influenced by the Static Stability Measured in Wind Tunnel Force Tests under Conditions of Constant Thrust and Constant Power. *NACA TN 2075*, 1950.

3.2 H. S. Ribner. Field of Flow about a Jet and Effects of Jets on Stability of Jet-Propelled Airplanes. *NACA Wartime Rept. L213*, 1946.

3.3 H. S. Ribner. Notes on the Propellor and Slip-stream in Relation to Stability. *NACA Wartime Rept. L25*, 1944.

3.4 H. S. Ribner. Formulas for Propellors in Yaw and Charts of the SideForce Derivative. *NACA Rept. 819*, 1945.

3.5 R. Smelt and H. Davies. Estimation of Increase in Lift Due to Slipstream. *ARC R&M 1788*, 1937.

3.6 J. Weil and W. C. Sleeman, Jr. Prediction of the Effects of Propellor Operation on the Static Longitudinal Stability of Single-Engine Tractor Monoplanes with Flaps Retracted. *NACA Rept. 941*, 1949.

3.7 E. Priestley. A General Treatment of Static Longitudinal Stability with Propellors, with Application to Single Engine Aircraft. *ARC R&M 2732*, 1953.

3.8 C. B. Millikan. The Influence of Running Propellors on Airplane Characteristics. *J. Aero. Sci.*, vol. 7, no. 3, 1940.

3.9 B. B. Klawans and H. I. Johnson. Some Effects of Fuselage Flexibility on Longitudinal Stability and Control. *NACA TN 3543*, 1956.

3.10 H. M. Lyon and J. Ripley. A General Survey of the Effects of Flexibility of the Fuselage, Tail Unit, and Control System on Longitudinal Stability and Control. *ARC R&M 2415* (1950).

3.11 M. J. Queijo. Theoretical Span Load Distributions and Rolling Moments for Sideslipping Wings of Arbitrary Plan Form in Incompressible Flow. *NACA Rept. 1269*, 1956.

3.12 E. C. Polhamus and K. P. Spreeman. Subsonic Wind-Tunnel Investigation of the Effect of Fuselage Afterbody on Directional Stability of Wing-Fuselage Combinations at High Angles of Attack. *NACA TN 3896*, 1956.

3.13 S. Katzoff and H. H. Sweberg. Ground Effect on Downwash Angles and Wake Location. *NACA TR 738*, 1943.

3.14 R. F. Goranson. A Method for Predicting the Elevator Deflection Required to Land. *NACA Wartime Rept. L-95*, 1944.

3.15 S. B. Gates. An Analysis of Static Longitudinal Stability in Relation to Trim and Control Forces. Part II, Engine On. *RAE Rept. BA 1549*, 1939.

3.16 S. B. Gates. Proposal for an Elevator Manoeuverability Criterion. *ARC R&M 2677*, 1942.

3.17 T. A. Toll. Summary of Lateral Control Research, *NACA TN 1245*, 1947.

3.18 H. C. Vetter. Effect of a Turbojet Engine on the Dynamic Stability of an Aircraft. *J. Aero. Sci.*, vol. 20, no. 11, 1953.

3.19 M. Brenckmann. Experimental Investigation of the Aerodynamics of a Wing in a Slipstream. *Univ. of Toronto UTIA Tech. Note 11*, 1957.

3.20 P. R. Owen and H. Hogg. Ground Effect on Downwash with Slipstream. *ARC R&M 2449*, 1952.

GENERAL EQUATIONS OF UNSTEADY MOTION

CHAPTER 4

4.1 GENERAL REMARKS

The analysis of the disturbed motions of airplanes rests upon the general equations of motion of the airplane and its control systems. These equations are derived in this chapter.

An airplane in flight is a very complicated dynamic system. It consists of an aggregate of elastic bodies so connected that both rigid and elastic relative motions can occur. For example, the propeller or jet-engine rotor rotates, the control surfaces move about their hinges, and bending and twisting of the various aerodynamic surfaces occur. The external forces that act on the airplane are also complicated functions of its shape and its motion. It seems clear that realistic analyses of engineering precision are not likely to be accomplished with a very simple mathematical model.

The equations are developed in this chapter in the following way. The airplane is first regarded as a single rigid body, and the equations of motion are derived with respect to a set of axes fixed in it. These are the classical Euler equations. Next the changes introduced by the presence of spinning rotors are evaluated. These are the gyroscopic effects. The equations of motion of the three control systems are then considered. Each of them is assumed to consist of a linkage of rigid elements with one degree of freedom relative to the body axes. The equations obtained are then discussed at some length, and in particular are reduced for the case of small disturbances, and put into nondimensional form. Finally the effects of elasticity are taken into account in the concluding section. The additional equations for the elastic degrees of freedom are derived, as are the changes that must be made in the basic rigid-body equations.

Free use is made of elementary vector analysis. A summary of this subject is contained in Appendix A.

4.2 THE RIGID-BODY EQUATIONS

In the interest of completeness, the rigid-body equations are derived from first principles.

Let δm (Fig. 4.1), an element of mass of the airplane, have the velocity $\mathbf{v}$ relative to inertial axes, and let $\delta\mathbf{F}$ be the resultant force that acts upon it.

Newton's second law then gives the equation of motion of δm; namely

$$\delta\mathbf{F} = \delta m \frac{d\mathbf{v}}{dt} \tag{4.2,1}$$

We sum Eq. 4.2,1 for all elements δm in the airplane to get

$$\Sigma\, \delta\mathbf{F} = \Sigma\, \delta m \frac{d\mathbf{v}}{dt} = \frac{d}{dt} \Sigma \mathbf{v}\, \delta m$$

The quantity $\Sigma\, \delta\mathbf{F}$ is a summation of all the forces that act upon all the elements. The internal forces, that is, those exerted by one element upon another, all occur in equal and opposite pairs, by Newton's third law, and hence contribute nothing to the summation. Thus $\Sigma\, \delta\mathbf{F} = \mathbf{F}$ is the resultant external force acting upon the airplane.

The velocity of δm is

$$\mathbf{v} = \mathbf{v}_c + \frac{d\mathbf{r}}{dt} \tag{4.2,2}$$

where $\mathbf{v}_c$ is the velocity of the mass center. Thus

$$\Sigma \mathbf{v}\, \delta m = \Sigma\left(\mathbf{v}_c + \frac{d\mathbf{r}}{dt}\right)\delta m$$

$$= m\mathbf{v}_c + \frac{d}{dt} \Sigma \mathbf{r}\, \delta m$$

Since C is the mass center, then $\Sigma \mathbf{r}\, \delta m = 0$, so that

$$\Sigma \mathbf{v}\, \delta m = m\mathbf{v}_c \tag{4.2,3}$$

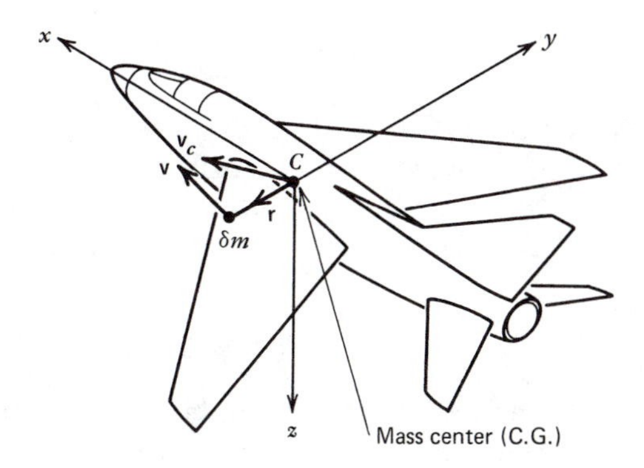

FIG. 4.1

where m is the total mass of the airplane. Thus Eq. 4.2,1 becomes

$$\mathbf{F} = m\frac{d\mathbf{v}_c}{dt} \tag{4.2,4}$$

This equation relates the external force on the airplane to the motion of the mass center. We need also the relation between the external moment and the rotation of the airplane. It is obtained from a consideration of the moment of momentum. The moment of momentum of δm is by definition $\delta\mathbf{h} = \mathbf{r} \times \mathbf{v}\,\delta m$. Consider

$$\begin{aligned}\frac{d}{dt}(\delta\mathbf{h}) &= \frac{d}{dt}(\mathbf{r} \times \mathbf{v})\,\delta m \\ &= \frac{d\mathbf{r}}{dt} \times \mathbf{v}\,\delta m + \mathbf{r} \times \frac{d\mathbf{v}}{dt}\,\delta m \end{aligned} \tag{4.2,5}$$

Now from Eq. 4.2,2,

$$\frac{d\mathbf{r}}{dt} = \mathbf{v} - \mathbf{v}_c$$

Also,

$$\mathbf{r} \times \frac{d\mathbf{v}}{dt}\,\delta m = \mathbf{r} \times \delta\mathbf{F} = \delta\mathbf{G} \tag{4.2,6}$$

Equation 4.2,6 follows from Eq. 4.2,1 and introduces $\delta\mathbf{G}$, the moment of $\delta\mathbf{F}$ about C. Equation 4.2,5 then becomes

$$\delta\mathbf{G} = \frac{d}{dt}(\delta\mathbf{h}) - (\mathbf{v} - \mathbf{v}_c) \times \mathbf{v}\,\delta m \tag{4.2,7}$$

Since $\mathbf{v} \times \mathbf{v} = 0$, Eq. 4.2,7 becomes

$$\delta\mathbf{G} = \frac{d}{dt}(\delta\mathbf{h}) + \mathbf{v}_c \times \mathbf{v}\,\delta m \tag{4.2,8}$$

Equation 4.2,8 is now summed for all elements, as was Eq. 4.2,1, and becomes

$$\Sigma\,\delta\mathbf{G} = \frac{d}{dt}\Sigma\,\delta\mathbf{h} + \mathbf{v}_c \times \Sigma\mathbf{v}\,\delta m \tag{4.2,9}$$

By an argument similar to that for $\Sigma\,\delta\mathbf{F}$, $\Sigma\,\delta\mathbf{G}$ is shown to be the resultant external moment about C, denoted $\mathbf{G}$. $\Sigma\,\delta\mathbf{h}$ is called the moment of momentum, or angular momentum of the airplane and is denoted $\mathbf{h}$. Formulas for $\mathbf{h}$ are derived in Sec. 4.3. Using Eq. 4.2,3 and noting that $\mathbf{v}_c \times \mathbf{v}_c = 0$, Eq. 4.2,9 reduces to

$$\mathbf{G} = \frac{d\mathbf{h}}{dt} \tag{4.2,10}$$

The reader should note that, in Eq. 4.2,10, both $\mathbf{G}$ and $\mathbf{h}$ are referred to a moving point, the mass center. For a moving reference point other than the mass center, the equation does *not* in general apply. The reader should also note that

Eqs. 4.2,4 and 4.2,10 are both valid when there is relative motion of parts of the airplane.

The two vector equations of motion of the airplane, equivalent to six scalar equations, are Eqs. 4.2,4 and 4.2,10.

$$\mathbf{F} = m\frac{d\mathbf{v}_c}{dt}$$

$$\mathbf{G} = \frac{d\mathbf{h}}{dt}$$

4.3 EVALUATION OF THE ANGULAR MOMENTUM h

From the preceding section we have the definition of **h**

$$\mathbf{h} = \Sigma\,\delta\mathbf{h} = \Sigma(\mathbf{r} \times \mathbf{v})\,\delta m$$

Let the angular velocity of the airplane be

$$\boldsymbol{\omega} = \mathbf{i}P + \mathbf{j}Q + \mathbf{k}R$$

where P, Q, R are the scalar components of $\boldsymbol{\omega}$, and **i**, **j**, **k** are unit vectors in the directions of x, y, z.

Now the velocity of a point in a rotating rigid body is given by[1]

$$\mathbf{v} = \mathbf{v}_c + \boldsymbol{\omega} \times \mathbf{r} \tag{4.3,1}$$

thus

$$\begin{aligned}\mathbf{h} &= \Sigma\mathbf{r} \times (\mathbf{v}_c + \boldsymbol{\omega} \times \mathbf{r})\,\delta m \\ &= \Sigma\mathbf{r} \times \mathbf{v}_c\,\delta m + \Sigma\mathbf{r} \times (\boldsymbol{\omega} \times \mathbf{r})\,\delta m \end{aligned} \tag{4.3,2}$$

Since $\mathbf{v}_c$ is constant with respect to the summation, and since $\Sigma\mathbf{r}\,\delta m = 0$, then the first sum in Eq. 4.3,2 is $(\Sigma\mathbf{r}\,\delta m) \times \mathbf{v}_c = 0$.

The second sum is conveniently expanded by the rule for a vector triple product.[2]

$$\begin{aligned}\mathbf{r} \times (\boldsymbol{\omega} \times \mathbf{r}) &= \boldsymbol{\omega}(\mathbf{r} \cdot \mathbf{r}) - \mathbf{r}(\boldsymbol{\omega} \cdot \mathbf{r}) \\ &= \boldsymbol{\omega} r^2 - \mathbf{r}(\boldsymbol{\omega} \cdot \mathbf{r})\end{aligned}$$

Since the vector **r** has the components (x, y, z) then Eq. 4.3,2 finally becomes

$$\mathbf{h} = \boldsymbol{\omega}\Sigma(x^2 + y^2 + z^2)\,\delta m - \Sigma\mathbf{r}(Px + Qy + Rz)\,\delta m \tag{4.3,3}$$

The scalar components of Eq. 4.3,3 are

$$\begin{aligned}h_x &= P\Sigma(y^2 + z^2)\,\delta m - Q\Sigma xy\,\delta m - R\Sigma xz\,\delta m \\ h_y &= -P\Sigma xy\,\delta m + Q\Sigma(x^2 + z^2)\,\delta m - R\Sigma yz\,\delta m \\ h_z &= -P\Sigma xz\,\delta m - Q\Sigma yz\,\delta m + R\Sigma(x^2 + y^2)\,\delta m\end{aligned}$$

[1] See Appendix A.

[2] See Appendix A.

The summations that occur in these equations are the moments and products of inertia of the airplane (see list of symbols), so that we have, upon substituting for them,

$$\begin{aligned} h_x &= AP - FQ - ER \\ h_y &= -FP + BQ - DR \\ h_z &= -EP - DQ + CR \end{aligned} \tag{4.3,4}$$

4.4 EULER'S EQUATIONS OF MOTION

It is noted from Eq. 4.2,10 that we require the derivatives with respect to time of Eqs. 4.3,4. If the axes of reference are nonrotating, then it is clear that, as the airplane rotates, the moments and products of inertia will vary, and derivatives of $A, B, \ldots, F$ will appear in the equations, as well as derivatives of P, Q, R. This is a most undesirable situation. It can be avoided, however, if the frame of reference $Cxyz$ is fixed to the airplane, and moves with it, for then $A, B, \ldots, F$ are constants. We are then in the position of requiring derivatives of vectors referred to a rotating frame of reference. This entails the inclusion of additional terms, but is certainly to be preferred to the variable inertia coefficients.

The derivative of a vector $\mathbf{A}$, referred to a frame of reference rotating with angular velocity $\boldsymbol{\omega}$ is[3]

$$\frac{d\mathbf{A}}{dt} = \frac{\delta \mathbf{A}}{\delta t} + \boldsymbol{\omega} \times \mathbf{A}$$

where

$$\frac{\delta \mathbf{A}}{\delta t} = \mathbf{i}\frac{dA_x}{dt} + \mathbf{j}\frac{dA_y}{dt} + \mathbf{k}\frac{dA_z}{dt} \tag{4.4,1}$$

The vector equations of motion 4.2,4 and 4.2,10 then become, when referred to the frame of reference $Cxyz$, fixed to the airplane,

$$\mathbf{F} = m\frac{\delta \mathbf{v}_c}{\delta t} + m\boldsymbol{\omega} \times \mathbf{v}_c$$

$$\mathbf{G} = \frac{\delta \mathbf{h}}{\delta t} + \boldsymbol{\omega} \times \mathbf{h} \tag{4.4,2}$$

These equations have the scalar components (see list of symbols):

$$\begin{aligned} F_x &= m(\dot{U} + QW - RV) \\ F_y &= m(\dot{V} + RU - PW) \\ F_z &= m(\dot{W} + PV - QU) \end{aligned} \tag{4.4,3}$$

$$\begin{aligned} L &= \dot{h}_x + Qh_z - Rh_y \\ M &= \dot{h}_y + Rh_x - Ph_z \\ N &= \dot{h}_z + Ph_y - Qh_x \end{aligned} \tag{4.4,4}$$

[3] See Appendix A.

where **h** is given by Eqs. 4.3,4. Equations 4.4,3, 4.4,4, and 4.3,4 are the Euler equations of motion of the airplane.

4.5 ORIENTATION AND POSITION OF THE AIRPLANE

Since the frame of reference adopted for the equations of motion is fixed to the airplane, and moves with it, the position and orientation of the airplane cannot be described relative to it.

For this purpose we introduce an Earth-fixed frame of reference $ox'y'z'$ (see Fig. 4.2). Let oz' be taken vertically downward, and ox' horizontal in the vertical

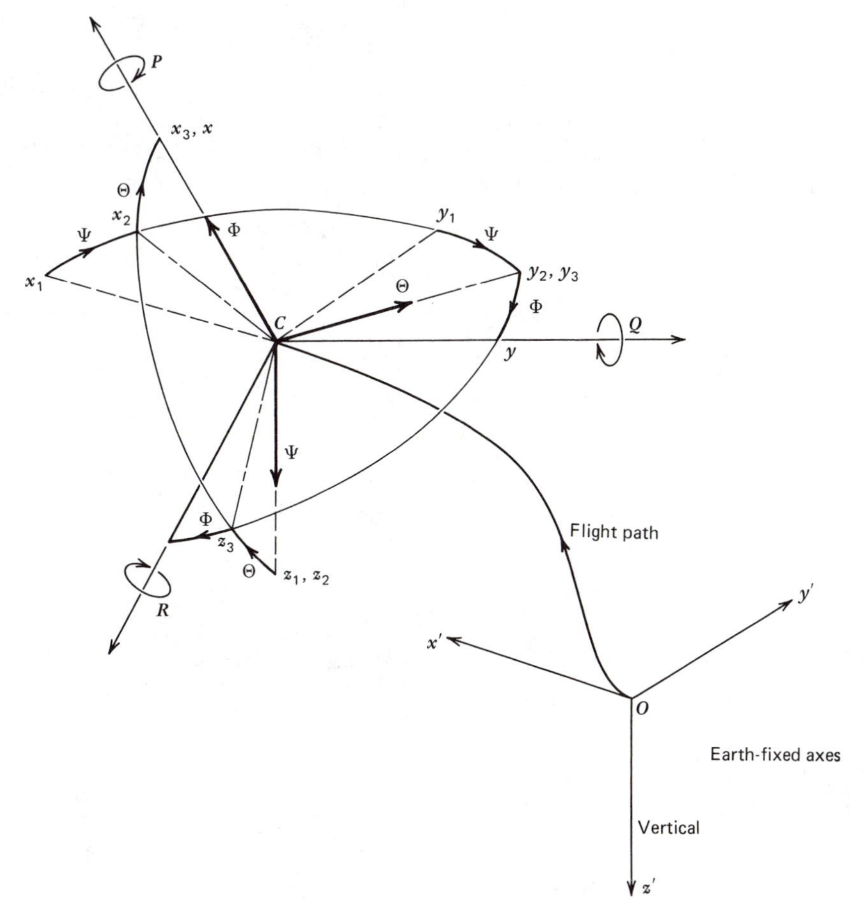

FIG. 4.2 Airplane orientation.

plane containing the initial velocity vector of the mass center. The origin O is assumed to coincide with C at $t = 0$.

The orientation of the airplane is then given by a series of three consecutive rotations, whose order is important. The airplane is imagined first to be oriented so that its axes are parallel to $ox'y'z'$. It is then in the position $Cx_1y_1z_1$. The following rotations are then applied.

1. A rotation $\boldsymbol{\Psi}$ about oz_1, carrying the axes to $Cx_2y_2z_2$ (bringing Cx to its final azimuth).
2. A rotation $\boldsymbol{\Theta}$ about oy_2, carrying the axes to $Cx_3y_3z_3$ (bringing Cx to its final elevation).
3. A rotation $\boldsymbol{\Phi}$ about ox_3, carrying the axes to their final position $Cxyz$ (giving the final angle of bank to the wings).

The Flight Path

We wish to have the coordinates of the flight path relative to the fixed frame $ox'y'z'$. To do this we require the velocity components in the directions x_1, y_1, z_1. Let these be denoted U_1, V_1, W_1, and similarly let subscripts 2 and 3 denote components in the directions of (x_2, y_2, z_2) and (x_3, y_3, z_3). We have then

$$\frac{dx'}{dt} = U_1$$

$$\frac{dy'}{dt} = V_1$$

$$\frac{dz'}{dt} = W_1$$

where

$$\begin{aligned} U_1 &= U_2 \cos \Psi - V_2 \sin \Psi \\ V_1 &= U_2 \sin \Psi + V_2 \cos \Psi \\ W_1 &= W_2 \end{aligned}$$

and

$$\begin{aligned} U_2 &= U_3 \cos \Theta + W_3 \sin \Theta \\ V_2 &= V_3 \\ W_2 &= -U_3 \sin \Theta + W_3 \cos \Theta \end{aligned}$$

and

$$\begin{aligned} U_3 &= U \\ V_3 &= V \cos \Phi - W \sin \Phi \\ W_3 &= V \sin \Phi + W \cos \Phi \end{aligned}$$

from this

$$\begin{aligned}
\frac{dx'}{dt} &= U \cos \Theta \cos \Psi + V(\sin \Phi \sin \Theta \cos \Psi - \cos \Phi \sin \Psi) \\
&\quad + W(\cos \Phi \sin \Theta \cos \Psi + \sin \Phi \sin \Psi) \\
\frac{dy'}{dt} &= U \cos \Theta \sin \Psi + V(\sin \Phi \sin \Theta \sin \Psi + \cos \Phi \cos \Psi) \\
&\quad + W(\cos \Phi \sin \Theta \sin \Psi - \sin \Phi \cos \Psi) \\
\frac{dz'}{dt} &= -U \sin \Theta + V \sin \Phi \cos \Theta + W \cos \Phi \cos \Theta
\end{aligned} \tag{4.5,1}$$

The coordinates (x', y', z') are then given by the integral of Eqs. 4.5,1. To integrate these equations in other than special simple cases is obviously a formidable problem. A case of engineering interest is that of small disturbances from steady flight, when the equations may be linearized and therefore much simplified This is dealt with in Sec. 4.14.

The Orientation of the Airplane

We wish to express the orientation of the airplane in terms of the angular-velocity components (P, Q, R). Let $(\mathbf{i}, \mathbf{j}, \mathbf{k})$ be unit vectors, with subscripts 1, 2, 3 denoting directions (x_1, y_1, z_1), etc.

Let the airplane experience, in time Δt, an infinitesimal rotation from the position defined by Ψ, Θ, Φ to that corresponding to $(\Psi + \Delta\Psi)$, $(\Theta + \Delta\Theta)$, $(\Phi + \Delta\Phi)$. The vector representing this rotation is approximately

$$\Delta\mathbf{n} \doteqdot \mathbf{i}\,\Delta\Phi + \mathbf{j}_3\,\Delta\Theta + \mathbf{k}_2\,\Delta\Psi$$

and the angular velocity is exactly

$$\boldsymbol{\omega} = \lim_{\Delta t \to 0} \frac{\Delta\mathbf{n}}{\Delta t} = \mathbf{i}\dot{\Phi} + \mathbf{j}_3\dot{\Theta} + \mathbf{k}_2\dot{\Psi} \tag{4.5,2}$$

If the components of $\boldsymbol{\omega}$ given in Eq. 4.5,2 be projected onto $Cxyz$, and remembering that $\boldsymbol{\omega} = \mathbf{i}P + \mathbf{j}Q + \mathbf{k}R$, we obtain the following relations:

$$\begin{aligned}
P &= \dot{\Phi} - \dot{\Psi} \sin \Theta \\
Q &= \dot{\Theta} \cos \Phi + \dot{\Psi} \cos \Theta \sin \Phi \\
R &= \dot{\Psi} \cos \Theta \cos \Phi - \dot{\Theta} \sin \Phi
\end{aligned} \tag{4.5,3}$$

In order to find the angles, (Ψ, Θ, Φ), in terms of (P, Q, R), the above equations must be solved. If they are regarded as algebraic equations in $\dot{\Psi}$, $\dot{\Theta}$, $\dot{\Phi}$, we obtain

$$\begin{aligned}
\dot{\Theta} &= Q \cos \Phi - R \sin \Phi \\
\dot{\Phi} &= P + Q \sin \Phi \tan \Theta + R \cos \Phi \tan \Theta \\
\dot{\Psi} &= (Q \sin \Phi + R \cos \Phi) \sec \Theta
\end{aligned} \tag{4.5,4}$$

From this,

$$\Theta = \int_0^t (Q \cos \Phi - R \sin \Phi)\, dt$$
$$\Phi = \int_0^t (P + Q \sin \Phi \tan \Theta + R \cos \Phi \tan \Theta)\, dt \tag{4.5,5}$$
$$\Psi = \int_0^t (Q \sin \Phi + R \cos \Phi) \sec \Theta\, dt$$

The exact integral Eqs. 4.5,5 can obviously be solved analytically only in simple special cases. In many cases, however, they can be linearized, and solutions obtained, (see Sec. 4.14).

Choice of Axes

The equations derived in the preceding sections are valid for any orthogonal axes fixed in the airplane, with origin at the mass center, and known as *body axes*. Since most aircraft are very nearly symmetrical, it is usual to assume exact symmetry, and to let Cxz be the plane of symmetry. Then Cx points "forward," Cz "downward," and Cy to the right. In this case, the two products of inertia D and F are zero, and Eqs. 4.3,4 are consequently simplified.

The directions of Cx and Cz in the plane of symmetry are conventionally fixed in one of three ways (see Fig. 4.3 and Appendix B.13).

Principal Axes. These are chosen to coincide with the principal axes of the machine, so that the remaining product of inertia E vanishes; Eqs. 4.3,4 then become

$$h_x = AP$$
$$h_y = BQ \tag{4.5,6}$$
$$h_z = CR$$

Stability Axes. These are chosen so that Cx points in the direction of motion of the airplane in a reference condition of steady symmetric flight. In this case, the

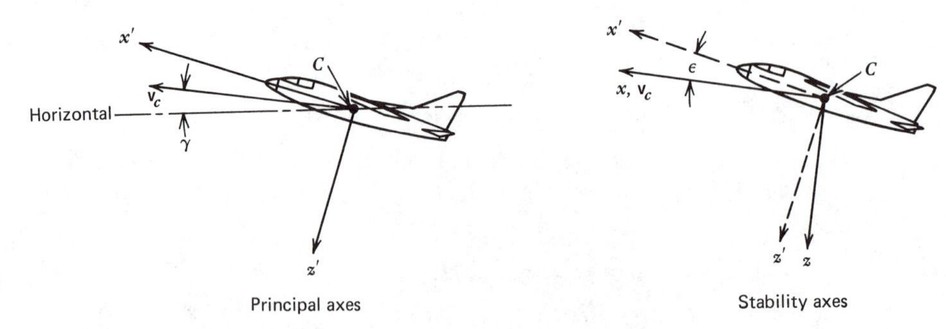

FIG. 4.3 Illustrating two choices of body axes.

reference values of V and W are zero, and the axes are termed *stability axes.* These are the axes that are adopted in this book, owing to the simplifications that result in the equations of motion, and in the expressions for the aerodynamic forces.

With this choice, it should be noted that for different initial flight conditions the axes are differently oriented in the airplane, and hence the values of A, C, and E will vary from problem to problem. The "stability axes," just as the principal axes, are body axes that remain fixed to the airplane during the motion considered in any one problem.

The following formulas are convenient for computing A, C, E when the values A' and C' are known for principal axes.

$$\begin{aligned} A &= A' \cos^2 \varepsilon + C' \sin^2 \varepsilon \\ C &= A' \sin^2 \varepsilon + C' \cos^2 \varepsilon \\ E &= \tfrac{1}{2}(A' - C') \sin 2\varepsilon \end{aligned} \tag{4.5,7}$$

where ε = angle between Cx and Cx' (see Fig. 4.3).

Body Axes. When the axes are neither principal axes nor stability axes, they are usually called simply *body axes.* In this case the x axis is usually fixed to a longitudinal reference line in the airplane. These axes may be the most convenient ones to use if the aerodynamic data have been measured by a wind-tunnel balance that resolves the forces and moments into body-fixed axes rather than tunnel-fixed axes. This is commonly so with the sting balances used in supersonic wind tunnels.

4.6 EFFECT OF SPINNING ROTORS ON THE EULER EQUATIONS

In evaluating the angular momentum **h** (Sec. 4.3) it was tacitly assumed that the airplane is a single rigid body. This is implied in the equation for the velocity of an element (Eq. 4.3,1). Let us now imagine that some portions of the airplane mass are spinning relative to the body axes: e.g., rotors of jet engines, or propellers. Each such rotor has an angular momentum relative to the body axes. This can be computed from Eqs. 4.3,4 by interpreting the moments and products of inertia therein as those of the rotor with respect to axes parallel to $Cxyz$ and origin at the rotor mass center. The angular velocities in Eqs. 4.3,4 are interpreted as those of the rotor relative to the airplane body axes. Let the resultant relative angular momentum of all rotors be $\mathbf{h}'$, with components (h'_x, h'_y, h'_z), which are assumed to be constant. It can be shown that the total angular momentum of an airplane with spinning rotors is obtained simply by adding $\mathbf{h}'$ to the $\mathbf{h}$ previously obtained in Sec. 4.3. The equations that correspond to Eqs. 4.3,4 are then

$$\begin{aligned} h_x &= AP - FQ - ER + h'_x \\ h_y &= -FP + BQ - DR + h'_y \\ h_z &= -EP - DQ + CR + h'_z \end{aligned} \tag{4.6,1}$$

Because of the additional terms in the angular momentum, certain extra terms appear in the right-hand side of the moment equations, Eqs. 4.4,4. These additional terms, known as the gyroscopic couples, are

$$\begin{aligned} &\text{In the } L \text{ equation:} \quad Qh'_z - Rh'_y \\ &\text{In the } M \text{ equation:} \quad Rh'_x - Ph'_z \\ &\text{In the } N \text{ equation:} \quad Ph'_y - Qh'_x \end{aligned} \tag{4.6,2}$$

As an example, suppose the rotor axes are parallel to Cx, with angular momentum $\mathbf{h}' = \mathbf{i}I\Omega$. Then the gyroscopic terms in the three equations are, respectively, 0, $I\Omega R$, and $-I\Omega Q$.

4.7 EQUATIONS OF MOTION OF THE CONTROL SYSTEMS

In this section we are concerned with finding the equations of motion of control systems that are either manually or power-operated. Each system is assumed to consist of a control surface or surfaces (elevator, rudder, or ailerons) assumed to be rigid, and a mechanical linkage of rigid elements that connects the control surface to the power source (pilot and/or hydraulic jack, electric motor, etc.). Relative to the body axes, each system is assumed to have only one degree of freedom, for which a convenient generalized coordinate is the control surface angle. We require the equations of motion in the relevant degrees of freedom.

Although under certain circumstances friction in a control system can be a significant factor, we shall assume for simplicity that the control systems are entirely free from friction.

General Formulation

The general method of attack is described first, and then applied to each control system separately. The required equations are obtained by the application of Lagrange's equation of motion in a moving frame of reference. Lagrange's equation is

$$\frac{d}{dt}\frac{\partial T}{\partial \dot{q}_k} - \frac{\partial T}{\partial q_k} = \mathscr{F} \tag{4.7,1}$$

where T = kinetic energy of the system relative to the chosen frame of reference

$\mathscr{F} = \partial W/\partial q_k$ = generalized force

W = work done on the system by the external forces that act upon it

q_k = generalized coordinate

In order to apply this equation in a non-Newtonian reference frame, it is necessary to include in the external forces the inertia-force field due to the acceleration and

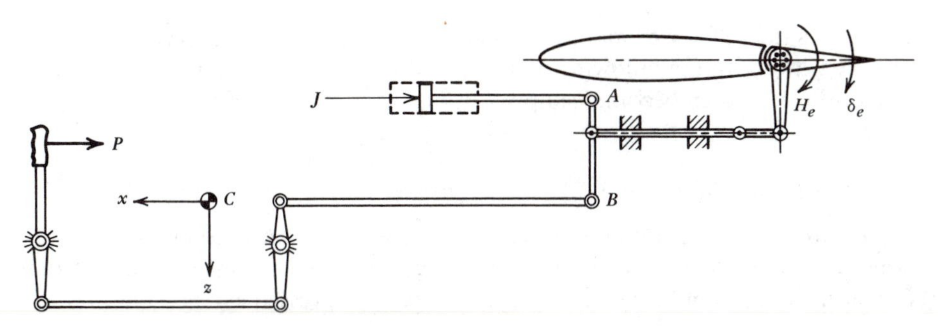

FIG. 4.4 A hypothetical elevator system.

rotation of the reference frame.[4] The determination of that part of $\mathscr{F}$ which arises from the inertia-force field requires a knowledge of the acceleration field in a moving frame of reference. A derivation of this is contained in Appendix A. In addition to the contribution of the inertia forces to $\mathscr{F}$, there are others associated with the control-surface aerodynamic hinge moment, the control force, and gravity.

The structure to which the control system is attached is assumed to be rigid, so that the reactions at the attachment points do not undergo any displacement when the control system is actuated. Therefore these reactions add nothing to W. The gravity contribution to $\mathscr{F}$ is neglected since it adds at most a rather small constant term.

4.8 THE ELEVATOR SYSTEM

In order to fix our ideas, we shall refer to the hypothetical control system of Fig. 4.4. The generalized coordinate of the system is the elevator angle δ_e, and the control force is supplied partially by the pilot (force P), and partially by a hydraulic jack (force J). In this system, the equilibrium of the lever AB requires that the forces P and J stand in a fixed constant ratio.[5] We must now apply Eq. 4.7,1 to find the equation of motion.

The Kinetic Energy T

The kinetic energy of the moving masses (elevator, levers, piston, etc.) may always be expressed in the form

$$T = \tfrac{1}{2} I_e \dot{\delta}_e^{\,2} \tag{4.8,1}$$

[4] See J. L. Synge and B. A. Griffith, *Principles of Mechanics*, Sec. 12.3, McGraw-Hill Book Co., New York, 1942.

[5] It is assumed that the control valve of the jack is so arranged that the necessary hydraulic pressure is automatically provided.

where I_e is the "effective" moment of inertia of the elevator, and may be constant or vary with δ_e. In the following it is assumed that I_e is a constant. The left-hand side of Eq. 4.7,1 then becomes simply $I_e\ddot{\delta}_e$.

The Generalized Force $\mathscr{F}$

In order to determine the generalized force $\mathscr{F}$ we must compute the work done by the external force system (including the inertia forces) during a virtual displacement $\delta(\delta_e)$ of the system. Let this work be expressed as

$$\delta W = H_e\,\delta(\delta_e) + F_e\,\delta(\delta_e) + \delta W_i \tag{4.8,2}$$

where F_e = generalized elevator control force

δW_i = work done by the inertia forces

The quantity F_e is related to the manual and power control forces as follows. Let the displacements of P and J be δs_P and δs_J, respectively. Then the work done by the control forces is

$$F_e\,\delta(\delta_e) = P\,\delta s_P + J\,\delta s_J \tag{4.8,3}$$

or

$$F_e = P\frac{ds_P}{d\delta_e} + J\frac{ds_J}{d\delta_e} \tag{4.8,4}$$

Equation 4.8,4 shows that F_e can be computed from a knowledge of P, J, and the "gearings" of the stick and jack relative to the elevator (c.f. Sec. 2.9).

Evaluation of δW_i. We come now to the evaluation of the term δW_i which arises from the acceleration and rotation of the airplane. The corresponding terms that appear in the equation of motion represent inertia couplings between motion of the airplane and motion of the control system. Such couplings are usually avoided as undesirable, although an exception to this was treated in Sec. 3.3; i.e., the bob weight. Let the acceleration of an element of mass of the system relative to inertial axes be **a**, and relative to the airplane body axes be **a**$'$. Then the acceleration due to the motion of the airplane is (**a** − **a**$'$). The relevant inertia force on the mass element dm is therefore

$$d\mathbf{F}_i = -(\mathbf{a} - \mathbf{a}')\,dm \tag{4.8,5}$$

From Appendix A.7, the scalar components of $d\mathbf{F}_i$ are found to be

$$\begin{aligned}
dF_{x_i} &= -[a_{Cx} + 2Q\dot{z} - 2R\dot{y} - x(Q^2 + R^2) + y(PQ - \dot{R}) \\
&\quad + z(PR + \dot{Q})]\,dm \\
dF_{y_i} &= -[a_{Cy} + 2R\dot{x} - 2P\dot{z} + x(\dot{R} + PQ) - y(P^2 + R^2) \\
&\quad + z(RQ - \dot{P})]\,dm \\
dF_{z_i} &= -[a_{Cz} + 2P\dot{y} - 2Q\dot{x} + x(PR - \dot{Q}) + y(RQ + \dot{P}) \\
&\quad - z(P^2 + Q^2)]\,dm
\end{aligned} \tag{4.8,6}$$

Now the terms containing $\dot{x}$, $\dot{y}$, and $\dot{z}$ are the components of the Coriolis force, $-2\boldsymbol{\omega} \times \mathbf{v}'$. This force is perpendicular to the relative velocity $\mathbf{v}'$, and hence perpendicular to the virtual displacement as well. It therefore does no work during the displacement, and the terms referred to may be dropped in the integration that follows.

A complete and exact calculation of δW_i would entail the computation of the work done by the force field of Eqs. 4.8,6 on every component of the elevator system during the virtual displacement $\delta(\delta_e)$. This can in principle be performed once the geometry is given, but is obviously a lengthy and involved process for any but the simplest of systems. We shall assume that we may neglect the contributions of all elements save that of the elevator itself, which we shall approximate as a lamina lying in the xy plane.

The relevant geometry is shown in Fig. 4.5*a*. The displacement of the element dm is in the direction of Cz, and of magnitude $r\,\delta(\delta_e)$. Hence the work done by the inertia forces is

$$\delta W_i = \int r\,\delta(\delta_e)\,dF_{z_i} \tag{4.8,7}$$

The value of dF_{z_i} is that given by Eq. 4.8,6, and the integration is to be carried out over both right and left elevators. Bearing in mind the remarks made above about

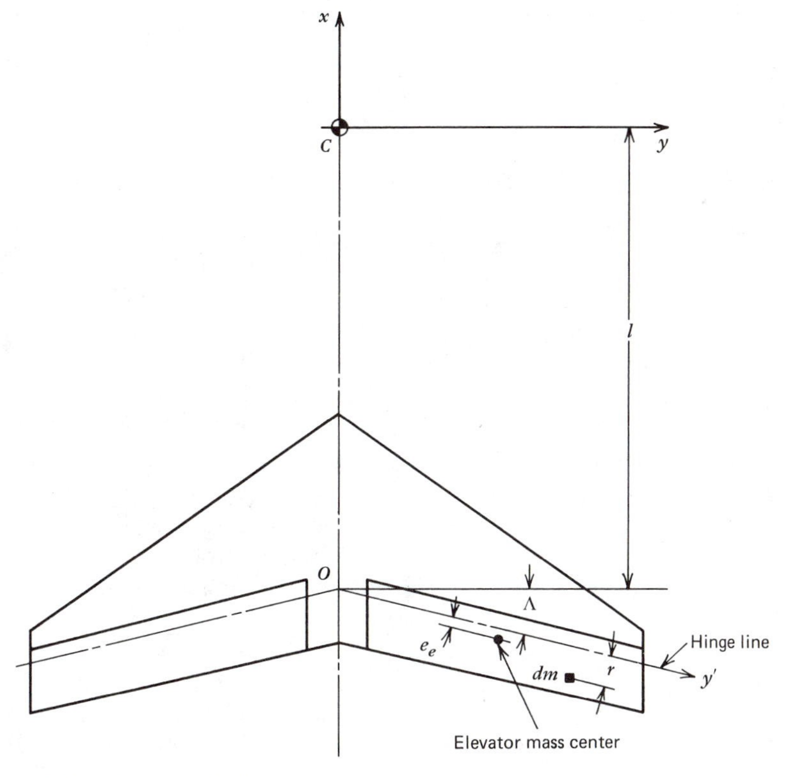

FIG. 4.5*a* Horizontal tail.

the Coriolis force, and that $z = 0$, we get from Eq. 4.8,7 that

$$\delta W_i = -\delta(\delta_e) \int [a_{Cz} + x(PR - \dot{Q}) + y(RQ + \dot{P})] r\, dm$$

or

$$\frac{\delta W_i}{\delta(\delta_e)} = -a_{Cz} \int r\, dm - (PR - \dot{Q}) \int xr\, dm - (RQ + \dot{P}) \int yr\, dm \tag{4.8,8}$$

From the definition of mass center, the first integral is

$$\int r\, dm = m_e \cdot e_e \tag{4.8,9}$$

where m_e = mass of both elevators

e_e = eccentricity of the elevator mass center (see Fig. 4.5)

The second integral in Eq. 4.8,8 is the product of inertia of the elevator with respect to the hinge line and the y axis. It is denoted

$$\int rx\, dm = P_{ex} \tag{4.8,10}$$

The third integral of Eq. 4.8,8 is zero by virtue of the symmetry about the x axis. Thus we get

$$\frac{\partial W_i}{\partial \delta_e} = \frac{\delta W_i}{\delta(\delta_e)} = -m_e e_e a_{Cz} - P_{ex}(PR - \dot{Q}) \tag{4.8,11}$$

Equation of Motion

Then the generalized force becomes, from Eq. 4.8,2,

$$\mathscr{F} = \frac{\partial W}{\partial \delta_e} = H_e + F_e - m_e e_e a_{Cz} - P_{ex}(PR - \dot{Q}) \tag{4.8,12}$$

Upon combining this result with Eqs. 4.7,1 and 4.8,1, we obtain the equation of motion of the elevator:

$$I_e \ddot{\delta}_e + m_e e_e a_{Cz} + P_{ex}(PR - \dot{Q}) = H_e + F_e \tag{4.8,13}$$

This equation of motion has been derived in a rather formal manner. Figure 4.5*b* provides a physical interpretation of it for a one-piece elevator with a straight hinge line. If the system of forces shown is supposed to be in statical equilibrium, then the given equation results. F_e is seen to be the control hinge moment, and H_e the aerodynamic hinge moment. $I_e \ddot{\delta}_e$ is the inertia moment associated with the angular acceleration $\ddot{\delta}_e$ about the hinge. $m_e a_{Cz}$ is the inertia force acting through the elevator mass center associated with the acceleration a_{Cz} of the airplane C.G. Finally, on each element dm there acts an inertia force as shown caused by the angular velocity and angular acceleration of the airplane. The integrated moment of these elementary inertia forces provides the remaining term $P_{ex}(PR - \dot{Q})$ of the equation.

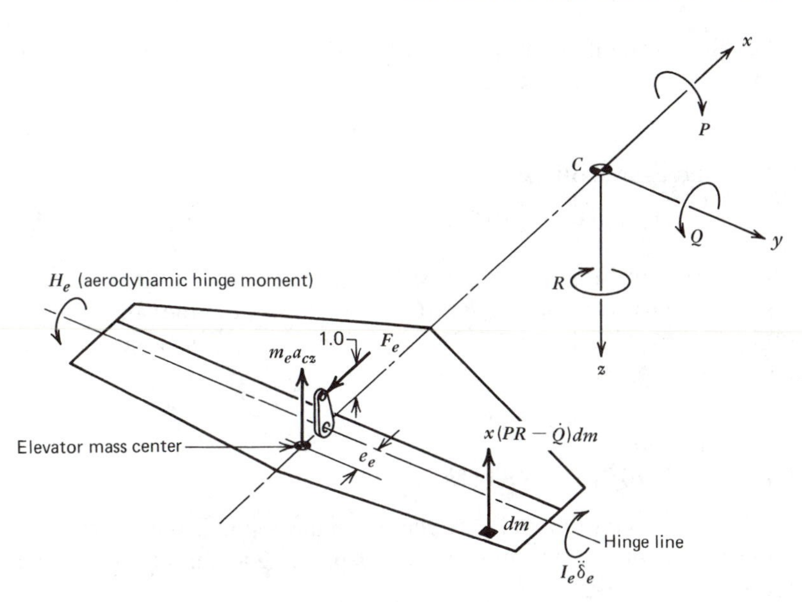

FIG. 4.5*b* An interpretation of the elevator equation of motion.

Dynamic Balance of the Elevator

The preceding equation reveals the requirements for dynamic balance of the elevator.[6] It shows that, if the eccentricity of the elevator C.G., e_e, is not zero, then acceleration of the airplane in the z direction (as in a pull-up, or banked turn) will tend to induce motion of the elevator. If the product of inertia P_{ex} is not zero, then a roll–yaw combination PR, or a pitching acceleration $\dot{Q}$ will tend to induce a response of the elevator. These effects are known as *inertial coupling* of the elevator with the relevant airplane degrees of freedom. When the inertial coupling is entirely eliminated, then the elevator is said to be *dynamically balanced.* Thus we find that for dynamic balance e_e and P_{ex} must both be zero. We must now inquire as to the possibility of satisfying these two conditions simultaneously. Consider first the case of a hinge line that is unswept; i.e., parallel to Cy. Then $x = -(l + r)$, where l is the distance shown on Fig. 4.5*a*, and

$$\begin{aligned} P_{ex} &= \int rx\,dm = -l\int r\,dm - \int r^2\,dm \\ &= -lm_e e_e - \int r^2\,dm \end{aligned} \tag{4.8,14}$$

It is clear that e_e and P_{ex} can both be zero only if $\int r^2\,dm$ is zero. The latter is, however, the moment of inertia of the elevator about its hinge, and cannot be zero. In this case, then, the elevator cannot be completely balanced.

[6] From the stability point of view, that is. The balancing requirements from the flutter point of view are similar, but not exactly the same.

When the hinge line is swept back at an angle Λ (Fig. 4.5*a*), then we find for the product of inertia the formula

$$P_{ex} = -lm_e e_e - \sin\Lambda \int r|y'|\,dm - \cos\Lambda \int r^2\,dm \tag{4.8,15}$$

Now, if P_{ex} and e_e are both zero, we have

$$\sin\Lambda \int r|y'|\,dm + \cos\Lambda \int r^2\,dm = 0 \tag{4.8,16}$$

It is theoretically possible, although perhaps practically difficult, to add balance weights in such a manner that Eq. 4.8,16 is satisfied while at the same time meeting the condition on e_e; i.e., $\int r\,dm = 0$. Hence an elevator with a swept hinge line can in principle be completely balanced.

4.9 THE RUDDER SYSTEM

The equation of motion of the rudder system is found in a manner completely analogous to that of Sec. 4.8, and with corresponding assumptions. The equation analogous to Eq. 4.8,1 is

$$T = \tfrac{1}{2} I_r \dot{\delta}_r^{\,2} \tag{4.9,1}$$

where I_r = effective moment of inertia of the rudder system

δ_r = rudder angle

The equation analogous to Eq. 4.8,2 is

$$\delta W = H_r \delta(\delta_r) + F_r \delta(\delta_r) + \delta W_i \tag{4.9,2}$$

where F_r = generalized rudder control force and is found in the same manner as F_e. The rudder is assumed to be a lamina in the xz plane, and hence the equation corresponding to Eq. 4.8,7 is

$$\delta W_i = -\int r\,\delta(\delta_r)\,dF_{y_i} \tag{4.9,3}$$

Upon applying Eq. 4.8,6, for dF_{y_i}, we get

$$\frac{\delta W_i}{\delta(\delta_r)} = a_{Cy} \int r\,dm + (\dot{R} + PQ) \int rx\,dm + (RQ - \dot{P}) \int rz\,dm$$

Hence the generalized force is

$$\mathscr{F} = H_r + F_r + a_{Cy} m_r e_r + (\dot{R} + PQ) P_{rx} + (RQ - \dot{P}) P_{rz}$$

where m_r = mass of the rudder

e_r = C.G. eccentricity of the rudder (cf. Fig. 4.5)

$P_{rx} = \int rx\,dm$ = product of inertia of the rudder with respect to its hinge line and the z axis

$P_{rz} = \int rz\,dm$ = product of inertia of the rudder with respect to its hinge line and the x axis

The equation of motion is then

$$I_r\ddot{\delta}_r - a_{C_y}m_r e_r - (\dot{R} + PQ)P_{rx} - (RQ - \dot{P})P_{rz} = H_r + F_r \tag{4.9,4}$$

4.10 THE AILERON SYSTEM

The aileron system differs from the other two controls in that two aerodynamic surfaces are involved. Although these surfaces are sometimes arranged so that the upgoing aileron moves through a larger angle than the downgoing one (aileron differential), this complication is omitted from the treatment given here. The deflection of the right-hand aileron is given by δ_a, positive downward, and the left-hand aileron by δ_a, positive upward. δ_a is taken to be the generalized coordinate of the system.

The equation for the kinetic energy is

$$T = \tfrac{1}{2}I_a\dot{\delta}_a^{\,2} \tag{4.10,1}$$

where I_a = effective moment of inertia of the whole aileron system (including both surfaces)

The expression for the work done in a virtual displacement $\delta(\delta_a)$ is

$$\delta W = 2H_a\,\delta(\delta_a) + F_a\,\delta(\delta_a) + \delta W_i \tag{4.10,2}$$

where H_a = the hinge moment on *one* aileron and

F_a = the generalized aileron control force

found in the same manner as F_e. The ailerons are assumed to be laminae lying in the xy plane, and hence the expression for δW_i is similar to Eq. 4.8,7, except that we must differentiate between the right- and left-hand ailerons.

$$\delta W_i = \int_{\text{right}} r\,\delta(\delta_a)\,dF_{z_i} - \int_{\text{left}} r\,\delta(\delta_a)\,dF_{z_i} \tag{4.10,3}$$

We substitute for dF_{z_i} from Eq. 4.8,6, and integrate to obtain

$$\begin{aligned}\frac{\delta W_i}{\delta(\delta_a)} = &-a_{Cz}\left\{\int_{\text{right}} r\,dm - \int_{\text{left}} r\,dm\right\}\\ &-(PR - \dot{Q})\left\{\int_{\text{right}} rx\,dm - \int_{\text{left}} rx\,dm\right\}\\ &-(RQ + \dot{P})\left\{\int_{\text{right}} ry\,dm - \int_{\text{left}} ry\,dm\right\}\end{aligned} \tag{4.10,4}$$

Because of symmetry the first two terms of Eq. 4.10,4 are zero, and

$$P_{ay} = \int_{\text{right}} ry\,dm = -\int_{\text{left}} ry\,dm \tag{4.10,5}$$

P_{ay} is the product of inertia of the right aileron with respect to its hinge line and the x axis. Then Eq. 4.10,4 becomes

$$\frac{\delta W_i}{\delta(\delta_a)} = -2P_{ay}(RQ + \dot{P}) \tag{4.10,6}$$

The generalized force is therefore

$$\mathscr{F} = 2H_a + F_a - 2P_{ay}(RQ + \dot{P}) \tag{4.10,7}$$

and the equation of motion is

$$I_a\ddot{\delta}_a + 2P_{ay}(RQ + \dot{P}) = 2H_a + F_a \tag{4.10,8}$$

4.11 THE APPLIED FORCES

The equations of motion derived in the foregoing sections contain the following forces and moments:

1. The external force acting on the airplane (F_x, F_y, F_z).
2. The external couple acting on the airplane (L, M, N).
3. The control surface hinge moments (H_e, H_r, H_a).
4. The generalized control forces (F_e, F_r, F_a).

The forces and moments in 1 to 3 above are the external actions upon the airplane. They are of two kinds, gravitational and aerodynamic, and are in general determined by the orientation, configuration, and motion of the airplane: i.e., by the variables (Θ, Φ, Ψ), (δ_e, δ_a, δ_r), (U, V, W), and (P, Q, R). The generalized control forces, on the other hand, are internal forces, and are arbitrary functions of time determined by the control commands of a human or automatic pilot.

The External Forces

From Fig. 4.6 the components of the weight in the directions of the axes are found to be

$$\begin{aligned} X_g &= -mg \sin \Theta \\ Y_g &= mg \cos \Theta \sin \Phi \\ Z_g &= mg \cos \Theta \cos \Phi \end{aligned} \tag{4.11,1}$$

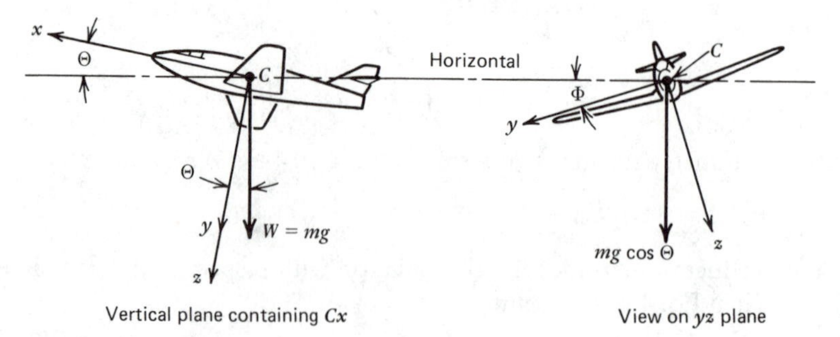

FIG. 4.6 The gravity force.

With the aerodynamic forces (including propulsive forces) denoted by (X, Y, Z), the resultant external forces are

$$
\begin{aligned}
F_x &= X - mg \sin \Theta \\
F_y &= Y + mg \cos \Theta \sin \Phi \qquad (4.11,2) \\
F_z &= Z + mg \cos \Theta \cos \Phi
\end{aligned}
$$

Since gravity does not contribute to the moments about the mass center, then (L, M, N) are entirely aerodynamic. The effect of mass unbalance of the control surfaces in producing a gravitational hinge moment will be assumed small and neglected, so that the hinge moments (H_e, H_r, H_a) are also purely aerodynamic.

4.12 THE EQUATIONS COLLECTED

The kinematical and dynamical equations derived in the foregoing are collected below for convenience. The assumption that Cxz is a plane of symmetry is used, so that $D = F = 0$, and Eqs. 4.3,4 are substituted into Eqs. 4.4,4 to give Eqs. 4.12,2.

$$
\begin{aligned}
X - mg \sin \Theta &= m(\dot{U} + QW - RV) && (a) \\
Y + mg \cos \Theta \sin \Phi &= m(\dot{V} + RU - PW) && (b) \quad (4.12,1) \\
Z + mg \cos \Theta \cos \Phi &= m(\dot{W} + PV - QU) && (c)
\end{aligned}
$$

$$
\begin{aligned}
L &= A\dot{P} - E\dot{R} + QR(C - B) - EPQ + Qh'_z - Rh'_y && (a) \\
M &= B\dot{Q} + RP(A - C) + E(P^2 - R^2) + Rh'_x - Ph'_z && (b) \quad (4.12,2) \\
N &= -E\dot{P} + C\dot{R} + PQ(B - A) + EQR + Ph'_y - Qh'_x && (c)
\end{aligned}
$$

$$
\begin{aligned}
P &= \dot{\Phi} - \dot{\Psi} \sin \Theta && (a) \\
Q &= \dot{\Theta} \cos \Phi + \dot{\Psi} \cos \Theta \sin \Phi && (b) \\
R &= \dot{\Psi} \cos \Theta \cos \Phi - \dot{\Theta} \sin \Phi && (c) \\
\dot{\Theta} &= Q \cos \Phi - R \sin \Phi && (d) \quad (4.12,3) \\
\dot{\Phi} &= P + Q \sin \Phi \tan \Theta + R \cos \Phi \tan \Theta && (e) \\
\dot{\Psi} &= (Q \sin \Phi + R \cos \Phi) \sec \Theta && (f)
\end{aligned}
$$

$$
\begin{aligned}
\frac{dx'}{dt} &= U \cos \Theta \cos \Psi + V(\sin \Phi \sin \Theta \cos \Psi - \cos \Phi \sin \Psi) && (a) \\
&\quad + W(\cos \Phi \sin \Theta \cos \Psi + \sin \Phi \sin \Psi) \\
\frac{dy'}{dt} &= U \cos \Theta \sin \Psi + V(\sin \Phi \sin \Theta \sin \Psi + \cos \Phi \cos \Psi) && (b) \quad (4.12,4) \\
&\quad + W(\cos \Phi \sin \Theta \sin \Psi - \sin \Phi \cos \Psi) \\
\frac{dz'}{dt} &= -U \sin \Theta + V \sin \Phi \cos \Theta + W \cos \Phi \cos \Theta && (c)
\end{aligned}
$$

$$
\begin{aligned}
H_e + F_e &= I_e\ddot{\delta}_e + m_e e_e a_{Cz} + P_{ex}(PR - \dot{Q}) && (a) \\
H_r + F_r &= I_r\ddot{\delta}_r - m_r e_r a_{Cy} - P_{rx}(PQ + \dot{R}) - P_{rz}(RQ - \dot{P}) && (b) \quad (4.12,5) \\
2H_a + F_a &= I_a\ddot{\delta}_a + 2P_{ay}(RQ + \dot{P}) && (c)
\end{aligned}
$$

The above equations are quite general, and, except for Eqs. 4.12,5, contain few assumptions. These are:

For Eqs. 4.12,1 to 4.12,4

1. The airplane is a rigid body, which may have attached to it any number of rigid spinning rotors.
2. Cxz is a plane of mirror symmetry.
3. The axes of any spinning rotors are fixed in direction relative to the body axes, and the rotors have constant angular speed relative to the body axes.

For Eqs. 4.12,5

1. Each control system consists of a linkage of rigid elements, attached to a rigid airplane.
2. Each control system has one degree of freedom relative to the body axes.
3. The control systems are frictionless.
4. For the purpose of calculating the inertial coupling, the control surfaces are approximated as laminae lying in the coordinate planes.
5. The contributions to the inertial coupling of all portions of the control systems other than the aerodynamic surfaces is neglected.
6. The elevator and ailerons are symmetrical with respect to the xz plane.
7. Hinge moments due to gravity are neglected.

4.13 DISCUSSION OF THE EQUATIONS

The equations of Sec. 4.12 are many in number and complex in nature. They can be identified as ordinary nonlinear differential equations in many variables, with the time t as independent variable. It is important to an understanding of the classes of problems that may arise to consider the number of dependent variables in relation to the number of equations.

Consider first Eqs. 4.12,4. The C.G. coordinates (x', y', z') that appear in them do not occur elsewhere. Hence we may, for the time being, drop these variables and these three equations from consideration. Furthermore, as remarked previously in Sec. 4.11, the nine aerodynamic forces and moments (X, Y, Z), (L, M, N), (H_e, H_r, H_a) are functions of the motion and configuration variables (U, V, W), (P, Q, R), $(\delta_e, \delta_r, \delta_a)$. This functional dependence reduces the number of independent variables by nine. Finally, it must be observed that only three of Eqs. 4.12,3 are independent.

The situation is then that we have *twelve* independent equations, Eqs. 4.12,1(a), (b),(c); Eqs. 4.12,2(a),(b),(c); Eqs. 4.12,3(a),(b),(c); and Eqs. 4.12,5(a),(b),(c), containing *fifteen* variables (U, V, W), (P, Q, R), (Θ, Φ, Ψ), $(\delta_e, \delta_r, \delta_a)$, (F_e, F_r, F_a). It is clear that three of the variables must somehow be specified in advance if the mathematical system is to be complete. The three actually selected are determined by the engineering problem to be solved, and the nature of the mathematical problem that ensues is strongly influenced by this choice. A discussion of the various possibilities follows.

Stability Problems—Controls Fixed

In these problems the airplane is considered to be disturbed from an initially steady flight condition, with the controls locked in position. The three control-system equations, Eqs. 4.12,5, are then dropped, as are the variables (F_e, F_r, F_a); the control-surface angles $(\delta_e, \delta_r, \delta_a)$ are constants. The result is a set of nine equations in nine variables. The equations are nonlinear and, consequently, extremely difficult to deal with analytically. In many problems of practical importance, it is satisfactory to linearize the equations by dealing only with small perturbations from the reference condition. In that case we obtain a set of homogeneous linear differential equations with constant coefficients, a type that is readily solved. Problems of this class are treated in Chaps. 6 and 7.

Stability Problems—Controls Free

These problems are derived from the fixed-control case by adding the three control-system equations, and the control angles $(\delta_e, \delta_r, \delta_a)$, while the forcing functions (F_e, F_r, F_a) are set equal to zero. The equations are again homogeneous and are treated in the same manner as those governing the fixed-control cases. Problems of this class also are treated in Chaps. 6 and 7.

Stability Problems—Automatic Controls

When the airplane has an automatic-control system, the controls are neither fixed nor free. Even though the pilot input (P_c, P_r, P_a) is zero, the generalized control forces are not. They are determined in general by a feedback of the airplane response into the input of the control system. All the twelve equations being discussed are in general involved, and the dynamics of the closed loop must be introduced. Problems of this class are treated in Chap. 10

Response to Controls

The effectiveness of an airplane's controls is conventionally studied by specifying the variation of $(\delta_e, \delta_r, \delta_a)$ with time arbitrarily, e.g., a step-function input of aileron angle. The airplane equations of motion then become inhomogeneous equations for (U, V, W), (P, Q, R), (Θ, Φ, Ψ) with the control angles as forcing functions. The three control-system equations are not required, unless it is desired to use them to compute the variation of the control force (F_e, F_r, F_a) consistent with the maneuver. Problems of this type are treated in Chapter 9.

Response to Atmospheric Turbulence

The response of an airplane to rough air is treated with the same equations as are used for stability. The nonuniform motion of the atmosphere, however, brings about the introduction of new forcing functions. The mathematical treatment then parallels that used for the response to controls. This case is treated at length in Ref. 1.10

Inverse Problems

A class of problems that has not received very much attention is that in which the specified variables are selected from among those that are normally treated as dependent: i.e., (U, V, W), (P, Q, R), (Θ, Φ, Ψ). This is the kind of problem that arises when we seek answers to questions of the type "Given the airplane motion, what pilot action is required to produce it?" Questions of this kind are frequently relevant to problems of control design and maneuvering loads.

The mathematical problem that results is generally simpler than those of stability and control. The equations to be solved are sometimes algebraic, sometimes differential. A decided advantage is the ability of this method to cope with the nonlinear equations of large disturbances.

4.14 THE SMALL-DISTURBANCE THEORY

As remarked in Sec. 4.13, the equations of motion are frequently linearized for use in stability and control analysis. It is assumed that the motion of the airplane consists of small deviations from a reference condition of steady flight. The use of the small-disturbance theory has been found in practice to give good results. It predicts with satisfactory precision the stability of unaccelerated flight, and it can be used, with sufficient accuracy for engineering purposes, for response calculations where the disturbances are not infinitesimal. There are, of course, limitations to the theory. It is not suitable for discussions of spinning or stalled flight, nor for solutions of problems in which large disturbances angles occur, for example $\Phi \doteqdot \pi/2$.

The reasons for the success of the method are twofold: (1) In many cases, the major aerodynamic effects are nearly linear functions of the disturbances, and (2) disturbed flight of considerable violence can occur with quite small values of the linear- and angular-velocity disturbances.

Notation for Small Disturbances

The reference values of all the variables are denoted by a subscript zero, and the small perturbations are indicated as follows:

1. Changes in (U, V, W), (P, Q, R), (Θ, Φ, Ψ) are denoted by lowercase letters; i.e., $U = u_0 + u$, $P = p_0 + p$, $\Theta = \theta_0 + \theta$ etc.
2. Changes in (X, Y, Z), (L, M, N), (H_e, H_r, H_a), (F_e, F_r, F_a) are denoted by a prefix Δ; i.e., $X = X_0 + \Delta X$, $L = L_0 + \Delta L$, etc.
3. Changes in the control-surface angles are denoted by (ξ, η, ζ), so that $\delta_a = \delta_{a_0} + \xi$, $\delta_e = \delta_{e_0} + \eta$, $\delta_r = \delta_{r_0} + \zeta$.

All the disturbance quantities and their derivatives are assumed to be small, so that their squares and products are negligible compared to first-order quantities.

The reference flight condition is assumed for convenience to be symmetric and with no angular velocity. Thus $v_0 = p_0 = q_0 = r_0 = \phi_0 = 0$. Furthermore, stability axes are selected as standard in this book, and thus $w_0 = 0$ for all the problems considered. u_0 is then equal to the reference flight speed, and θ_0 to the reference angle of climb (not assumed to be small). In dealing with the trigonometric functions in the equations the following relations are used:

$$\begin{aligned} \sin(\theta_0 + \theta) &= \sin\theta_0 \cos\theta + \cos\theta_0 \sin\theta \\ &\doteq \sin\theta_0 + \theta\cos\theta_0 \\ \cos(\theta_0 + \theta) &= \cos\theta_0 \cos\theta - \sin\theta_0 \sin\theta \\ &\doteq \cos\theta_0 - \theta\sin\theta_0 \end{aligned} \tag{4.14,1}$$

Further Assumptions

The small-disturbance equations will be slightly restricted by the adoption of two more assumptions, which correspond to current practice. These are

1. The effects of spinning rotors are negligible. This is the case when the airplane is in gliding flight with power off, when the symmetrical engines have opposite rotation, or when the rotor angular momentum is small.
2. All controls are dynamically balanced. This is almost invariably the case, since the controls must be closely balanced to avoid flutter difficulties.

The Linear Equations

When the small-disturbance notation is introduced into the equations of Sec. 4.12, the additional assumptions noted above are incorporated, and only the first-order terms in disturbance quantities are kept, then the following linear equations are obtained.

$$\begin{aligned} X_0 + \Delta X - mg(\sin\theta_0 + \theta\cos\theta_0) &= m\dot{u} & (a) \\ Y_0 + \Delta Y + mg\phi\cos\theta_0 &= m(\dot{v} + u_0 r) & (b) \\ Z_0 + \Delta Z + mg(\cos\theta_0 - \theta\sin\theta_0) &= m(\dot{w} - u_0 q) & (c) \end{aligned} \tag{4.14,2}$$

$$\begin{aligned} L_0 + \Delta L &= A\dot{p} - E\dot{r} & (a) \\ M_0 + \Delta M &= B\dot{q} & (b) \\ N_0 + \Delta N &= -E\dot{p} + C\dot{r} & (c) \end{aligned} \tag{4.14,3}$$

$$\begin{aligned} H_{e_0} + \Delta H_e + F_{e_0} + \Delta F_e &= I_e\ddot{\eta} & (a) \\ H_{r_0} + \Delta H_r + F_{r_0} + \Delta F_r &= I_r\ddot{\zeta} & (b) \\ 2H_{a_0} + 2\Delta H_a + F_{a_0} + \Delta F_a &= I_a\ddot{\xi} & (c) \end{aligned} \tag{4.14,4}$$

$$\begin{aligned} &\dot{\theta} = q & (a) \\ &\dot{\phi} = p + r\tan\theta_0, \qquad p = \dot{\phi} - \dot{\psi}\sin\theta_0 & (b) \\ &\dot{\psi} = r\sec\theta_0 & (c) \end{aligned} \tag{4.14,5}$$

$$\frac{dx'}{dt} = (u_0 + u)\cos\theta_0 - u_0\theta\sin\theta_0 + w\sin\theta_0 \qquad (a)$$

$$\frac{dy'}{dt} = u_0\psi\cos\theta_0 + v \qquad (b) \qquad (4.14,6)$$

$$\frac{dz'}{dt} = -(u_0 + u)\sin\theta_0 - u_0\theta\cos\theta_0 + w\cos\theta_0 \qquad (c)$$

Reference Steady State

If all the disturbance quantities are set equal to zero in the foregoing equations, then they apply to the reference flight condition. When this is done, we get the following relations, which may be used to eliminate all the initial forces and moments from the equations:

$$\begin{aligned} X_0 - mg\sin\theta_0 &= 0 \\ Y_0 &= 0 \\ Z_0 + mg\cos\theta_0 &= 0 \\ L_0 = M_0 = N_0 &= 0 \\ H_{e_0} + F_{e_0} &= 0 \\ H_{r_0} + F_{r_0} &= 0 \\ 2H_{a_0} + F_{a_0} &= 0 \end{aligned} \qquad (4.14,7)$$

$$\left(\frac{dx'}{dt}\right)_0 = u_0\cos\theta_0, \qquad \left(\frac{dy'}{dt}\right)_0 = 0, \qquad \left(\frac{dz'}{dt}\right)_0 = -u_0\sin\theta_0$$

The Linear Air Reactions

The method of treating the aerodynamic force and moment perturbations given below is essentially that introduced by Bryan (refs. 1.2 and 1.3). It leads to a form for the equations of motion that has been used with great success.

Despite this success, however, the method is not (mathematically speaking) sound. Indeed it does not give correct answers in certain cases where the aerodynamic forces change very rapidly, as in the penetration of a sharp-edged gust at high speed, or when a control is very rapidly displaced. An alternative method has been proposed that is not subject to these objections (ref. 4.2). This method is explained in Sec. 4.16, and it is applied to the formulation of an alternative set of small-disturbance equations.

The method of Bryan is based on the assumption that the aerodynamic forces and moments are functions of the instantaneous values of the disturbance velocities, control angles, and their derivatives. Expressions for the forces and moments are obtained in the form of a Taylor series in these variables, which is linearized by discarding all the higher-order terms. Thus, if A represents a typical aerodynamic reaction, then

$$\Delta A = A_u u + A_{\dot{u}} \dot{u} + \cdots + A_\zeta \zeta + A_{\ddot{\zeta}} \ddot{\zeta} \tag{4.14,8}$$

where

$$A_u = \left(\frac{\partial A}{\partial u} \right)_0$$

The subscript zero indicates that the derivative is to be evaluated at the reference flight condition. The derivatives such as A_u are called the *stability derivatives* of the airplane. There has now been amassed a great volume of information about these parameters for a wide variety of configurations and flight conditions. Chapter 5 is devoted to a discussion of them.

For a truly symmetric airplane, it is evident that the side force Y, the rolling moment L, the yawing moment N, and the rudder and aileron hinge moments H_r and H_a will all be identically zero in any condition of symmetric flight. Thus the derivatives of the *asymmetric* forces and moments with respect to the *symmetric* variables u, w, q, η are all zero. In writing out the complete linear expressions for the forces and moments, we use this fact, and in addition make these further approximations:

1. We may neglect the derivatives of the symmetric forces and moments X, Z, M, H_e with respect to the asymmetric variables v, p, r, ξ, ζ.
2. We may neglect all derivatives with respect to acceleration quantities, except for $M_{\dot{w}}$, $Z_{\dot{w}}$ and $H_{e_{\dot{w}}}$.
3. The following derivatives are negligibly small: X_q, X_η, $X_{\dot{\eta}}$, $Z_{\dot{\eta}}$, Y_ξ, $Y_{\dot{\xi}}$, $Y_{\dot{\zeta}}$, $L_{\dot{\zeta}}$, $N_{\dot{\xi}}$, H_{r_ξ}, $H_{r_{\dot{\xi}}}$, H_{a_v}, H_{a_ζ}, $H_{a_{\dot{\zeta}}}$.

It should be emphasized that none of these assumptions is basically necessary for the solution of airplane dynamics problems. They are made as a matter of experience and convenience. When it appears necessary to do so, any of the terms dropped can be restored into the equations. With these assumptions, however, the linear forces and moments are:

$$\begin{aligned} \Delta X &= X_u u + X_w w \\ \Delta Y &= Y_v v + Y_p p + Y_r r + Y_\zeta \zeta \\ \Delta Z &= Z_u u + Z_w w + Z_{\dot{w}} \dot{w} + Z_q q + Z_\eta \eta \end{aligned} \tag{4.14,9}$$

$$\begin{aligned} \Delta L &= L_v v + L_p p + L_r r + L_\xi \xi + L_{\dot{\xi}} \dot{\xi} + L_\zeta \zeta \\ \Delta M &= M_u u + M_w w + M_{\dot{w}} \dot{w} + M_q q + M_\eta \eta + M_{\dot{\eta}} \dot{\eta} \\ \Delta N &= N_v v + N_p p + N_r r + N_\xi \xi + N_\zeta \zeta + N_{\dot{\zeta}} \dot{\zeta} \end{aligned} \tag{4.14,10}$$

$$\begin{aligned} \Delta H_e &= H_{e_u} u + H_{e_w} w + H_{e_{\dot{w}}} \dot{w} + H_{e_q} q + H_{e_\eta} \eta + H_{e_{\dot{\eta}}} \dot{\eta} \\ \Delta H_r &= H_{r_v} v + H_{r_p} p + H_{r_r} r + H_{r_\zeta} \zeta + H_{r_{\dot{\zeta}}} \dot{\zeta} \\ \Delta H_a &= H_{a_p} p + H_{a_r} r + H_{a_\xi} \xi + H_{a_{\dot{\xi}}} \dot{\xi} \end{aligned} \tag{4.14,11}$$

When Eqs. 4.14,7,9,10 and 11 are substituted into the linear equations 4.14,2 to 4.14,4, these can be recast into Eqs. 4.14,12 and 4.14,13.

$$
\begin{aligned}
&\left(m\frac{d}{dt}-X_u\right)u - (X_w)w + (mg\cos\theta_0)\theta = 0\\
&(-Z_u)u + \left[(m-Z_{\dot{w}})\frac{d}{dt}-Z_w\right]w - \left[(mu_0+Z_q)\frac{d}{dt}-mg\sin\theta_0\right]\theta - (Z_\eta)\eta = 0\\
&(-M_u)u - \left(M_{\dot{w}}\frac{d}{dt}+M_w\right)w + \left(B\frac{d^2}{dt^2}-M_q\frac{d}{dt}\right)\theta - \left(M_{\dot{\eta}}\frac{d}{dt}+M_\eta\right)\eta = 0\\
&(-H_{e_u})u - \left(H_{e_{\dot{w}}}\frac{d}{dt}+H_{e_w}\right)w - \left(H_{e_q}\frac{d}{dt}\right)\theta + \left(I_e\frac{d^2}{dt^2}-H_{e_{\dot{\eta}}}\frac{d}{dt}-H_{e_\eta}\right)\eta = \Delta F_e\\
&q-\dot{\theta} = 0
\end{aligned}
\tag{4.14,12}
$$

$$
\begin{aligned}
&\left(m\frac{d}{dt}-Y_v\right)v - (Y_p)p + (mu_0-Y_r)r - (mg\cos\theta_0)\phi - (Y_\zeta)\zeta = 0\\
&(-L_v)v + \left(A\frac{d}{dt}-L_p\right)p - \left(E\frac{d}{dt}+L_r\right)r - \left(L_{\dot{\xi}}\frac{d}{dt}+L_{\xi}\right)\xi - (L_\zeta)\zeta = 0\\
&(-N_v)v - \left(E\frac{d}{dt}+N_p\right)p + \left(C\frac{d}{dt}-N_r\right)r - (N_\xi)\xi - \left(N_{\dot{\zeta}}\frac{d}{dt}+N_\zeta\right)\zeta = 0\\
&-(2H_{a_p})p - (2H_{a_r})r + \left(I_a\frac{d^2}{dt^2}-2H_{a_{\dot{\xi}}}\frac{d}{dt}-2H_{a_\xi}\right)\xi = \Delta F_a\\
&(-H_{r_v})v - (H_{r_p})p - (H_{r_r})r + \left(I_r\frac{d^2}{dt^2}-H_{r_{\dot{\zeta}}}\frac{d}{dt}-H_{r_\zeta}\right)\zeta = \Delta F_r\\
&p + r\tan\theta_0 - \dot{\phi} = 0\\
&r\sec\theta_0 - \dot{\psi} = 0
\end{aligned}
\tag{4.14,13}
$$

Separation into Two Groups

As a consequence of the simplifying assumptions made in their derivation, the preceding equations have the exceedingly useful property of splitting into two independent groups. Suppose that ϕ, v, p, r, ζ, ξ, ΔF_r and ΔF_a are identically zero. Then Eqs. 4.14,13 are all identically satisfied. The remaining five equations (4.14,12) form a complete set for the five homogeneous variables u, w, q, θ, η. Thus we may conclude that modes of motion are possible in which *only* these five variables differ from zero. Such motions are called *longitudinal* or *symmetric*, and the corresponding equations and variables are likewise named. Conversely, if the longitudinal variables are set equal to zero, the remaining seven equations (4.14,13) form a complete set for the determination of the variables ϕ, ψ, v, p, r, ξ, ζ. These are known as the *lateral* variables, the corresponding equations and motions being likewise named.

It is worth while recording here the specific assumptions upon which this separation depends. A study of the various steps that have led to the final equations reveals these facts—the existence of the pure longitudinal motions depends on only two assumptions:

1. The existence of a plane of symmetry.
2. The absence of rotor gyroscopic effects.

The existence of the pure lateral motions, however, depends on more restrictive approximations; namely

1. The linearization of the equations.
2. The absence of rotor gyroscopic effects.
3. The neglect of all aerodynamic cross-coupling (approximation 1 p. 109).

If the equations were not linearized, then there would be inertial cross-coupling between the longitudinal and lateral modes, as evidenced by terms such as mPV in Eq. 4.12,1*c* and $(A - C)RP$ in Eq. 4.12,2*b*. That is, motion in the lateral modes would induce longitudinal motion.

4.15 THE NONDIMENSIONAL SYSTEM

In most aerodynamics problems, the maximum of convenience and generality is achieved by the use of nondimensional coefficients to represent the aerodynamic parameters involved. In this way the major effects of speed, size, and air density are automatically accounted for. The equations of airplane dynamics are no exception. However, in addition to introducing nondimensional forms for aerodynamic forces, we shall find it desirable to express the mass of the airplane and the time scale of its motions in nondimensional form as well. The reason for this is made clear by the application of dimensional analysis.[7] Imagine a class of

[7] See, for example, Henry L. Langhaar, *Dimensional Analysis and Theory of Models*, Chap. 2, John Wiley & Sons, New York, 1951.

geometrically similar airplanes of various sizes and masses in steady unaccelerated flight at various heights and speeds. Suppose that one of these airplanes is subjected to a disturbance. After the disturbance, some typical nondimensional variable π varies with time. For example, π may be the angle of yaw, the load factor, or the helix angle in roll. Thus, for this one airplane, under one particular set of conditions we shall have

$$\pi = f(t) \tag{4.15,1}$$

Let it be assumed that this equation can be generalized to cover the whole class of airplanes, under all flight conditions. That is, we shall assume that π is a function not of t alone, but also of

$$u_0, \rho, m, l, g, \mathbf{M}, RN$$

where m is the airplane mass and l is a characteristic length. Instead of Eq. 4.15,1, then, we write

$$\pi = f(u_0, \rho, m, l, g, \mathbf{M}, RN, t) \tag{4.15,2}$$

Buckingham's π theorem tells us that, since there are nine quantities in Eq. 4.15,2 containing three fundamental dimensions, L, M, and T, then there are $9 - 3 = 6$ independent dimensionless combinations of the nine quantities. These six so-called π functions are to be regarded as the meaningful physical variables of the equation, instead of the original nine. By inspection, we can easily form the following six independent nondimensional combinations:

$$\pi, \mathbf{M}, RN, \frac{m}{\rho l^3}, \frac{u_0 t}{l}, \frac{u_0^2}{lg}$$

Following the π theorem, we write as the symbolic solution to our problem

$$\pi = f\left(\mathbf{M}, RN, \frac{m}{\rho l^3}, \frac{u_0 t}{l}, \frac{u_0^2}{lg}\right) \tag{4.15,3}$$

The effects of the six variables m, ρ, l, g, u_0, and t are thus seen to be compressed into the three combinations: $m/\rho l^3$, u_0^2/lg, and $u_0 t/l$. We replace l^3 by Sl, where S is a characteristic area, without changing its dimensions, and denote the resulting nondimensional quantity $m/\rho Sl$ by μ. The quantity l/u_0 has the dimensions of time and is denoted t^*. The quantity u_0^2/lg is the Froude number (FN). Equation 4.15,3 then becomes

$$\pi = f(\mathbf{M}, RN, FN, \mu, t/t^*) \tag{4.15,4}$$

The significance of Eq. 4.15,4 is that it shows π to be a function of only five variables, instead of the original eight. The result is of sufficient importance that it is customary to elaborate on it still further. Since μ is the ratio of the airplane mass to the mass of a volume Sl of air, it is called the *relative mass parameter* or *relative density parameter*. It is smallest at sea level and increases with altitude. t^* is called the unit of aerodynamic time, and $\hat{t} = t/t^*$ is the aerodynamic time, in *airsecs*. It is equal to the distance traveled in units of l. The symbol D is used for derivatives with respect to $\hat{t}$, i.e.,

$$D = \frac{d}{d\hat{t}} = t^* \frac{d}{dt} \tag{4.15,5}$$

The significance of the Froude number, as it applies to airplanes, is seen by combining it with μ. Thus $\mu/FN = mg/\rho S u_0^2 = C_{L_0}/2$, where C_{L_0} is the lift coefficient in the initial steady flight condition. Equation 4.15,4 may therefore alternatively be written

$$\pi = f(\mathbf{M}, RN, C_{L_0}, \mu, \hat{t}) \tag{4.15,6}$$

Complete dynamical similarity exists between different members of the class of geometrically similar aircraft when all the nondimensional parameters (the *class* parameters) on the right-hand side of Eq. 4.15,6 are the same for each member.

Nondimensional Systems

One of the difficulties that must be faced by the student of this subject is acquiring familiarity with the system of notation. The equations themselves become meaningful only as the symbols contained in them acquire individual "personalities," through an appreciation of their physical connotations. The acquisition of the required familiarity and appreciation is hampered by the unfortunate fact that there are a number of nondimensional systems in use. The differences among the various systems are centered about three main points:

1. The definition of the time constant (here t^*).
2. The choice of characteristic length l.
3. The definitions of the nondimensional stability derivatives.

The main systems in use are the British, as given by Duncan (ref. 1.9), and the NACA system.[8] There are variations in practice within the NACA system itself, as sometimes the time constant is l/u_0, and sometimes it is the same as that used in Britain; i.e., $\tau = m/\rho S u_0$. Further confusion is added by the use of certain symbols such as l_p with two different definitions. The NACA definition is $l_p = C_{l_p}/4K_x^2$ while the British is $l_p = C_{l_p}$, with C_{l_p} having the identical meaning in the two definitions. Under the circumstances there are only two things the author can do. The first is to warn the student (and indeed the practicing engineer) to exercise great care when reading the literature, and to note in detail the definitions of the symbols used. The second is to present the system used herein as clearly as possible, and to give the reasons for choosing it.

The system adopted in this book (Table 4.1) uses the NACA stability derivatives, with time constant $t^* = l/u_0$. The characteristic length in the longitudinal equations is $l = \frac{1}{2}\bar{c}$, and in the lateral equations, $l = \frac{1}{2}b$. The main reasons for this choice are:

1. The bulk of the experimental and theoretical data on stability derivatives is published in the NACA notation.
2. The NACA stability derivatives are virtually self-explanatory, and do not require constant reference to a key.

[8] As used in NACA Reports, e.g., ref. 5.22.

TABLE 4.1
The Nondimensional System

(1) Dimensional Quantity	(2) Divisor[a]	(3) Nondimensional Quantity
$X \quad Y \quad Z$	$\frac{1}{2}\rho {v_c}^2 S$	$C_x \quad C_y \quad C_z$
$L \quad M \quad N$	$\rho {v_c}^2 S l$	$C_l \quad C_m \quad C_n$
$H_a \quad H_e \quad H_r$	$\frac{1}{2}\rho {v_c}^2 S_f {c_f}^b$	$C_{ha} \quad C_{he} \quad C_{hr}$
$F_a \quad F_e \quad F_r$	$\frac{1}{2}\rho {v_c}^2 S_f {c_f}^b$	$C_{fa} \quad C_{fe} \quad C_{fr}$
$u \quad v \quad w$	u_0	$\hat{u} \quad \beta \quad \alpha^c$
$p \quad q \quad r$	$1/t^*$	$\hat{p} \quad \hat{q} \quad \hat{r}^d$
$\dot{\beta} \quad \dot{\alpha}$	$1/t^*$	$D\beta \quad D\alpha$
$\dot{\xi} \quad \dot{\eta} \quad \dot{\zeta}$	$1/t^*$	$D\xi \quad D\eta \quad D\zeta^d$
m	$\rho S l$	μ
$A \quad B \quad C \quad E$	$\rho S l^3$	$i_A \quad i_B \quad i_C \quad i_E$
$I_a \quad I_e \quad I_r$	$\frac{1}{2}\rho S_f c_f l^2$	$i_a \quad i_e \quad i_r$
t	$t^* = l/u_0$	$\hat{t}$

[a] The items of col. 1 are divided by those of col. 2 to obtain col. 3.

[b] The subscript f denotes *flap control*. It may be a, e, or r for aileron, elevator, or rudder, respectively.

[c] Let α be the angle of attack of the axis Cx. Then $\alpha = \tan^{-1}(W/U) = \tan^{-1}[w/(u_0 + u)]$. To the first order in w and u this is equivalent to $\alpha = w/u_0$. Similarly the sideslip angle is $\beta = \sin^{-1}(v/v_c)$, or, to the first order, $\beta = v/u_0$. (In Chaps. 2 and 3, α denotes the angle of attack of the zero lift line of the airplane. In this and succeeding chapters it is used for the angle of attack of the x axis. It is then simply the perturbation angle of attack.)

[d] Thus the nondimensional rates of rotation are

$$\hat{p} = \frac{pb}{2u_0}, \qquad \hat{q} = \frac{qc}{2u_0}, \qquad \hat{r} = \frac{rb}{2u_0}, \qquad D\xi = \frac{\dot{\xi}b}{2u_0}, \qquad \text{etc.}$$

3. It is convenient to use the time constant l/u_0, for then all inertia terms in the equations are readily identified by the presence of one of the mass or moment-of-inertia parameters.
4. The choice of the *semichord* and *semispan* for the characteristic lengths makes the time constant used in nondimensionalizing the motion variables the same as that used in nondimensionalizing the stability derivatives. This reduces the number of numerical constants in the equations.

Nondimensional Stability Derivatives. The nondimensional stability derivatives are exemplified by C_{x_u}, C_{n_p}, $C_{m\dot{\alpha}}$, C_{ha_ξ}. In every case these represent partial derivatives of the force and moment coefficients with respect to the nondimensional form of the variable indicated by the subscript. Thus

$$C_{x_u} = \frac{\partial C_x}{\partial \hat{u}}, \quad C_{n_p} = \frac{\partial C_n}{\partial \hat{p}}, \quad C_{m\dot{\alpha}} = \frac{\partial C_m}{\partial(\dot{\alpha}c/2u_0)}, \quad C_{ha_\xi} = \frac{\partial C_{ha}}{\partial \xi}, \quad \text{etc.}$$

The Nondimensional Equations

The application of the definitions of Table 4.1 to Eqs. 4.14,12 and 4.14,13 leads to the desired nondimensional equations. By way of illustration, the procedure is given below for one force and one moment equation.

The Z-Force Equation. The relevant equation is the second of Eqs. 4.14,12. From Table 4.1, $Z = C_z \frac{1}{2}\rho v_c^2 S$ where $v_c^2 = U^2 + v^2 + w^2$ and $U = u_0 + u$. Hence

$$Z_u = \left(\frac{\partial Z}{\partial u}\right)_0 = C_{z_0}\rho u_0 S\left(\frac{\partial v_c}{\partial u}\right)_0 + \tfrac{1}{2}\rho u_0^2 S\left(\frac{\partial C_z}{\partial u}\right)_0$$

where the subscript zero indicates the reference flight condition. But $2v_c(\partial v_c/\partial u) = 2U(\partial U/\partial u)$, and hence $(\partial v_c/\partial u)_0 = 1$. Also

$$\left(\frac{\partial C_z}{\partial u}\right)_0 = \frac{1}{u_0}\left(\frac{\partial C_z}{\partial \hat{u}}\right)_0 = \frac{1}{u_0} C_{z_u}$$

Because of the choice of axes, C_z is initially normal to the flight path, and Z_0 is the negative of the initial lift. Hence $C_{z_0} = -C_{L_0}$, and

$$Z_u = -\rho u_0 S C_{L_0} + \tfrac{1}{2}\rho u_0 S C_{z_u}$$

Also, since $(\partial v_c/\partial w)_0 = 0$, then

$$Z_w = \left(\frac{\partial Z}{\partial w}\right)_0 = \tfrac{1}{2}\rho u_0^2 S\left(\frac{\partial C_z}{\partial w}\right)_0$$

But $w = u_0\alpha$, so that

$$Z_w = \tfrac{1}{2}\rho u_0 S\left(\frac{\partial C_z}{\partial \alpha}\right)_0 = \tfrac{1}{2}\rho u_0 S C_{z_\alpha}$$

In a similar way,

$$Z_{\dot{w}} = \left(\frac{\partial Z}{\partial \dot{w}}\right)_0 = \tfrac{1}{2}\rho u_0^2 S\left(\frac{\partial C_z}{\partial \dot{w}}\right) = \tfrac{1}{2}\rho u_0 S\left(\frac{\partial C_z}{\partial \dot{\alpha}}\right)_0$$

But

$$\left(\frac{\partial C_z}{\partial \dot{\alpha}}\right)_0 = t^* C_{z_{\dot{\alpha}}}$$

Hence

$$Z_{\dot{w}} = \tfrac{1}{2}t^*\rho u_0 S C_{z_{\dot{\alpha}}}$$

Also,

$$Z_q = \left(\frac{\partial Z}{\partial q}\right)_0 = \tfrac{1}{2}\rho u_0^2 S\left(\frac{\partial C_z}{\partial q}\right)_0$$

But

$$q = \frac{1}{t^*}\hat{q}$$

Hence

$$Z_q = \tfrac{1}{2}t^*\rho u_0^2 S C_{z_q}$$

Also

$$Z_\eta = \left(\frac{\partial Z}{\partial \eta}\right)_0 = \tfrac{1}{2}\rho u_0{}^2 S\left(\frac{\partial C_z}{\partial \eta}\right)_0 = \tfrac{1}{2}\rho u_0{}^2 S C_{z_\eta}$$

The remaining elements of the Z-force equation are transformed as follows:

$$m = \rho S l \mu$$

$$\frac{d}{dt} = \frac{1}{t^*} D$$

$$\begin{aligned} mg \sin\theta_0 &= mg\cos\theta_0 \tan\theta_0 \\ &= -Z_0 \tan\theta_0 \qquad \text{(from Eq. 4.14,7)} \\ &= L_0 \tan\theta_0 \end{aligned}$$

If the above relations are substituted into the equation and then $\frac{1}{2}\rho u_0{}^2 S$ is divided out, we get

$$(2C_{L_0} - C_{z_u})\hat{u} + (2\mu D - C_{z_{\dot{\alpha}}} D - C_{z_\alpha})\alpha - (2\mu D + C_{z_q} D - C_{L_0}\tan\theta_0)\theta - C_{z_\eta}\eta = 0$$

The L-Moment Equation. The relevant equation is the second of Eqs. 4.14,3. From Table 4.1, $L = C_l \rho v_c{}^2 S l$. Hence

$$\begin{aligned} L_v &= \left(\frac{\partial L}{\partial v}\right)_0 = \rho u_0{}^2 S l\left(\frac{\partial C_l}{\partial v}\right)_0 = \rho u_0 S l\left(\frac{\partial C_l}{\partial \beta}\right)_0 \\ &= \rho u_0 S l C_{l_\beta} \end{aligned}$$

Also

$$\begin{aligned} L_p &= \left(\frac{\partial L}{\partial p}\right)_0 = \rho u_0{}^2 S l\left(\frac{\partial C_l}{\partial p}\right)_0 = t^*\rho u_0{}^2 S l\left(\frac{\partial C_l}{\partial \hat{p}}\right)_0 \\ &= t^*\rho u_0{}^2 S l C_{l_p} \end{aligned}$$

Similarly

$$\begin{aligned} L_r &= \left(\frac{\partial L}{\partial r}\right)_0 = \rho u_0{}^2 S l\left(\frac{\partial C_l}{\partial r}\right)_0 \\ &= t^*\rho u_0{}^2 S l C_{l_r} \end{aligned}$$

and

$$\begin{aligned} L_{\dot{\xi}} &= t^*\rho u_0{}^2 S l C_{l_{\dot{\xi}}} \\ L_\xi &= \rho u_0{}^2 S l C_{l_\xi} \\ L_\zeta &= \rho u_0{}^2 S l C_{l_\zeta} \end{aligned}$$

The moments of inertia are

$$\begin{aligned} A &= i_A \rho S l^3 \\ E &= i_E \rho S l^3 \end{aligned}$$

The above substitutions are made, and the equation divided through by $\rho u_0{}^2 S l$ to get

$$-C_{l_\beta}\beta + (i_A D - C_{l_p})\hat{p} - (i_E D + C_{l_r})\hat{r} - (C_{l_{\dot{\xi}}} D + C_{l_\xi})\xi - C_{l_\zeta}\zeta = 0$$

By applying the method illustrated above to all of the Eqs. 4.14,12 and 4.14,13 Eqs. 4.15,7 and 4.15,8 are obtained.

Longitudinal Equations (Eqs. 4.15,7)

$$(2\mu D - 2C_{L_0}\tan\theta_0 - C_{x_u})\hat{u} - C_{x_\alpha}\alpha + C_{L_0}\theta + 0 = 0 \quad (a)$$

$$(2C_{L_0} - C_{z_u})\hat{u} + (2\mu D - C_{z_{\dot{\alpha}}}D - C_{z_\alpha})\alpha - [(2\mu + C_{z_q})D - C_{L_0}\tan\theta_0]\theta - C_{z_\eta}\eta = 0 \quad (b)$$

$$-C_{m_u}\hat{u} - (C_{m_{\dot{\alpha}}}D + C_{m_\alpha})\alpha + (i_B D^2 - C_{m_q}D)\theta - (C_{m_{\dot{\eta}}}D + C_{m_\eta})\eta = 0 \quad (c)$$

$$-(2C_{he} + C_{he_u})\hat{u} - (C_{he_{\dot{\alpha}}}D + C_{he_\alpha})\alpha - C_{he_q}D\theta + (i_e D^2 - C_{he_{\dot{\eta}}}D - C_{he_\eta})\eta = \Delta C_{fe} \quad (d)$$

$$\mu = \frac{m}{\rho S l}, \qquad t^* = \frac{l}{u_0}, \qquad l = \frac{\bar{c}}{2} \qquad\qquad \hat{q} - D\theta = 0 \quad (e)$$

Lateral Equations (Eqs. 4.15,8)

$$(2\mu D - C_{y_\beta})\beta - C_{y_p}\hat{p} + (2\mu - C_{y_r})\hat{r} - C_{L_0}\phi + 0 - C_{y_\zeta}\zeta = 0 \quad (a)$$

$$-C_{l_\beta}\beta + (i_A D - C_{l_p})\hat{p} - (i_E D + C_{l_r})\hat{r} + 0 - (C_{l_{\dot{\xi}}}D + C_{l_\xi})\xi - C_{l_\zeta}\zeta = 0 \quad (b)$$

$$-C_{n_\beta}\beta - (i_E D + C_{n_p})\hat{p} + (i_C D - C_{n_r})\hat{r} + 0 - C_{n_\xi}\xi - (C_{n_{\dot{\zeta}}}D + C_{n_\zeta})\zeta = 0 \quad (c)$$

$$0 - 2C_{ha_p}\hat{p} - 2C_{ha_r}\hat{r} + 0 + (i_a D^2 - 2C_{ha_{\dot{\xi}}}D - 2C_{ha_\xi})\xi + 0 = \Delta C_{fa} \quad (d)$$

$$-C_{hr_\beta}\beta - C_{hr_p}\hat{p} - C_{hr_r}\hat{r} + 0 + 0 + (i_r D^2 - C_{hr_{\dot{\zeta}}}D - C_{hr_\zeta})\zeta = \Delta C_{fr} \quad (e)$$

$$\hat{p} + \hat{r}\tan\theta_0 - D\phi = 0 \quad (f)$$

$$\hat{r}\sec\theta_0 - D\psi = 0 \quad (g)$$

$$\mu = \frac{m}{\rho S l}, \qquad t^* = \frac{l}{u_0}, \qquad l = \frac{b}{2}$$

4.16 AERODYNAMIC TRANSFER FUNCTIONS

It was pointed out in Sec. 4.14 that the conventional equations of motion, based on the Bryan representation of the aerodynamic forces, are subject to certain theoretical and practical objections. These can be overcome by the application of the Laplace transform and the transfer-function concept. These mathematical tools are explained in Chap. 8; an understanding of them is prerequisite to reading this section.

We assume to begin with that the differential equations governing the unsteady pressure distribution over the surfaces of the airplane are linear in the usual sense of aerodynamic theory. This means that, if we impose a particular perturbation, say $\alpha(\hat{t})$, and obtain the pitching-moment response $C_m(\hat{t})$, then the functions $\alpha(\hat{t})$ and $C_m(\hat{t})$ are related by a linear differential or integral equation. This equation may or may not be known. The *aerodynamic transfer function* relating α and C_m is defined as the ratio of the Laplace transforms[9] of C_m and α

$$G_{m\alpha}(s) = \frac{\bar{C}_m(s)}{\bar{\alpha}(s)} \tag{4.16,1}$$

This transfer function is unique if $\alpha(\hat{t}) = C_m(\hat{t}) = 0$ for $\hat{t} < 0$. This requires that we deal with motions for which the initial conditions are those of the reference flight condition.

Because of the assumed linearity, the C_m corresponding to several perturbations, e.g., $\hat{u}$, α, $\hat{q}$, η is obtained by adding the separate values: i.e.,

$$\bar{C}_m(s) = G_{mu}\bar{u} + G_{m\alpha}\bar{\alpha} + G_{mq}\bar{q} + G_{m\eta}\bar{\eta} \tag{4.16,2}$$

where $\bar{u}$, $\bar{\alpha}$, $\bar{q}$, and $\bar{\eta}$ are the Laplace transforms of $\hat{u}(\hat{t})$, $\alpha(\hat{t})$, $\hat{q}(\hat{t})$ and $\eta(\hat{t})$. The transfer functions are illustrated by the block diagram of Fig. 4.7, which also shows a comparison with the stability-derivative representation. It is evident that the use of stability derivatives implies an approximation to the transfer function in the form of a power series

$$G_{m\alpha}(s) \doteq C_{m_\alpha} + C_{m_{\dot{\alpha}}}s + C_{m_{\ddot{\alpha}}}s^2 + \cdots \tag{4.16,3}$$

Although an infinite series such as this can sometimes give a correct value for $G_{m\alpha}$, there are other times when the series is not convergent (ref. 4.2).

The Transformed Equations

The small-disturbance equations, Eqs. 4.15,7 and 4.15,8 can be modified to incorporate the transfer functions instead of the stability derivatives.

The terms

$$(C_{z_{\dot{\alpha}}}D + C_{z_\alpha})\alpha \tag{4.16,4}$$

[9] $\bar{C}_m(s) = \int_{\hat{t}=0}^{\infty} e^{-s\hat{t}} C_m(\hat{t})\, d\hat{t}$, $\quad \bar{q}(s) = \int_{\hat{t}=0}^{\infty} e^{-s\hat{t}} \hat{q}(\hat{t})\, d\hat{t}$, $\quad$ etc.

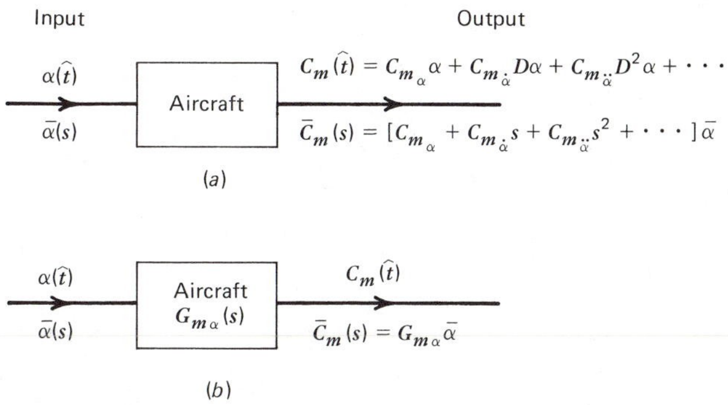

Input Output

$\alpha(\hat{t})$ → Aircraft → $C_m(\hat{t}) = C_{m_\alpha}\alpha + C_{m_{\dot\alpha}}D\alpha + C_{m_{\ddot\alpha}}D^2\alpha + \cdots$

$\bar\alpha(s)$ → $\bar C_m(s) = [C_{m_\alpha} + C_{m_{\dot\alpha}}s + C_{m_{\ddot\alpha}}s^2 + \cdots]\bar\alpha$

(*a*)

$\alpha(\hat{t})$ → Aircraft $G_{m\alpha}(s)$ → $C_m(\hat{t})$

$\bar\alpha(s)$ → $\bar C_m(s) = G_{m\alpha}\bar\alpha$

(*b*)

FIG. 4.7 Comparison of transfer function and stability derivatives. (*a*) Conventional representation by means of stability derivatives. (*b*) Representation by means of aerodynamic transfer functions.

are typical of the expressions giving the aerodynamic forces in the preceding equations. They represent the change in C_z from the reference flight condition. In the present scheme, this change is represented by

$$\Delta\bar{C}_z = G_{z\alpha}\bar{\alpha} \tag{4.16,5}$$

By taking the Laplace transform[10] of the equations, and replacing expressions such as (4.16,4) by expressions such as (4.16,5), Eqs. 4.16,6 and 4.16,7 are obtained.

No identifying subscript is used on the transfer functions for the elevator hinge moment, Eq. 4.16,6*d*, since there can be no ambiguity. Subscripts *a* and *r* are used to identify the aileron and rudder hinge moments in Eqs. 4.16,7*d* and *e*.

The aerodynamic transfer functions defined here may be considered as generalizations of the conventional derivatives. Methods of determining them are given in Chap. 5.

4.17 ELASTIC DEGREES OF FREEDOM

In the preceding sections of this chapter, the airplane was treated as though it were a rigid body, and the equations of motion were derived on that basis. Now it is known that the stability and control characteristics of high-speed flexible airplanes may be significantly influenced by the distortions of the structure under transient loading conditions (refs. 4.6 to 4.8). In this section are presented the changes that must be made to the system of rigid-body equations to allow for this. The distortions, and the rigid-body motions except for U, are all assumed to be small.

[10] Because of the requirement that the system be initially quiescent (see after Eq. 4.16,1), all initial values, such as $\hat{q}(0)$ are set equal to zero when forming the Laplace transforms.

Longitudinal Equations (Eqs. 4.16,6)

$$\begin{aligned}
(2\mu s - 2C_{L_0}\tan\theta_0 - G_{xu})\bar{u} - G_{x\alpha}\bar{\alpha} + C_{L_0}\bar{\theta} + 0 &= 0 && (a)\\
(2C_{L_0} - G_{zu})\bar{u} + (2\mu s - G_{z\alpha})\bar{\alpha} - [(2\mu + G_{zq})s - C_{L_0}\tan\theta_0]\bar{\theta} - G_{z\eta}\bar{\eta} &= 0 && (b)\\
-G_{mu}\bar{u} - G_{m\alpha}\bar{\alpha} + (i_B s^2 - G_{mq}s)\bar{\theta} - G_{m\eta}\bar{\eta} &= 0 && (c)\\
-(2C_{he_0} + G_{hu})\bar{u} - G_{h\alpha}\bar{\alpha} - G_{hq}s\bar{\theta} + (i_e s^2 - G_{h\eta})\bar{\eta} &= \Delta\bar{C}_{fe} && (d)\\
\bar{q} - s\bar{\theta} &= 0 && (e)
\end{aligned}$$

Lateral Equations (Eqs. 4.16,7)

$$\begin{aligned}
(2\mu s - G_{y\beta})\bar{\beta} - G_{yp}\bar{p} + (2\mu - G_{yr})\bar{r} - C_{L_0}\bar{\phi} + 0 - G_{y\zeta}\bar{\zeta} &= 0 && (a)\\
-G_{l\beta}\bar{\beta} + (i_A s - G_{lp})\bar{p} - (i_E s + G_{lr})\bar{r} + 0 - G_{l\xi}\bar{\xi} - G_{l\zeta}\bar{\zeta} &= 0 && (b)\\
-G_{n\beta}\bar{\beta} - (i_E s + G_{np})\bar{p} + (i_C s - G_{nr})\bar{r} + 0 - G_{n\xi}\bar{\xi} - G_{n\zeta}\bar{\zeta} &= 0 && (c)\\
0 - 2G_{hp_a}\bar{p} - 2G_{hr_a}\bar{r} + 0 + (i_a s^2 - 2G_{h\xi_a})\bar{\xi} + 0 &= \Delta\bar{C}_{fa} && (d)\\
-G_{h\beta_r}\bar{\beta} - G_{hp_r}\bar{p} - G_{hr_r}\bar{r} + 0 + 0 - (i_r s^2 - G_{h\zeta_r})\bar{\zeta} &= \Delta\bar{C}_{fr} && (e)\\
\bar{p} + \bar{r}\tan\theta_0 - s\bar{\phi} &= 0 && (f)\\
\bar{r}\sec\theta_0 - s\bar{\psi} &= 0 && (g)
\end{aligned}$$

The Method of Quasistatic Deflections

Many of the important effects of distortion can be accounted for simply by altering the aerodynamic derivatives. The assumption is made that the changes in aerodynamic loading take place slowly enough that the structure is at all times in static equilibrium. (This is equivalent to assuming that the natural frequencies of vibration of the structure are much higher than the frequencies of the rigid-body motions). Thus a change in load produces a proportional change in the shape of the airplane, which in turn influences the load. Examples of this kind of analysis are given in Sec. 3.6 (effect of fuselage bending on the location of the neutral point) and Sec. 5.3 (effect on the u derivatives).

The Method of Normal Modes

When the separation in frequency between the elastic degrees of freedom and the rigid-body motions is not large, then significant coupling can occur between the two. In that case a dynamic analysis is required, which takes account of the time dependence of the elastic motions.

The method that is described here for accomplishing this purpose is based on the representation of the deformation of the elastic airplane in terms of its normal modes of free vibration. Let the position of an element δm of the airplane in the reference steady state be (x_0, y_0, z_0) and its position at any time t be (x, y, z). (The coordinate axes are those previously used, with origin always at the mass center, and with linear and angular velocities $\mathbf{v}_c$ and $\boldsymbol{\omega}$). Thus $(x - x_0)$, $(y - y_0)$, and $(z - z_0)$ are the elastic displacements of δm. They are assumed to be given by the following expressions

$$
\begin{aligned}
x - x_0 &= \sum_{n=1}^{\infty} f_n(x_0, y_0, z_0)\varepsilon_n(t) \\
y - y_0 &= \sum_{n=1}^{\infty} g_n(x_0, y_0, z_0)\varepsilon_n(t) \qquad (4.17,1) \\
z - z_0 &= \sum_{n=1}^{\infty} h_n(x_0, y_0, z_0)\varepsilon_n(t)
\end{aligned}
$$

in which the f_n, g_n, h_n functions give the shape of the nth normal mode of free vibration, and $\varepsilon_n(t)$ is the generalized coordinate giving the displacement in that mode. In this way the general elastic deformation is constructed as a superposition of normal modes.

An important property of the normal modes may be obtained from the following consideration. Let the elastic body (the airplane) be at rest relative to an inertial frame of reference. Then let it be deformed by a set of external forces so arranged that the resultant force and moment are zero. If these forces be suddenly released, a complex vibration of the body will result, containing in general all the modes, each with its own natural frequency. Let the linear and angular momenta corresponding to the nth mode be $m\mathbf{v}_{c_n}$ and $\mathbf{h}_n$, respectively. Because of the periodicity of the motion in this mode, these momenta, if nonzero, must be periodic with frequency ω_n. Hence the total linear and angular momenta, relative to the inertial frame of reference, are each of the form

$$\sum_{n=1}^{\infty} (a_n \sin \omega_n t + b_n \cos \omega_n t)$$

But the system has at all times been free of resultant external force and moment, and was initially at rest. Hence it follows that the summations giving the momenta are zero. This can only be true when all the amplitudes a_n and b_n are individually zero. The important conclusion then is that in each normal mode of free vibration the resultant linear and angular momenta are zero.

Influence on the Six Euler Equations

The six Euler equations of motion were derived in Sec. 4.4 from the fundamental Eqs. 4.2,4 and 4.2,10, together with the assumption that the velocity field is given by

$$\mathbf{v} = \mathbf{v}_c + \boldsymbol{\omega} \times \mathbf{r}$$

With the addition of the elastic motions, the velocity field is altered to

$$\mathbf{v} = \mathbf{v}_c + \boldsymbol{\omega} \times \mathbf{r} + \mathbf{v}_e$$

where

$$\mathbf{v}_e = \mathbf{i}\dot{x} + \mathbf{j}\dot{y} + k\dot{z} \tag{4.17,2}$$

and where $\dot{x}$, $\dot{y}$, $\dot{z}$ are obtained from Eqs. 4.17,1. The addition of $\mathbf{v}_e$ to the velocity field, however, does not change the resultant linear and angular momenta of the system,[11] by virtue of the property of the free normal modes noted above. This being so, the six Euler equations remain unchanged in form, and are still given by Eqs. 4.4,3 and 4.4,4. This is equivalent to saying that there is no inertia coupling between the elastic degrees of freedom and the rigid-body degrees of freedom. When applied to the airplane problem, the elastic and rigid-body equations are not nevertheless entirely uncoupled. The deformations of the structure will in general produce perturbations in the aerodynamic forces and moments. These may be introduced into the linearized equations of motion by the addition of appropriate derivatives to the expressions for the aerodynamic forces given by Eqs. 4.14,9 to 4.14,11. For example, the added terms in the pitching moment associated with the nth elastic degree of freedom would be

$$M_{\varepsilon_n}\varepsilon_n + M_{\dot{\varepsilon}_n}\dot{\varepsilon}_n + M_{\ddot{\varepsilon}_n}\ddot{\varepsilon}_n \tag{4.17,3}$$

Similar expressions appear for each of the added degrees of freedom, and in each of the aerodynamic force and moment equations. An example of the elastic stability

[11] The small elastic distortions cause small perturbations in the inertia coefficients $A, B, \cdots F$. These are all multiplied in the moment equations by the small angular velocities P, Q, R; so to the first order only the values of the inertias in the reference state appear in the equations.

derivatives is given in Sec. 5.16. Alternatively, the aerodynamic forces may be formulated in the form of transfer functions, along the lines indicated in Sec. 4.16.

The Additional Equations of Motion

The additional equations are found by applying Lagrange's equation, with the ε_n as generalized coordinates. The influence of the motion of the frame of reference is accounted for by introducing the associated inertia-force field in the computation of the generalized forces just as in Sec. 4.7. When the system considered is flexible, and has a strain energy associated with it, the appropriate form of Lagrange's equation is

$$\frac{d}{dt}\frac{\partial T}{\partial \dot{q}_k} - \frac{\partial T}{\partial q_k} + \frac{\partial U}{\partial q_k} = \mathscr{F}_k \tag{4.17,4}$$

where U is the strain energy, and the remaining symbols have the same meaning as in Eq. 4.7,1. Since normal modes have been chosen as the degrees of freedom, then the individual equations of motion are independent of one another insofar as elastic and inertia forces are concerned (this is a property of the normal modes), although the equations will be coupled through the aerodynamic contributions to the $\mathscr{F}$'s. The lack of elastic and inertia coupling permits the left-hand side of Eq. 4.17,4 to be evaluated by considering only a single elastic degree of freedom to be excited. Let its generalized coordinate be ε_n. The kinetic energy is given by

$$T = \tfrac{1}{2}\int(\dot{x}^2 + \dot{y}^2 + \dot{z}^2)\,dm$$

where the integration is over all elements of mass of the body. From Eqs. 4.17,1, this becomes (with only ε_n excited)

$$T = \tfrac{1}{2}\dot{\varepsilon}_n^{\,2}\int(f_n^{\,2} + g_n^{\,2} + h_n^{\,2})\,dm$$

The integral is the *generalized inertia* in the nth mode, and is denoted by

$$I_n = \int(f_n^{\,2} + g_n^{\,2} + h_n^{\,2})\,dm \tag{4.17,5}$$

so that

$$T = \tfrac{1}{2}I_n\dot{\varepsilon}_n^{\,2} \tag{4.17,6}$$

The first term of Eq. 4.17,4 is therefore $I_n\ddot{\varepsilon}_n$, and the second term is zero.

The strain-energy term is conveniently evaluated in terms of the natural frequency of the nth mode by applying Rayleigh's method. This uses the fact that, when the system vibrates in an undamped normal mode, the maximum strain energy occurs when all elements are simultaneously at the extreme position, and the kinetic energy is zero. This maximum strain energy must be equal to the maximum kinetic energy that occurs when all elements pass simultaneously through their equilibrium position, where the strain energy is zero. Hence, if $\varepsilon_n = a \sin \omega_n t$, then the maximum kinetic energy is, from Eq. 4.17,6

$$T_{\max} = \tfrac{1}{2}I_n\omega_n^{\,2}a$$

Since the stress-strain relation is assumed to be linear, the strain energy[12] is a quadratic function of ε_n; i.e., $U = \frac{1}{2}k\varepsilon_n{}^2$. Hence

$$U_{\max} = \tfrac{1}{2}ka^2 = T_{\max} = \tfrac{1}{2}I_n\omega_n{}^2a$$

It follows that $k = I_n\omega_n{}^2$, and that

$$U = \tfrac{1}{2}I_n\omega_n{}^2\varepsilon_n{}^2$$

and hence $\partial U/\partial\varepsilon_n = I_n\omega_n{}^2\varepsilon_n$. The left side of Eq. 4.17,4 is therefore as follows:

$$I_n\ddot{\varepsilon}_n + I_n\omega_n{}^2\varepsilon_n = \mathscr{F}_n \tag{4.17,7}$$

Evaluation of $\mathscr{F}_n$

The generalized force is calculated from the work done during a virtual displacement,

$$\mathscr{F}_n = \frac{\partial W}{\partial \varepsilon_n} \tag{4.17,8}$$

where W is the work done by all the external forces, including the inertia forces associated with nonuniform motion of the frame of reference. The latter are given by Eqs. 4.8,6, and the work done by them is

$$\delta W_i = \int (\delta x\, dF_{x_i} + \delta y\, dF_{y_i} + \delta z\, dF_{z_i})$$

where the integration is over the whole body. Introducing Eqs. 4.17,1, this becomes

$$\delta W_i = \sum_{n=1}^{\infty} \delta\varepsilon_n \int (f_n\, dF_{x_i} + g_n\, dF_{y_i} + h_n\, dF_{z_i})$$

thus

$$\mathscr{F}_{n_i} = \frac{\partial W_i}{\partial \varepsilon_n} = \int (f_n\, dF_{x_i} + g_n\, dF_{y_i} + h_n\, dF_{z_i}) \tag{4.17,9}$$

When the inertia-force expressions of Eqs. 4.8,6 are linearized to small disturbances, and substituted into Eq. 4.17,9, all the remaining terms contain integrals of the following types:

$$\int f_n\, dm, \qquad \int (y_0 h_n - z_0 g_n)\, dm$$

The first of these is zero because the origin is the mass center, and the second is zero because the angular momentum associated with the elastic mode vanishes. The net result is that $\mathscr{F}_{n_i} = 0$. This result simply verifies what was found above; i.e., that there is no inertial coupling between the elastic and rigid-body degrees of freedom.

The remaining contribution to $\mathscr{F}_n$ is that of the aerodynamic forces. Let the local normal-pressure perturbation at an element dS of the airplane's surface be

[12] For example, in a spring of stiffness k and stretch x, the strain energy is $U = \frac{1}{2}kx^2$.

$p(x_0, y_0, z_0)$, and let the local outward normal be $\mathbf{n}(n_x, n_y, n_z)$. Then the work done by the aerodynamic forces in a virtual displacement is

$$\delta W_a = -\int p\mathbf{n} \cdot (\mathbf{r} - \mathbf{r}_0)\, dS$$

where the integral is over the whole surface of the airplane, and $(\mathbf{r} - \mathbf{r}_0)$ is the vector displacement at dS. It is given by

$$\mathbf{r} - \mathbf{r}_0 = \sum_{n=1}^{\infty} (\mathbf{i}f_n + \mathbf{j}g_n + \mathbf{k}h_n)\, \delta\varepsilon_n$$

hence

$$\delta W_a = -\sum_{n=1}^{\infty} \delta\varepsilon_n \int p(n_x f_n + n_y g_n + n_z h_n)\, dS$$

and

$$\mathscr{F}_n = \frac{\partial W_a}{\partial \varepsilon_n} = -\int p(n_x f_n + n_y g_n + n_z h_n)\, dS \tag{4.17,10}$$

Each of the variables inside the integral is a function of (x_0, y_0, z_0), i.e., of position of the surface, and moreover, p is in the most general case a function of all the generalized coordinates, of their derivatives, and of the control-surface angles. The result is that $\mathscr{F}_n$ is a linear function of all these variables, which may be expressed in terms of a set of generalized aerodynamic derivatives (or alternatively aerodynamic transfer functions), namely,

$$\begin{aligned} \mathscr{F}_n = {} & A_{nu}u + A_{n\dot{u}}\dot{u} + \cdots + A_{np}p + \cdots + A_{n\zeta}\zeta + \cdots \\ & + \sum_{m=1}^{\infty} a_{nm}\varepsilon_m + \sum_{m=1}^{\infty} b_{nm}\dot{\varepsilon}_m + \sum_{m=1}^{\infty} c_{nm}\ddot{\varepsilon}_m \end{aligned} \tag{4.17,11}$$

In application, only the important derivatives would be retained in any given case, and the equation could be made nondimensional along the lines of Sec. 4.15. The values of the derivatives kept would be computed by application of Eq. 4.17,10. An example of this computation is given in Sec. 5.16.

Résumé

The effects of structural dynamics on the stability and control equations can be incorporated by adding structural degrees of freedom based on free normal modes. For an exact representation, an infinite number of such modes are required; however, in practice only a few of the lowest modes need be employed. The six rigid-body equations are altered only to the extent of additional aerodynamic terms of the type given in Eq. 4.17,3. One additional equation is required for each elastic degree of freedom (Eq. 4.17,7). The generalized forces appearing in the added equations contain only aerodynamic contributions, which are computed from Eq. 4.17,10 and expressed as in Eq. 4.17,11.

4.18 ADDITIONAL SYMBOLS INTRODUCED IN CHAPTER 4

$\mathbf{a}_c$	acceleration vector of airplane mass center
(a_{cx}, a_{cy}, a_{cz})	scalar components of $\mathbf{a}_c$
A, B, C	moments of inertia about (x, y, z) axes
D	product of inertia $\int yz\,dm$
E	product of inertia $\int xz\,dm$
F	product of inertia $\int xy\,dm$
e_a, e_e, e_r	mass eccentricities of control surfaces (see Fig. 4.5)
$\mathbf{F}$	resultant external force vector
(F_x, F_y, F_z)	scalar components of $\mathbf{F}$
F_a, F_e, F_r	generalized control forces (see Sec. 4.8)
$\mathscr{F}$	generalized force in Lagrange's equation
$\mathbf{G}$	resultant external moment vector, about the mass center
$G_{av}(s)$	aerodynamic transfer function, C_a output, v input
$\mathbf{h}$	angular momentum vector of the airplane
$\mathbf{h}'$	angular momentum vector of spinning rotors
(h_x, h_y, h_z)	scalar components of $\mathbf{h}$
(h'_x, h'_y, h'_z)	scalar components of $\mathbf{h}'$
H_a, H_e, H_r	hinge moments on aileron, elevator, rudder
I_a, I_e, I_r	effective moments of inertia of control systems
(L, M, N)	scalar components of G
m_a, m_e, m_r	masses of aileron, elevator, rudder
P_{ay}	aileron product of inertia (see Sec. 4.10)
P_{ex}	elevator product of inertia (Eq. 4.8,10)
P_{rx}, P_{rz}	rudder products of inertia (see Sec. 4.9)
(P, Q, R)	scalar components of $\boldsymbol{\omega}$, radians/sec
(p, q, r)	perturbation of (P, Q, R)
s	Laplace transform variable
(U, V, W)	scalar components of $\mathbf{v}_c$
(u, v, w)	perturbations of (U, V, W)
$\mathbf{v}_c$	velocity vector of airplane mass center
(X, Y, Z)	components of resultant aerodynamic force acting on the airplane
(x', y', z')	coordinates of airplane mass center relative to fixed axes (see Fig. 4.2)
ε_n	generalized coordinate of the nth elastic mode

$\delta_e, \delta_r, \delta_a$	angles of elevator, rudder, and aileron
$\boldsymbol{\omega}$	angular velocity vector of the airplane
(Ψ, Θ, Φ)	Euler angles, radians (see Sec. 4.5)
(ψ, θ, ϕ)	perturbations of (Ψ, Θ, Φ)
(ξ, η, ζ)	perturbations of $(\delta_a, \delta_e, \delta_r)$

See also Secs. 4.15 and 4.16, and Table 4.1.

See also Secs. 2.11 and 3.11.

4.19 BIBLIOGRAPHY

4.1 E. J. Routh. *Dynamics of a System of Rigid Bodies* 6th ed. Macmillan & Co., London, 1905.

4.2 B. Etkin. Aerodynamic Transfer Functions: An Improvement on Stability Derivatives for Unsteady Flight. *Univ. Toronto UTIA Rept. 42*, 1956.

4.3 M. J. Abzug. Effects of Certain Steady Motions on Small-Disturbance Airplane Dynamics. *J. Aero. Sci.*, vol. 21, no. 11, 1954.

4.4 R. L. Nelson. The Motions of Rolling Symmetrical Missiles Referred to a Body Axis System. *NACA TN 3737*, 1956.

4.5 A. C. Charters. The Linearized Equations of Motion Underlying the Dynamic Stability of Aircraft, Spinning Projectiles, and Symmetrical Missiles. *NACA TN 3350*, 1955.

4.6 W. P. Rodden. An Aeroelastic Parameter for Estimation of the Effect of Flexibility on the Lateral Stability and Control of Aircraft. *J. Aero. Sci.*, vol. 23, no. 7, 1956.

4.7 B. B. Klawans and H. I. Johnson. Some Effects of Fuselage Flexibility on Longitudinal Stability and Control. *NACA TN 3543*, 1956.

4.8 M. D. McLaughlin. A Theoretical Investigation of the Short-Period Dynamic Longitudinal Stability of Airplanes Having Elastic Wings of 0° to 60° Sweepback. *NACA TN 3251*, 1954.

THE STABILITY DERIVATIVES

CHAPTER 5

5.1 GENERAL REMARKS

We saw in Chap. 4 how the aerodynamic actions on the airplane can be represented approximately by means of stability derivatives, or more exactly by aerodynamic transfer functions. Indeed, all the aerodynamics involved in airplane dynamics is concentrated in this section of the subject: i.e., in the determination of these derivatives or transfer functions. Each of the stability derivatives contained in the equations of motion is discussed in the following sections. Wherever possible, formulas for them are given in terms of the more elementary parameters used in static stability and performance. Where this is not feasible, it is shown in a qualitative way how the particular force or moment is related to the relevant perturbation quantity. No data for estimation are contained in this chapter; these are all in Appendix B.

Expressions for C_x and C_z

For convenience, we shall want the derivatives of C_x and C_z expressed in terms of lift, drag, and thrust coefficients. The relevant forces are shown in Fig. 5.1. As shown, the thrust line does not necessarily lie on the x axis. However, the angle between them is generally small, and we shall assume it to be zero. With this assumption, and for small α, we get[1]

$$\begin{aligned} C_x &= C_T + C_L\alpha - C_D \\ C_z &= -(C_L + C_D\alpha) \end{aligned} \tag{5.1,1}$$

where C_T is the coefficient of thrust, $T/\frac{1}{2}\rho v_c{}^2 S$.

[1] Since X and Z are the *aerodynamic* forces acting on the airplane, there are no weight components in Eqs. 5.1,1.

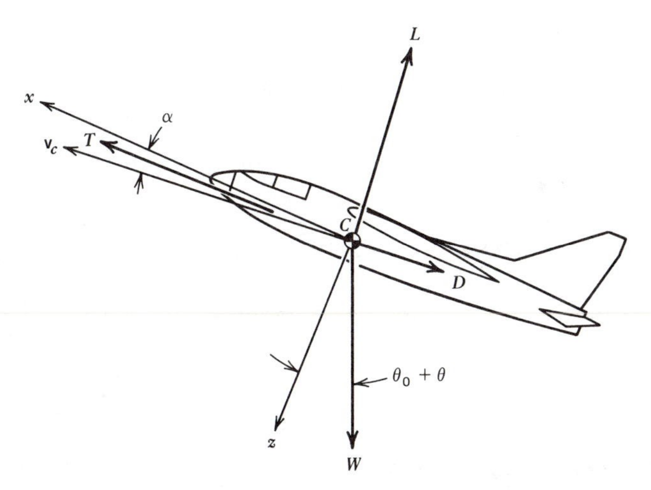

FIG. 5.1 Forces in symmetric flight.

5.2 THE α DERIVATIVES (C_{x_α}, C_{z_α}, C_{m_α}, C_{he_α})

The α derivatives describe the changes that take place in the forces and moments when the angle of attack of the airplane is increased. They are normally an increase in the lift, an increase in the drag, a negative pitching moment, and a negative elevator hinge moment. The contents of Chap. 2 are relevant to these derivatives.

The Derivative C_{x_α}

By definition, $C_{x_\alpha} = (\partial C_x/\partial \alpha)_0$, where the subscript zero indicates that the derivative is evaluated when the disturbance quantities are zero. From Eq. 5.1,1

$$\frac{\partial C_x}{\partial \alpha} = \frac{\partial C_T}{\partial \alpha} + C_L + \alpha \frac{\partial C_L}{\partial \alpha} - \frac{\partial C_D}{\partial \alpha}$$

We may assume that the thrust coefficient is sensibly independent of α so that $\partial C_T/\partial \alpha = 0$, and hence

$$C_{x_\alpha} = \left(\frac{\partial C_x}{\partial \alpha}\right)_0 = C_{L_0} - \left(\frac{\partial C_D}{\partial \alpha}\right)_0 \tag{5.2,1}$$

where the subscript zero again indicates the reference flight condition, in which, with stability axes, $\alpha = 0$. When the drag is given by a parabolic polar in the form $C_D = C_{D_{\min}} + C_L{}^2/\pi Ae$, then

$$\left(\frac{\partial C_D}{\partial \alpha}\right) = \frac{2C_{L_0}}{\pi Ae} C_{L_\alpha} \tag{5.2,2}$$

The Derivative C_{z_α}

By definition, $C_{z_\alpha} = (\partial C_z/\partial\alpha)_0$. From Eq. 5.1,1, we get

$$\frac{\partial C_z}{\partial \alpha} = -\left(C_{L_\alpha} + C_D + \alpha\frac{\partial C_D}{\partial \alpha}\right)$$

Therefore

$$C_{z_\alpha} = -(C_{L_\alpha} + C_{D_0}) \tag{5.2,3}$$

C_{D_0} will frequently be negligible compared to C_{L_α}, and consequently $C_{z_\alpha} \doteqdot -C_{L_\alpha}$.

The Derivative C_{m_α}

C_{m_α} is the static stability derivative, which was treated at some length in Chap. 2. It is conveniently expressed in terms of the stick-fixed neutral point (Eq. 2.3,3):

$$C_{m_\alpha} = a(h - h_n) \tag{5.2,4}$$

For statically stable airplanes, $h < h_n$, and C_{m_α} is negative.

The Derivative C_{he_α}

C_{he_α} gives the rate of change of elevator hinge moment with airplane angle of attack. If the airplane is tailless, then C_{he_α} is one of the primary parameters to be estimated or found by wind-tunnel testing. When the airplane has a tail, the primary parameter is $C_{h_{\alpha_t}}$ (see Sec. 2.6). Then we have

$$\frac{\partial C_{he}}{\partial \alpha} = \frac{\partial C_{he}}{\partial \alpha_t}\frac{\partial \alpha_t}{\partial \alpha}$$

But $\partial\alpha_t/\partial\alpha = (1 - \partial\varepsilon/\partial\alpha)$, where ε is the downwash at the tail; hence

$$C_{he_\alpha} = \left(1 - \frac{\partial \varepsilon}{\partial \alpha}\right)C_{he_{\alpha_t}} \tag{5.2,5}$$

5.3 THE u DERIVATIVES (C_{x_u}, C_{z_u}, C_{m_u}, C_{he_u})

The u derivatives give the effect on the forces and moments of an increase in the forward speed, while the angle of attack, the elevator angle, and the throttle position remain fixed. If the *coefficients* of lift, drag, and hinge moment did not change, then this would imply an increase in these forces and moments in accordance with the speed-squared law, i.e.,

$$\frac{\text{Force or moment}}{\text{Initial force or moment}} = \frac{(u_0 + u)^2}{{u_0}^2} \doteqdot 1 + 2\hat{u}$$

Since the pitching moment is initially zero, then, so long as C_m does not change with u, it will remain zero. The situation is actually more complicated than this, for the nondimensional coefficients are in general functions of Mach number and Reynolds number, both of which increase with increasing u. The variation with Reynolds number is usually neglected, but the effect of Mach number must be included.

The presence of the thrust term in Eq. 5.1,1 indicates that there will also be some contribution from the propulsive system, depending on how the thrust varies with speed. This is governed mainly by the type of power plant (e.g., pure jet, propeller turbine, rocket). The propulsive effect also depends, of course, on the flight condition; i.e., C_T is zero in gliding flight, and largest in slow-speed maximum power climbs.

Finally, the increased loading on the airframe due to the speed increase may induce significant structural distortion. This is a static aeroelastic effect. For example, the tail lift coefficient may be influenced appreciably by the loading (see Sec. 3.6).

The Derivative C_{x_u}

C_{x_u} is the *speed damping* derivative, since it gives the resistance due to an increase in speed. By definition $C_{x_u} = (\partial C_x/\partial \hat{u})_0 = u_0(\partial C_x/\partial u)_0$. From Eq. 5.1,1 (with $\alpha = 0$),

$$\frac{\partial C_x}{\partial u} = \frac{\partial C_T}{\partial u} - \frac{\partial C_D}{\partial u} \tag{5.3,1}$$

We shall neglect the aeroelastic effect on C_D, and assume that $C_D = f(\alpha, \mathbf{M})$. Then

$$\frac{\partial C_D}{\partial u} = \frac{\partial C_D}{\partial \mathbf{M}} \frac{\partial \mathbf{M}}{\partial u}$$

Since $\mathbf{M} = (u_0 + u)/a$, where a is the speed of sound, then $\partial \mathbf{M}/\partial u = 1/a$, and

$$\frac{\partial C_D}{\partial u} = \frac{1}{a} \frac{\partial C_D}{\partial \mathbf{M}} \tag{5.3,2}$$

Since $C_T = T/\frac{1}{2}\rho v_c^2 S$, then

$$\frac{\partial C_T}{\partial u} = \frac{\partial T/\partial u}{\frac{1}{2}\rho v_c^2 S} - \frac{2T}{\frac{1}{2}\rho v_c^3 S} \frac{\partial v_c}{\partial u}$$

In the undisturbed reference condition $v_c = u_0$, and $\partial v_c/\partial u = 1$, and thus

$$\left(\frac{\partial C_T}{\partial u}\right)_0 = \frac{(\partial T/\partial u)_0}{\frac{1}{2}\rho u_0^2 S} - \frac{2}{u_0} C_{T_0} \tag{5.3,3}$$

When Eqs. 5.3,2 and 5.3,3 are substituted into Eq. 5.3,1, the result obtained for C_{x_u} is

$$C_{x_u} = u_0 \left(\frac{\partial C_x}{\partial u}\right)_0 = \frac{(\partial T/\partial u)_0}{\frac{1}{2}\rho u_0 S} - 2C_{T_0} - \mathbf{M} \frac{\partial C_D}{\partial \mathbf{M}} \tag{5.3,4}$$

In the above equations $(\partial T/\partial u)_0$ is the derivative of the thrust with forward speed for the given speed, height, and throttle position. We shall consider three particular cases for this quantity:

1. Gliding flight: $T \equiv 0$; therefore $(\partial T/\partial u)_0 = 0$, and $C_{T_0} = 0$.
2. Jet and rocket propulsion: For jet and rocket power plants, the thrust variation with speed is usually quite small. A fair approximation is to assume $(\partial T/\partial u)_0 = 0$.
3. Variable-pitch propeller plus piston engine; power plants of this type may be approximated with the assumption of constant thrust horsepower: i.e., $T(u_0 + u) = \text{const}$. It follows that

$$T\,du + (u_0 + u)\,dT = 0$$

or

$$\left(\frac{\partial T}{\partial u}\right)_0 = -\frac{T}{u_0}$$

and hence

$$C_{x_u} = -3C_{T_0} - \mathbf{M}\frac{\partial C_D}{\partial \mathbf{M}}$$

The reference value of the thrust is related to the reference flight conditions as follows (see Fig. 5.1):

$$T_0 - D_0 = W \sin\theta_0 = L_0 \tan\theta_0$$

or

$$C_{T_0} = C_{D_0} + C_{L_0}\tan\theta_0 \tag{5.3,5}$$

When this value of C_{T_0} is substituted into Eq. 5.3,4, we have the general formula for C_{x_u} in the convenient form:

$$C_{x_u} = \frac{(\partial T/\partial u)_0}{\frac{1}{2}\rho u_0 S} - 2(C_{D_0} + C_{L_0}\tan\theta_0) - \mathbf{M}\frac{\partial C_D}{\partial \mathbf{M}}$$

The values appropriate to the three special cases are now summarized below:

1. Gliding flight:

$$C_{x_u} = -\mathbf{M}\frac{\partial C_D}{\partial \mathbf{M}}$$

2. Jet and rocket engines:

$$C_{x_u} = -2(C_{D_0} + C_{L_0}\tan\theta_0) - \mathbf{M}\frac{\partial C_D}{\partial \mathbf{M}}$$

3. Variable-pitch propeller and reciprocating engine:

$$C_{x_u} = -3(C_{D_0} + C_{L_0}\tan\theta_0) - \mathbf{M}\frac{\partial C_D}{\partial \mathbf{M}}$$

The Derivative C_{z_u}

From Eq. 5.1,1 C_{z_u} is given by

$$C_{z_u} = u_0 \left(\frac{\partial C_z}{\partial u}\right)_0 = -u_0 \left(\frac{\partial C_L}{\partial u}\right)_0$$

As in the case of the drag derivative (Eq. 5.3,2), we neglect aeroelastic and Reynolds-number effects, so that C_{z_u} depends entirely on the variation of C_L with Mach number:

$$C_{z_u} = -\mathbf{M} \frac{\partial C_L}{\partial \mathbf{M}} \tag{5.3,6a}$$

This derivative tends to be small except at transonic speeds. Theoretical values are easily calculated for high-aspect-ratio straight wings. At subsonic speeds, the Prandtl–Glauert rule[2] gives the lift coefficient for two-dimensional flow as

$$C_L = \frac{a_i \alpha}{\sqrt{1 - \mathbf{M}^2}}$$

where a_i is the lift-curve slope in incompressible flow. Upon differentiation with respect to **M**, we get

$$\frac{\partial C_L}{\partial \mathbf{M}} = \frac{\mathbf{M}}{1 - \mathbf{M}^2} C_L$$

and hence

$$C_{z_u} = -\frac{\mathbf{M}^2}{1 - \mathbf{M}^2} C_{L_0} \tag{5.3,6b}$$

In level flight with the lift equal to the weight, $\mathbf{M}^2 C_{L_0}$ is constant, and hence $C_{z_u} \sim 1/(1 - \mathbf{M}^2)$. At supersonic speeds, the two-dimensional lift is given by[3]

$$C_L = \frac{4\alpha}{\sqrt{\mathbf{M}^2 - 1}}$$

After differentiation with respect to **M**, we get exactly the same result as for subsonic speeds. That is, Eq. 5.3,6*b* applies over the whole Mach-number range, except of course near $\mathbf{M} = 1$ where the cited airfoil theories do not apply. Equation 5.3,6*b* gives an extreme value of C_{z_u}. Highly swept wings and low-aspect-ratio wings are less sensitive to changes in **M**.

The Derivative C_{m_u}

Significant contributions to C_{m_u} may come from aeroelastic or compressibility effects. In order to include both in a formal manner, let the dynamic pressure

[2] Reference 1.17, Sec. 12.3.
[3] Reference 1.17, Sec. 12.6.

be denoted by

$$p_d = \tfrac{1}{2}\rho v_c^{\,2} \tag{5.3,7}$$

Then, since the airframe distortions, and the load distribution, are dependent on p_d, we express C_m as

$$C_m = f(\alpha, \mathbf{M}, p_d)$$

Then

$$C_{m_u} = u_0\left(\frac{\partial C_m}{\partial u}\right)_0 = u_0\,\frac{\partial C_m}{\partial \mathbf{M}}\,\frac{\partial \mathbf{M}}{\partial u} + u_0\,\frac{\partial C_m}{\partial p_d}\left(\frac{\partial p_d}{\partial u}\right)_0 \tag{5.3,8}$$

We had previously that $\partial\mathbf{M}/\partial u = 1/a$, and, from Eq. 5.3,7,

$$\left(\frac{\partial p_d}{\partial u}\right)_0 = \rho u_0$$

Hence

$$C_{m_u} = \mathbf{M}\,\frac{\partial C_m}{\partial \mathbf{M}} + \rho u_0^{\,2}\,\frac{\partial C_m}{\partial p_d} \tag{5.3,9}$$

Values of $\partial C_m/\partial\mathbf{M}$ can be found from wind-tunnel tests on a rigid model. They are largest at transonic speeds and are strongly dependent on the wing planform. The main factor that contributes to this derivative is the backward shift of the wing center of pressure that occurs in the transonic range. On two-dimensional symmetrical wings, for example, the center of pressure moves from approximately $0.25c$ to approximately $0.50c$ as the Mach number increases from subsonic to supersonic values. Thus an increase in $\mathbf{M}$ in this range produces a diving-moment increment; i.e., C_{m_u} is negative. For wings of very low aspect ratio, the center of pressure movement is much less, and the values of C_{m_u} are correspondingly smaller.

To find $\partial C_m/\partial p_d$ requires either an aeroelastic analysis or tests on a flexible model. As an example of this phenomenon, let us consider an airplane with a tail and a flexible fuselage.[4] We found in Sec. 3.6 that the tail lift coefficient is given by

$$C_{L_t} = \frac{a_t}{1 + k a_t p_d S_t}\,(\alpha_{wb} - \varepsilon - i_t) \tag{5.3,10}$$

The pitching moment contributed by the tail is (Eq. 2.2,8)

$$C_{m_t} = -V_H C_{L_t}$$

Hence

$$\left(\frac{\partial C_m}{\partial p_d}\right)_{\text{tail}} = -V_H\,\frac{\partial C_{L_t}}{\partial p_d} \tag{5.3,11}$$

When Eq. 5.3,10 is differentiated with respect to p_d and simplified, and the resulting expression is substituted into Eq. 5.3,11, we obtain the result

$$\left(\frac{\partial C_m}{\partial p_d}\right)_{\text{tail}} = -C_{m_t}\,\frac{k a_t S_t}{1 + k a_t p_d S_t} \tag{5.3,12}$$

[4] It is not meant to imply that fuselage bending is the only important aeroelastic contribution to C_{m_u}. Distortion of the wing and tail may also be important.

The corresponding contribution to C_{m_u} is (see Eq. 5.3,9)

$$(C_{m_u})_{\text{tail}} = -C_{m_t} \frac{2p_d k a_t S_t}{1 + k a_t p_d S_t} \tag{5.3,13}$$

All the factors in this expression are positive, except for C_{m_t}, which may be of either sign. The contribution of the tail to C_{m_u} may therefore be either positive or negative. The tail pitching moment is usually positive at high speeds and negative at low speeds. Therefore its contribution to C_{m_u} is usually negative at high speeds and positive at low speeds. Since the dynamic pressure occurs as a multiplying factor in Eq. 5.3,13, then the aerolastic effect on C_{m_u} goes up with speed and down with altitude.

The Derivative C_{he_u}

As with the pitching moment, the effects of speed on C_{he} which we take into account are those associated with the Mach number and flexibility; i.e.,

$$\begin{aligned} C_{he_u} &= u_0 \frac{\partial C_{he}}{\partial \mathbf{M}} \frac{\partial \mathbf{M}}{\partial u} + u_0 \frac{\partial C_{he}}{\partial p_d} \left(\frac{\partial p_d}{\partial u} \right)_0 \\ &= \mathbf{M} \frac{\partial C_{he}}{\partial \mathbf{M}} + \rho u_0^2 \frac{\partial C_{he}}{\partial p_d} \end{aligned} \tag{5.3,14}$$

The terms of Eq. 5.3,14 would have to be determined by wind-tunnel testing and aeroelastic analysis.

5.4 THE q DERIVATIVES (C_{z_q}, C_{m_q}, C_{he_q})

These derivatives represent the aerodynamic effects that accompany rotation of the airplane about a spanwise axis through the C.G. while α remains zero. An example of this kind of motion was treated in Sec. 3.1 (i.e., the steady pull-up). Figure 5.2*b* shows the general case in which the flight path is arbitrary. This should be contrasted with the situation illustrated in Fig. 5.2*a*, where $q = 0$ while α is changing.

Both the wing and the tail are affected by the rotation, although, when the airplane has a tail, the wing contribution to C_{z_q} and C_{m_q} is often negligible in comparison with that of the tail. In such cases it is common practice to increase the tail effect by an arbitrary amount, of the order of 10%, to allow for the wing and body.

Contributions of a Tail

As was illustrated in Fig. 3.3, the main effect of q on the tail is to increase its angle of attack by (ql_t/u_0)radians, where u_0 is the flight speed. It is this change in α_t that accounts for the changed forces on the tail. The assumption is implicit in the

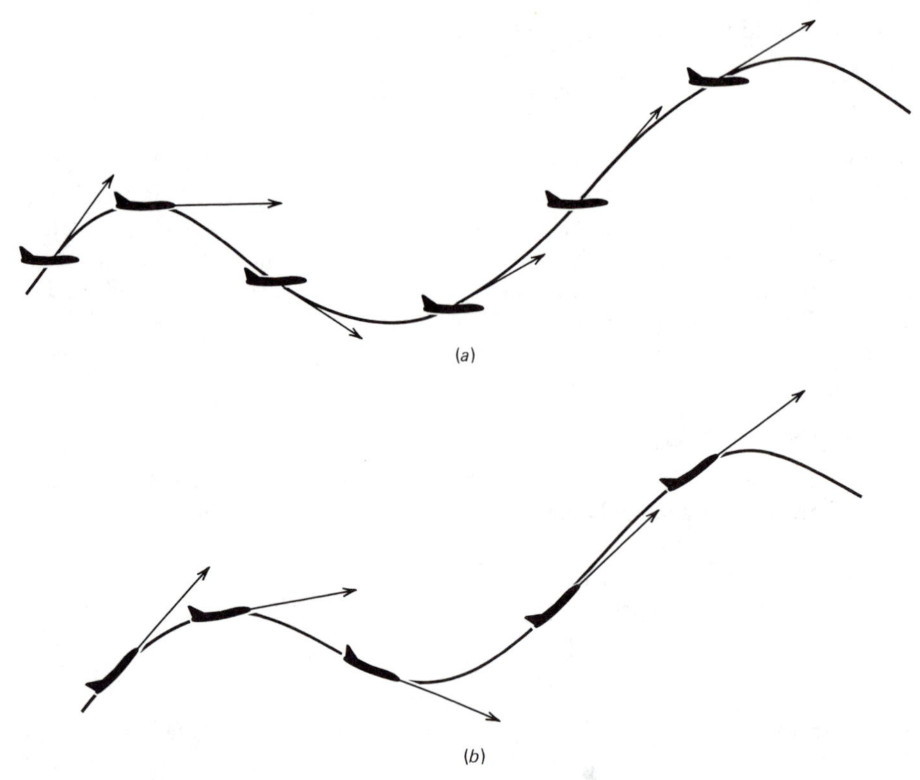

FIG. 5.2 (*a*) Motion with zero q, but varying α. (*b*) Motion with zero α, but varying q.

following derivations that the instantaneous forces on the tail correspond to its instantaneous angle of attack; i.e., no account is taken of the fact that it takes a finite time for the tail lift to build up to its steady-state value following a sudden change in q. (A method of including this refinement has recently been given by Tobak, ref. 5.1.) The derivatives obtained are therefore *quasistatic.*

C_{z_q} of the Tail

By definition, $C_{z_q} = (\partial C_z/\partial \hat{q})_0 = (2u_0/\bar{c})(\partial C_z/\partial q)_0$, and, from Eq. 5.1,1, $(\partial C_z/\partial q)_0 = -(\partial C_L/\partial q)_0$. The change in tail lift coefficient caused by the rotation q is

$$\Delta C_{L_t} = a_t \Delta\alpha_t = a_t \frac{ql_t}{u_0} \tag{5.4,1}$$

and the corresponding change in airplane lift coefficient is

$$\Delta C_L = \frac{S_t}{S}\Delta C_{L_t} = \frac{S_t}{S} a_t \frac{ql_t}{u_0}$$

Therefore

$$\frac{\partial C_L}{\partial q} = a_t \frac{S_t}{S} \frac{l_t}{u_0}$$

and

$$(C_{z_q})_{\text{tail}} = -\frac{2u_0}{\bar{c}} a_t \frac{S_t l_t}{S u_0} = -2a_t V_H \tag{5.4,2}$$

C_{m_q} of the Tail

The increment in pitching moment that corresponds to ΔC_{L_t} is (see Eq. 2.2,8)

$$\Delta C_m = -V_H \Delta C_{L_t} = -a_t V_H \frac{q l_t}{u_0}$$

Hence

$$\frac{\partial C_m}{\partial q} = -a_t V_H \frac{l_t}{u_0}$$

and

$$(C_{m_q})_{\text{tail}} = \frac{2u_0}{\bar{c}} \left(\frac{\partial C_m}{\partial q}\right)_0 = -2a_t V_H \frac{l_t}{\bar{c}} \tag{5.4,3}$$

C_{he_q} of the Tail

The change in α_t produces a change in the elevator hinge moment of amount

$$\Delta C_{he} = C_{he_{\alpha_t}} \frac{q l_t}{u_0}$$

Hence

$$C_{he_q} = \frac{2u_0}{\bar{c}} \left(\frac{\partial C_{he}}{\partial q}\right)_0 = 2 \frac{l_t}{\bar{c}} C_{he_{\alpha_t}} \tag{5.4,4}$$

Contributions of a Wing

As previously remarked, on airplanes with tails the wing contributions to the q derivatives are frequently negligible. However, if the wing is highly swept or of low aspect ratio, it may have significant values of C_{z_q} and C_{m_q}; and of course, on tailless airplanes, the wing supplies the major contribution. The q derivatives of wings alone are therefore of great engineering importance.

Unfortunately, no simple formulas can be given, because of the complicated dependence on the wing planform and the Mach number. However, the following discussion of the physical aspects of the flow indicates how linearized wing theory can be applied to the problem. Consider a plane lifting surface, at zero α, with forward speed u_0 and angular velocity q about a spanwise axis (Fig. 5.3). Each point in the wing has a velocity component, relative to the resting atmosphere, of qx normal to the surface. This velocity distribution is shown in the figure for the central and tip chords. Now there is an equivalent cambered wing that would have

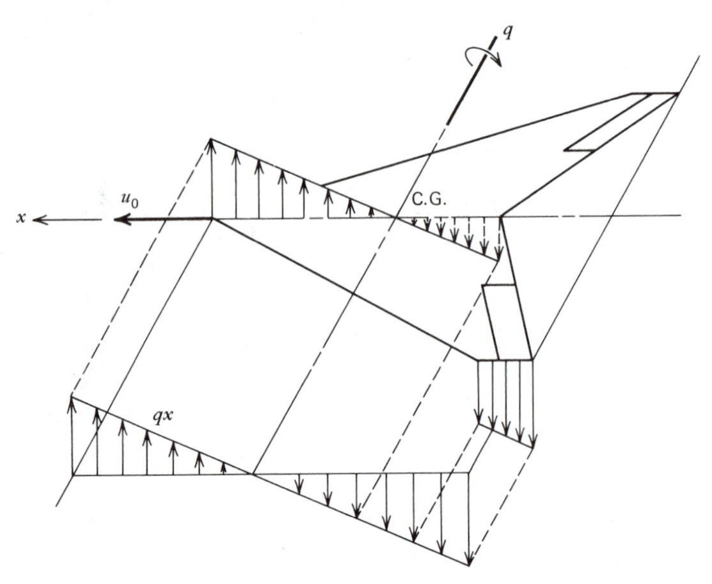

FIG. 5.3 Wing velocity distribution due to pitching.

the identical distribution of velocities normal to its surface when in rectilinear translation at speed u_0. This is illustrated in Fig. 5.4*a*. The cross section of the curved surface S is shown in (*b*). The normal velocity distribution will be the same as in Fig. 5.3 if

$$u_0 \frac{\partial z}{\partial x} = qx \quad \text{or} \quad \frac{\partial z}{\partial x} = \frac{q}{u_0} x$$

Hence

$$z = \frac{1}{2} \frac{q}{u_0} x^2 \tag{5.4,5}$$

and the cross section of S is a parabolic arc. In linearized wing theory, both subsonic and supersonic, the boundary condition is the same for the original plane wing with rotation q and the equivalent curved wing in rectilinear flight. The problem of finding the q derivatives then is reduced to that of finding the pressure distribution over the equivalent cambered wing. Because of the form of Eq. 5.4,5, the pressures are proportional to q/u_0. From the pressure distribution, C_{z_q}, C_{m_q}, and C_{he_q} can all be calculated. The derivatives can in principle also be found by experiment, by testing a model of the equivalent wing.

The values obtained by this approach are quasistatic; i.e., they are steady-state values corresponding to $\alpha = 0$ and a small constant value of q. This implies that the flight path is a circle (as in Fig. 3.1), and hence that the vortex wake is not rectilinear. Now both the linearized theory and the wind-tunnel measurement apply to a straight wake, and to this extent are approximate. Since the values of the derivatives obtained are in the end applied to arbitrary flight paths, as in Fig. 5.2*b*, there is little point in correcting them for the curvature of the wake.

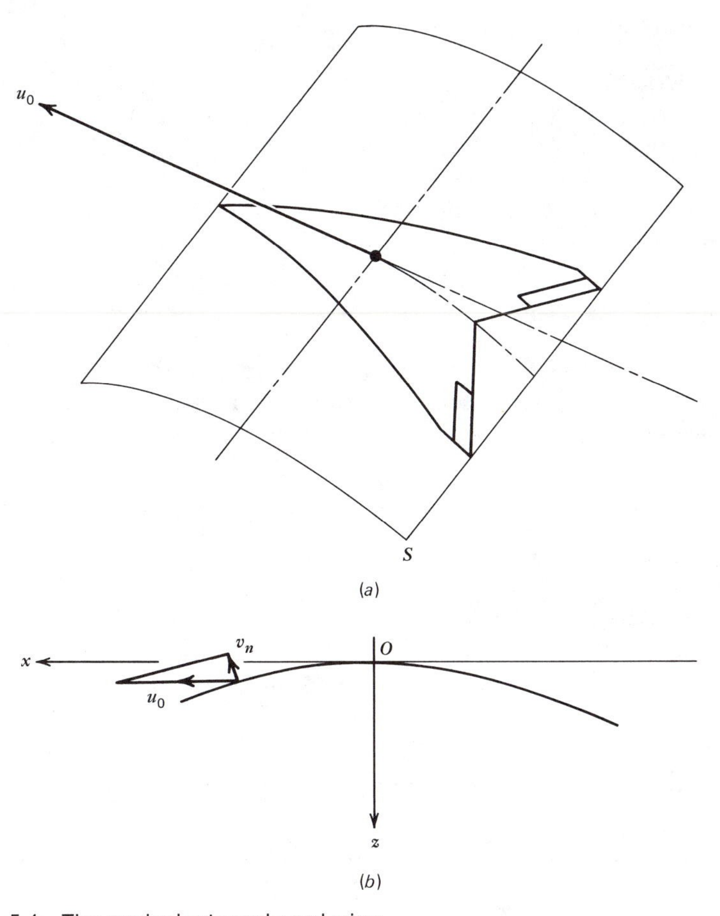

FIG. 5.4 The equivalent cambered wing.

The error involved in the application of the quasistatic derivatives to unsteady flight is not as great as might be expected. It has been shown that, when the flight path is a sine wave, the quasistatic derivatives apply so long as the reduced frequency is small, that is,

$$k = \frac{\omega \bar{c}}{2u_0} \ll 1 \tag{5.4,6}$$

where ω is the circular frequency of the pitching oscillation. If l is the wavelength of the flight path, then

$$k = \pi \frac{\bar{c}}{l}$$

so that the condition $k \ll 1$ implies that the wavelength must be long compared to the chord, e.g., $l > 60\bar{c}$ for $k < 0.05$.

5.5 THE $\dot{\alpha}$ DERIVATIVES ($C_{z_{\dot{\alpha}}}$, $C_{m_{\dot{\alpha}}}$, $C_{he_{\dot{\alpha}}}$)

The $\dot{\alpha}$ derivatives owe their existence to the fact that the pressure distribution on a wing or tail does not adjust itself instantaneously to its equilibrium value when the angle of attack is suddenly changed. The calculation of this effect, or its measurement, involves unsteady flow. In this respect, the $\dot{\alpha}$ derivatives are very different from those discussed previously, which can all be determined on the basis of steady-state aerodynamics.

Contributions of a Wing

Consider a wing in horizontal flight at zero α. Let it be subjected to a downward impulse, so that it suddenly acquires a constant downward velocity component. Then, as shown in Fig. 5.5, its angle of attack undergoes a step increase. The lift

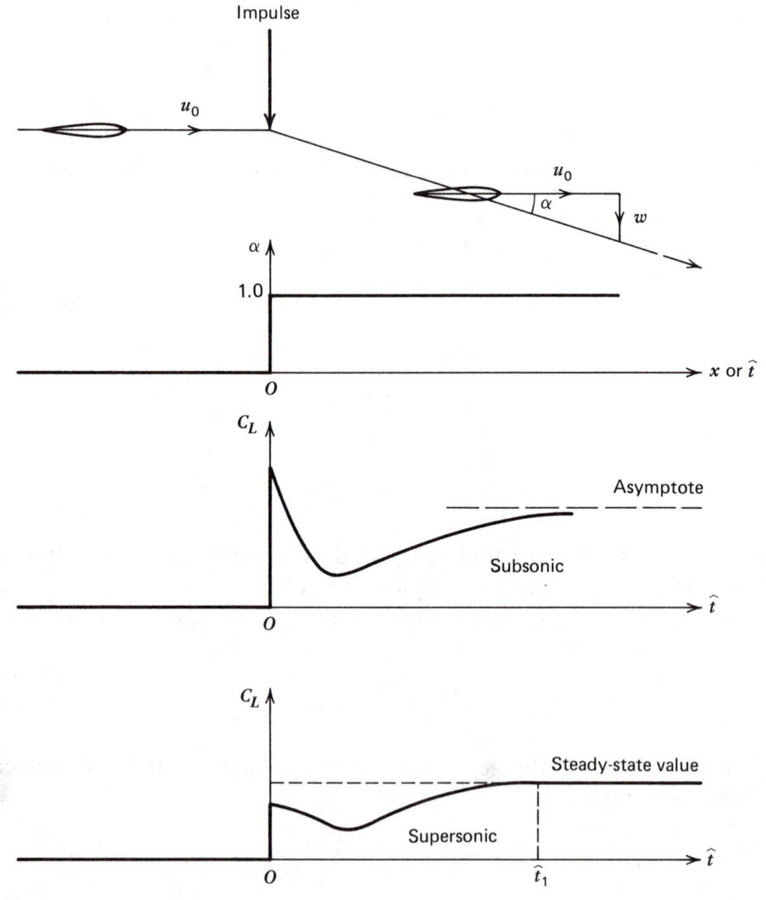

FIG. 5.5 Lift response to step change in α.
(After Tobak, *NACA Rept. 1188*.)

then responds in a transient manner (the *indicial* admittance, see Sec. 8.5) the form of which depends on whether **M** is greater or less than 1. In subsonic flight, the vortices the wing leaves behind it can influence it at all future times, so that the steady state is approached only asymptotically. In supersonic flight, the upstream traveling disturbances move more slowly than the wing, so that it outstrips the disturbance field of the initial impulse in a finite time t_1. From that time on the lift remains constant.

In order to find the lift associated with $\dot{\alpha}$, let us consider the motion of an airfoil with a small constant $\dot{\alpha}$, but with $q = 0$. The motion, and the angle of attack, are shown in Fig. 5.6. The method used follows that introduced by Tobak (ref. 5.1).

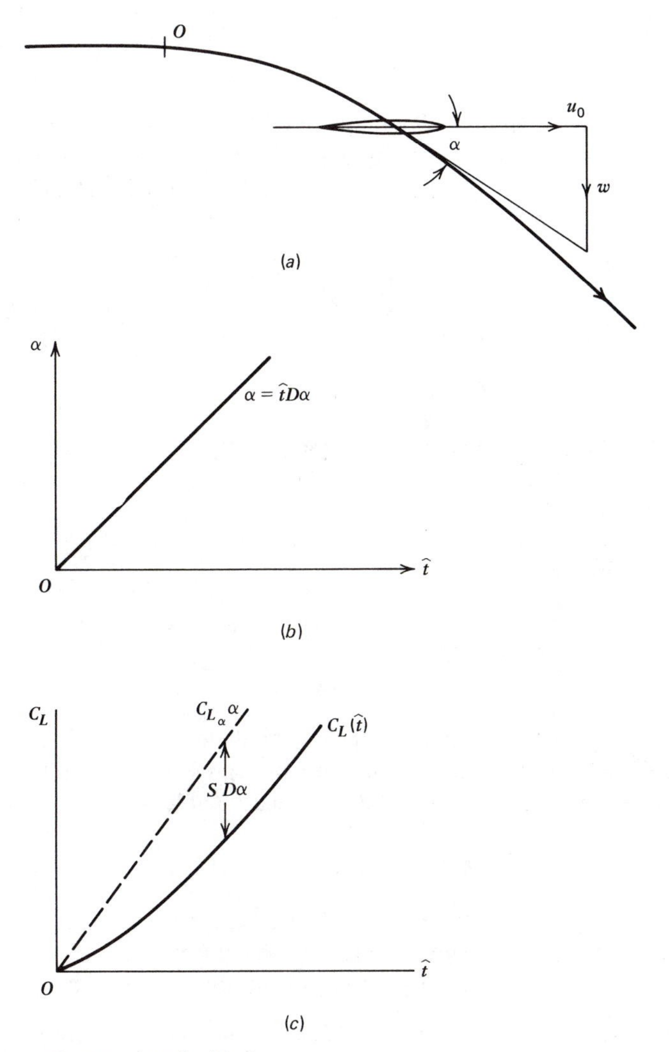

FIG. 5.6 Lift associated with $\dot{\alpha}$.

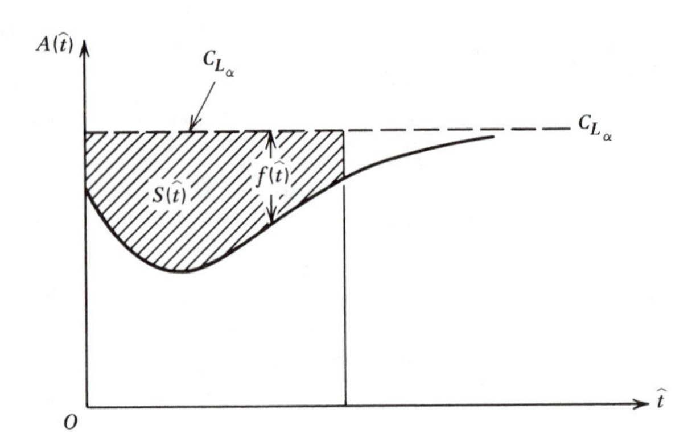

FIG. 5.7

We assume that the differential equation that relates $C_L(\hat{t})$ with $\alpha(\hat{t})$ is linear. Hence the method of superposition may be used to derive the response to a linear $\alpha(\hat{t})$ from the indicial admittance (see Sec. 8.6). Let the admittance be $A(\hat{t})$. Then, by Eq. 8.6,5*b*, the lift coefficient at time $\hat{t}$ is

$$C_L(\hat{t}) = \int_{\tau=0}^{\hat{t}} A(\hat{t} - \tau)\alpha'(\tau)\,d\tau$$

Since $\alpha'(\tau) = D\alpha =$ constant, then

$$C_L(\hat{t}) = D\alpha \int_{\tau=0}^{\hat{t}} A(\hat{t} - \tau)\,d\tau \tag{5.5,1}$$

The ultimate C_L response to a unit step α input is C_{L_α}. Let the lift defect (see Fig. 5.7) be $f(\hat{t})$; i.e.,

$$f(\hat{t}) = C_{L_\alpha} - A(\hat{t})$$

Then Eq. 5.5,1 becomes

$$\begin{aligned} C_L(\hat{t}) &= D\alpha\, C_{L_\alpha}\hat{t} - D\alpha \int_{\tau=0}^{\hat{t}} f(\hat{t} - \tau)\,d\tau \\ &= C_{L_\alpha}\alpha - S\,D\alpha \end{aligned} \tag{5.5,2}$$

where $S(\hat{t}) = \int_{\tau=0}^{\hat{t}} f(\hat{t} - \tau)\,d\tau$ is the area shown hatched on Fig. 5.7. The term $S\,D\alpha$ is also shown on Fig. 5.6. Now, if the idea of representing the lift by means of aerodynamic derivatives is to be valid, we must be able to write, for the motion in question,

$$C_L(\hat{t}) = C_{L_\alpha}\alpha(\hat{t}) + C_{L_{\dot{\alpha}}} D\alpha \tag{5.5,3}$$

where C_{L_α} and $C_{L_{\dot{\alpha}}}$ are constants. Comparing Eqs. 5.5,2 and 5.5,3, we find that $C_{L_{\dot{\alpha}}} = -S(\hat{t})$, a function of time. Hence, during the initial part of the motion, the *derivative concept is invalid.* However, for all finite wings,[5] the area $S(\hat{t})$ con-

[5] For two-dimensional incompressible flow, the area $S(\hat{t})$ diverges as $\hat{t} \to \infty$. That is, the derivative concept is definitely *not* applicable to that case.

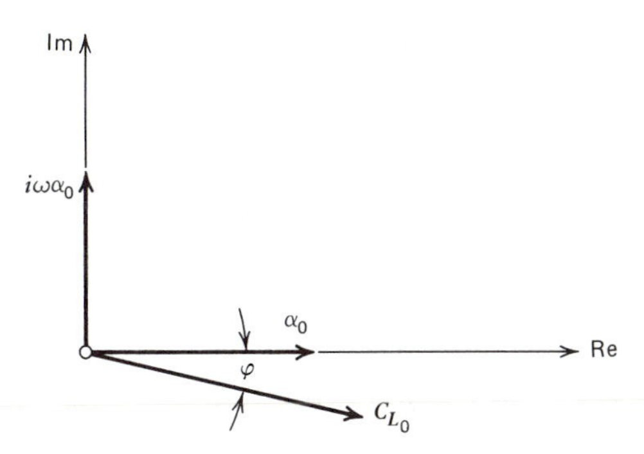

FIG. 5.8 Vector diagram of lift response to oscillatory α.

verges to a finite value as $\hat{t}$ increases indefinitely. In fact, for supersonic wings, S reaches its limiting value in a finite time, as is evident from Fig. 5.5. Thus Eq. 5.5,3 is valid,[6] with constant $C_{L_{\dot{\alpha}}}$, for values of $\hat{t}$ greater than a certain minimum. This minimum is not large, being the time required for the wing to travel a few chord lengths. In the time range where S is constant, or differs only infinitesimally from its asymptotic value, the $C_L(\hat{t})$ curve of Fig. 5.6c is parallel to $C_{L_\alpha}\alpha$. A similar situation exists with respect to C_m and C_{he}.

There is a second useful approach to the $\dot{\alpha}$ derivatives, and that is via the theory of oscillating wings. Extensive treatments of wings in oscillatory motion are available in the literature,[7] primarily in relation to flutter problems. Because of the time lag previously noted, the amplitude and phase of the oscillatory lift will be different from the quasisteady values. Let us represent the periodic angle of attack and lift coefficient by the complex numbers

$$\alpha = \alpha_0 e^{i\omega t} \quad \text{and} \quad C_L = C_{L_0} e^{i\omega t} \tag{5.5,4}$$

where α_0 is the amplitude (real) of α, and C_{L_0} is a complex number such that $|C_{L_0}|$ is the amplitude of the C_L response, and arg C_{L_0} is its phase angle. The relation between C_{L_0} and α_0 appropriate to the low frequencies characteristic of dynamic stability is illustrated in Fig. 5.8. In terms of these vectors, we may derive the value of $C_{L_{\dot{\alpha}}}$ as follows. The $\dot{\alpha}$ vector is

$$\dot{\alpha} = i\omega\alpha_0 e^{i\omega t}$$

Thus C_L may be expressed as

$$\begin{aligned} C_L &= R[C_{L_0}]e^{i\omega t} + iI[C_{L_0}]e^{i\omega t} \\ &= R[C_{L_0}]\frac{\alpha}{\alpha_0} + I[C_{L_0}]\frac{\dot{\alpha}}{\omega\alpha_0} \end{aligned}$$

Hence

$$C_{L_{\dot{\alpha}}} = \frac{\partial C_L}{\partial(\dot{\alpha}\bar{c}/2u_0)} = \frac{I[C_{L_0}]}{k\alpha_0} \tag{5.5,5}$$

[6] Exactly for supersonic wings, and approximately for subsonic wings.
[7] See bibliography at the end of the chapter.

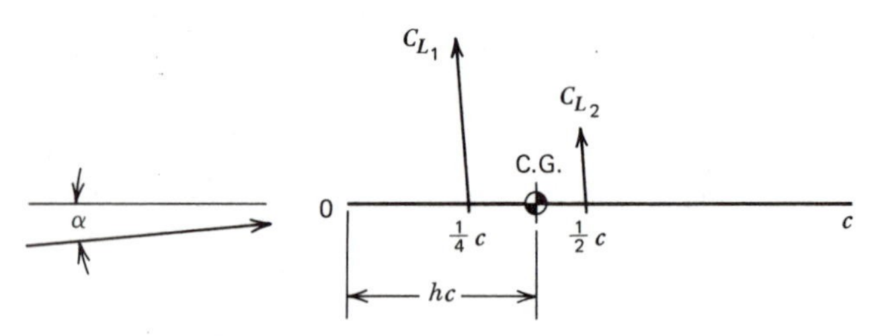

FIG. 5.9 Lift on oscillating two-dimensional airfoil.

or, if the amplitude α_0 is unity, $C_{L_{\dot{\alpha}}} = I[C_{L_0}]/k$, where k is the reduced frequency $\omega\bar{c}/2u_0$.

To assist in forming a physical picture of the behavior of a wing under these conditions, we give here the results for a two-dimensional airfoil in incompressible flow. The motion of the airfoil is a plunging oscillation; i.e., it is like that shown in Fig. 5.2*a*, except that the flight path is a sine wave. The instantaneous lift on the airfoil is given in two parts (see Fig. 5.9):

$$C_L = C_{L_1} + C_{L_2}$$

where

$$C_{L_1} = 2\pi\left[\alpha F(k) + \left(\frac{\dot{\alpha}c}{2u_0}\right)\frac{G(k)}{k}\right] \tag{5.5,6}$$

$$C_{L_2} = \pi\left(\frac{\dot{\alpha}c}{2u_0}\right)$$

and $F(k)$ and $G(k)$ are the real and imaginary parts of the Theodorsen function $C(k)$.[8] The lift that acts at the midchord is proportional to $\dot{\alpha} = \ddot{z}/u_0$, where z is the translation (vertically downward) of the airfoil. That is, it represents a force opposing the downward acceleration of the airfoil. This force is exactly that which is required to impart an acceleration $\ddot{z}$ to a mass of air contained in a cylinder, the diameter of which equals the chord c. This is known as the "apparent additional mass." It is as though the mass of the airfoil were increased by this amount. Except in cases of very low relative density $\mu = 2m/\rho S\bar{c}$, this added mass is small compared to that of the airplane itself, and hence the force C_{L_2} is relatively unimportant. Physically, the origin of this force is in the reaction of the air associated with its downward acceleration. The other component, C_{L_1}, which acts at the $\frac{1}{4}$ chord point, is associated with the circulation around the airfoil, and is a consequence of the imposition of the Kutta–Joukowski condition at the trailing edge. It is seen that it contains one term proportional to α and another proportional to $\dot{\alpha}$. From Fig. 5.9, the pitching-moment coefficient about the C.G. is obtained as

$$C_m = C_{L_1}(h - \tfrac{1}{4}) + C_{L_2}(h - \tfrac{1}{2}) \tag{5.5,7}$$

[8] A table and graph of $F(k)$ and $G(k)$ are given in ref. 1.19, pp. 214, 215.

From Eqs. 5.5,6 and 5.5,7, the following derivatives are found for frequency k.

$$\begin{aligned} C_{L_\alpha} &= 2\pi F(k) \\ C_{L_{\dot\alpha}} &= \pi + 2\pi \frac{G(k)}{k} \\ C_{m_\alpha} &= 2\pi F(k)(h - \tfrac{1}{4}) \\ C_{m_{\dot\alpha}} &= \pi(h - \tfrac{1}{2}) + 2\pi \frac{G(k)}{k}(h - \tfrac{1}{4}) \end{aligned} \tag{5.5,8}$$

The awkward situation is evident, from Eqs. 5.5,8, that the derivatives are frequency-dependent. That is, in free oscillations one does not know the value of the derivative until the solution to the motion (i.e., the frequency) is known. In cases of forced oscillations at a given frequency, this difficulty is not present. Since the entire linearized theory of airplane dynamics is based on small perturbations from uniform flight, it is consistent to compute the derivatives in question for the limiting case $k \to 0$. The $\lim_{k\to 0} F(k) = 1$, so that $C_{L_\alpha} = 2\pi$ and $C_{m\alpha} = 2\pi(h - \frac{1}{4})$, the theoretical steady flow values. This conclusion (i.e., that C_{L_α} and C_{m_α} are the quasistatic values) also holds for finite wings at all Mach numbers. The result for the limiting values of $C_{L_{\dot\alpha}}$ and $C_{m_{\dot\alpha}}$ is not so clear, since $\lim_{k\to 0} G(k)/k = \infty$. However this anomaly is entirely a result of the assumption of two-dimensional flow in the derivation of Eqs. 5.5,6. For wings of finite aspect ratio the limiting value of the corresponding quantity is finite (see ref. 5.2). This leads to definite values of $C_{L_{\dot\alpha}}$ and $C_{m_{\dot\alpha}}$. If the airfoil has a control flap, the hinge moment associated with $\dot\alpha$ behaves like $C_{L_{\dot\alpha}}$ and $C_{m_{\dot\alpha}}$. The limiting values described above can be obtained from a first-order-in-frequency analysis of an oscillating wing.

In summary, the $\dot\alpha$ *derivatives of wing alone may be computed from the indicial admittances of lift, pitching moment, and hinge moment, or from first-order-in-frequency analyses of harmonically plunging wings.*

Contributions of a Tail

There is an approximate method for evaluating the contributions of a tail surface, which is satisfactory in many cases. This is based on the concept of the *lag of the downwash.* It neglects entirely the nonstationary character of the lift response of the tail to changes in tail angle of attack, and attributes the result entirely to the fact that the downwash at the tail does not respond instantaneously to changes in wing angle of attack. The downwash is assumed to be dependent primarily on the strength of the wing's trailing vortices in the neighborhood of the tail. Since the vorticity is convected with the stream, then a change in the circulation at the wing will not be felt as a change in downwash at the tail until a time $\Delta t = l_t/u_0$ has elapsed, where l_t is the tail length. It is therefore assumed that the instantaneous downwash at the tail, $\varepsilon(t)$, corresponds to the wing α at time $(t - \Delta t)$. The corrections to the quasistatic downwash and tail angle of attack are therefore

$$\Delta\varepsilon = -\frac{\partial\varepsilon}{\partial\alpha}\dot{\alpha}\,\Delta t = -\frac{\partial\varepsilon}{\partial\alpha}\dot{\alpha}\frac{l_t}{u_0}$$

$$= -\Delta\alpha_t \tag{5.5,9}$$

$C_{z_{\dot{\alpha}}}$ of a Tail

The correction to the tail lift coefficient for the downwash lag is

$$\Delta C_{L_t} = a_t\,\Delta\alpha_t = a_t\dot{\alpha}\frac{l_t}{u_0}\frac{\partial\varepsilon}{\partial\alpha} \tag{5.5,10}$$

The correction to the airplane lift is therefore

$$\Delta C_L = a_t\dot{\alpha}\frac{l_t}{u_0}\frac{\partial\varepsilon}{\partial\alpha}\frac{S_t}{S}$$

Therefore

$$\frac{\partial C_z}{\partial\dot{\alpha}} = -\frac{\partial C_L}{\partial\dot{\alpha}} = -a_t\frac{\partial\varepsilon}{\partial\alpha}\frac{l_t}{u_0}\frac{S_t}{S}$$

and

$$(C_{z_{\dot{\alpha}}})_{\text{tail}} = \frac{\partial C_z}{\partial\left(\dfrac{\dot{\alpha}\bar{c}}{2u_0}\right)} = -2a_t V_H\frac{\partial\varepsilon}{\partial\alpha} \tag{5.5,11}$$

$C_{m_{\dot{\alpha}}}$ of a Tail

The correction to the pitching moment is obtained from ΔC_{L_t} as

$$\Delta C_m = -V_H\,\Delta C_{L_t} = -a_T\dot{\alpha}\frac{\partial\varepsilon}{\partial\alpha}\frac{l_T}{u_0}V_H$$

Therefore

$$\frac{\partial C_m}{\partial\dot{\alpha}} = -a_t V_H\frac{l_t}{u_0}\frac{\partial\varepsilon}{\partial\alpha}$$

and

$$(C_{m_{\dot{\alpha}}})_{\text{tail}} = \frac{\partial C_m}{\partial\left(\dfrac{\dot{\alpha}\bar{c}}{2u_0}\right)} = -2a_t V_H\frac{l_t}{\bar{c}}\frac{\partial\varepsilon}{\partial\alpha} \tag{5.5,12}$$

$C_{he_{\dot{\alpha}}}$ of a Tail

The correction to α_t produces a change in the elevator hinge moment

$$\Delta C_{he} = C_{he_{\alpha_t}}\Delta\alpha_t = C_{he_{\alpha_t}}\frac{\partial\varepsilon}{\partial\alpha}\dot{\alpha}\frac{l_t}{u_0}$$

Therefore

$$\frac{\partial C_{he}}{\partial\dot{\alpha}} = C_{he_{\alpha_t}}\frac{\partial\varepsilon}{\partial\alpha}\frac{l_t}{u_0}$$

and

$$C_{he_{\dot{\alpha}}} = \frac{\partial C_{he}}{\partial\left(\dfrac{\dot{\alpha}\bar{c}}{2u_0}\right)} = 2C_{he_{\alpha_t}}\frac{l_t}{\bar{c}}\frac{\partial \varepsilon}{\partial \alpha} \tag{5.5,13}$$

5.6 THE η DERIVATIVES (C_{z_η}, C_{m_η}, C_{he_η})

These three derivatives give the changes in the aerodynamic forces and moments that accompany a deflection of the elevator. They are simply related to the static control parameters introduced in Chap. 2.

The Derivative C_{z_η}

From Eq. 5.1,1 we get

$$C_{z_\eta} = \left(\frac{\partial C_z}{\partial \eta}\right)_0 = -\left(\frac{\partial C_L}{\partial \eta}\right)_0$$

$\partial C_L/\partial \eta$ is identical with the control parameter C_{L_δ} (see Secs. 2.4 and 2.5). Hence

$$C_{z_\eta} = -C_{L_\delta} \tag{5.6,1}$$

On airplanes with a tail, from Eq. 2.5,1,

$$C_{L_\delta} = a_e \frac{S_t}{S}$$

On tailless airplanes, C_{L_δ} is one of the primary parameters to be estimated or found from tunnel testing.

The Derivative C_{m_η}

As with C_{z_η} above, we find that

$$C_{m_\eta} = C_{m_\delta} \tag{5.6,2}$$

which, on airplanes with a tail, is (Eq. 2.5,3*b*)

$$C_{m_\eta} = -a_e V_H$$

The Derivative C_{he_η}

Likewise, we have that (Eq. 2.6,1)

$$C_{he_\eta} = C_{h_\delta} = b_2$$

which is the rate of change of elevator hinge-moment coefficient with elevator angle.

5.7 THE $\dot{\eta}$ DERIVATIVES ($C_{m_{\dot{\eta}}}$, $C_{he_{\dot{\eta}}}$)

The two derivatives of this section are closely related to those of Sec. 5.5, in that they involve oscillating-wing theory. The discussions of the nonstationary aspects of wing theory contained in that section are applicable to the $\dot{\eta}$ derivatives, with the difference that it is the control-flap angle which oscillates harmonically, while the main surface angle of attack remains zero. The required derivatives can be obtained from oscillating-flap studies carried to the first order in the reduced frequency.

Although the effect of $\dot{\eta}$ on C_m may often be unimportant, the effect on the hinge moment, (i.e., $C_{he_{\dot{\eta}}}$) is generally not. This is because $C_{he_{\dot{\eta}}}$ is frequently the major part of the control-system damping.

5.8 THE β DERIVATIVES (C_{y_β}, C_{l_β}, C_{n_β}, C_{hr_β})

These derivatives all are obtainable from wind-tunnel tests on yawed models (ref. 7.1). Generally speaking, estimation methods do not give completely reliable results, and testing is a necessity.

The Derivative C_{y_β}

This is the side-force derivative, giving the force that acts in the y direction (right) when the airplane has a positive β or v (i.e., a sideslip to the right, see Fig. 3.12). C_{y_β} is usually negative, and frequently small enough to be neglected entirely. The main contributions are those of the body and the vertical tail, although the wing, and wing-body interference, may modify it significantly. Of these, only the tail effect is readily estimated. It may be expressed in terms of the vertical-tail lift-curve slope and the sidewash factor (see Sec. 3.9). (In this and the following sections the fin velocity ratio V_F/V is assumed to be unity.)

$$(C_y)_{\text{tail}} = -a_F(\beta - \sigma)\frac{S_F}{S}$$

or

$$(C_{y_\beta})_{\text{tail}} = -a_F\left(1 - \frac{\partial\sigma}{\partial\beta}\right)\frac{S_F}{S} \tag{5.8,1}$$

The most troublesome component of this equation is the sidewash derivative $\partial\sigma/\partial\beta$, which is difficult to estimate because of its dependence on the wing and fuselage geometry (see Sec. 3.9).

The Derivative C_{l_β}

C_{l_β} is the dihedral effect, which was discussed at some length in Sec. 3.10.

The Derivative C_{n_β}

C_{n_β} is the weathercock stability derivative, dealt with in Sec. 3.9.

The Derivative C_{hr_β}

This derivative gives the rudder hinge moment due to sideslip. It is analogous to the elevator hinge moment due to angle of attack. It is given by

$$\frac{\partial C_{hr}}{\partial \beta} = C_{hr_{\alpha_F}} \frac{\partial \alpha_F}{\partial \beta}$$

where $C_{hr_{\alpha_F}}$ is the C_{h_α} of the rudder. By using Eq. 3.9,2 we get

$$C_{hr_\beta} = -C_{hr_{\alpha_F}}\left(1 - \frac{\partial \sigma}{\partial \beta}\right) \tag{5.8,2}$$

5.9 THE p DERIVATIVES (C_{y_p}, C_{l_p}, C_{n_p}, C_{ha_p}, C_{hr_p})

When an airplane rolls with angular velocity p about its x axis (the flight direction), its motion is instantaneously like that of a screw. This motion affects the airflow (local angle of attack) at all stations of the wing and tail surfaces. This is illustrated in Fig. 5.10 for two points: a wing tip and the fin tip. It should be noted that the nondimensional rate of roll, $\hat{p} = pb/2u_0$ is, for small p, the angle (in radians) of the helix traced by the wing tip. These angle-of-attack changes

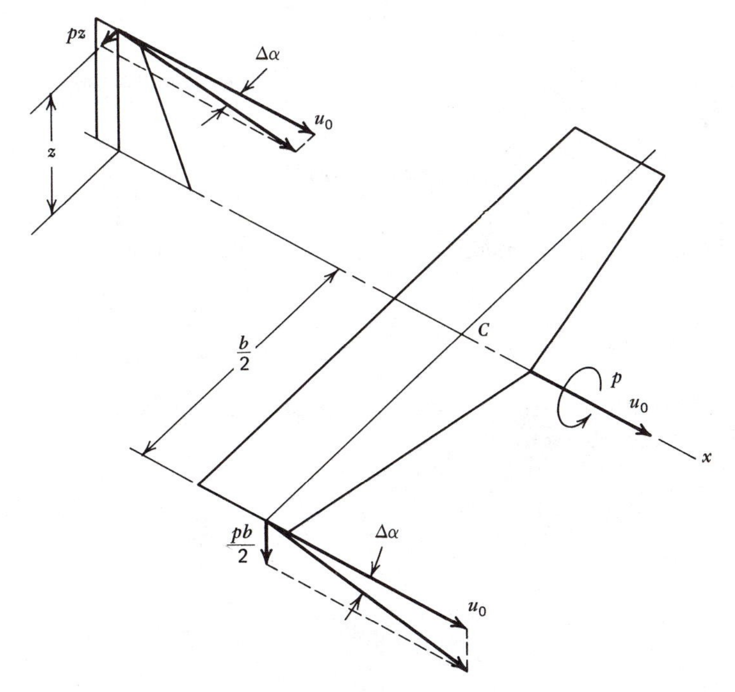

FIG. 5.10 Angle of attack changes due to p.

bring about alterations in the aerodynamic load distribution over the surfaces, and thereby introduce perturbations in the forces and moments. The change in the wing load distribution also causes a modification to the trailing vortex sheet. The vorticity distribution in it is no longer symmetrical about the x axis, and a sidewash (positive, i.e., to the right) is induced at a vertical tail conventionally placed. This further modifies the angle-of-attack distribution on the vertical-tail surface. This sidewash due to rolling is characterized by the derivative $\partial\sigma/\partial\hat{p}$. It has been studied theoretically and experimentally by Michael (ref. 5.3), who has shown its importance in relation to correct estimation of the tail contributions to the rolling derivatives. Finally, the helical motion of the wing produces a trailing vortex sheet that is not flat, but helical. For the small rates of roll admissible in a linear theory, this effect may be neglected with respect to both wing and tail forces.

The Derivative C_{y_p}

The side force due to rolling is often negligible. When it is not, the contributions that need to be considered are those from the wing[9] and from the vertical tail. The vertical-tail effect may be estimated in the light of its angle-of-attack change (Fig. 5.10) as follows. Let the mean change in α_F (Fig. 3.13) due to the rolling velocity be

$$\Delta\alpha_F = -\frac{pz_F}{u_0} + p\frac{\partial\sigma}{\partial p}$$

where z_F is an appropriate mean height of the fin. Introducing the nondimensional rate of roll, we may rewrite this as

$$\Delta\alpha_F = -\hat{p}\left(2\frac{z_F}{b} - \frac{\partial\sigma}{\partial\hat{p}}\right) \tag{5.9,1}$$

The incremental side-force coefficient on the fin is obtained from $\Delta\alpha_F$,

$$\Delta C_{y_F} = a_F\,\Delta\alpha_F = -a_F\hat{p}\left(2\frac{z_F}{b} - \frac{\partial\sigma}{\partial\hat{p}}\right) \tag{5.9,2}$$

where a_F is the lift-curve slope of the vertical tail. The incremental side force on the airplane is then given by

$$\Delta C_y = \frac{S_F}{S}\,\Delta C_{y_F} = -a_F\hat{p}\frac{S_F}{S}\left(2\frac{z_F}{b} - \frac{\partial\sigma}{\partial\hat{p}}\right)$$

thus

$$(C_{Y_p})_{\text{tail}} = -a_F\frac{S_F}{S}\left(2\frac{z_F}{b} - \frac{\partial\sigma}{\partial\hat{p}}\right) \tag{5.9,3}$$

[9] For the effect of the wing at low speeds, see ref. 7.1.

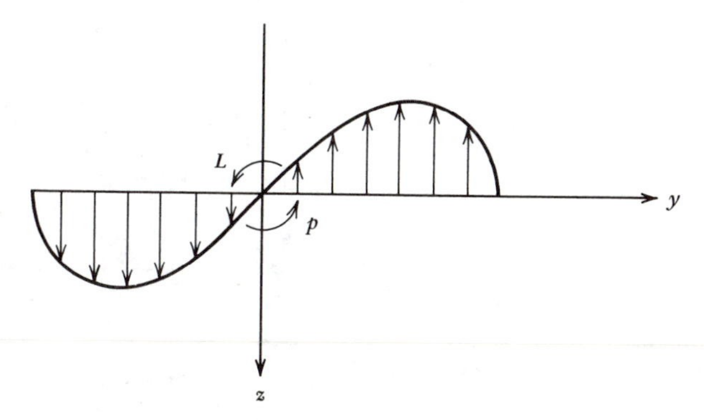

FIG. 5.11 Spanwise lift distribution due to rolling.

The Derivative C_{l_p}

C_{l_p} is known as the *damping-in-roll* derivative. It expresses the resistance of the airplane to rolling. Except in unusual circumstances, only the wing contributes significantly to this derivative. As can be seen from Fig. 5.10, the angle of attack due to p varies linearly across the span, from the value $pb/2u_0$ at the right wing tip to $-pb/2u_0$ at the left tip. This antisymmetric α distribution produces an antisymmetric increment in the lift distribution as shown in Fig. 5.11. In the linear range this is superimposed on the symmetric lift distribution associated with the wing angle of attack in undisturbed flight. The large rolling moment L produced by this lift distribution is proportional to the tip angle of attack $\hat{p}$, and C_{l_p} is a negative constant, so long as the local angle of attack remains below the local stalling angle.[10]

The Derivative C_{n_p}

The yawing moment produced by the rolling motion is one of the so called *cross derivatives*. It is the existence of these cross derivatives that causes the rolling and yawing motions to be so closely coupled. The wing and tail both contribute to C_{n_p}.

The wing contribution is in two parts. The first comes from the change in profile drag associated with the change in wing angle of attack. The wing α is increased on the right-hand side and decreased on the left-hand side. These changes will normally be accompanied by an increase in profile drag on the right side, and a decrease on the left side, combining to produce a *positive* (nose-right) yawing moment. The second wing effect is associated with the fore-and-aft inclination of the lift vector caused by the rolling in subsonic flight and in supersonic flight when the leading edge is subsonic. It depends on the leading-edge suction. The physical situation is illustrated in Fig. 5.12. The directions of motion of two typical wing

[10] Should the downgoing wing angle of attack exceed the stalling angle, the local lift-curve slope may fall to zero or even reverse its sign. C_{l_p} may then be zero or become positive. This is the situation when a wing *autorotates*, as in spinning.

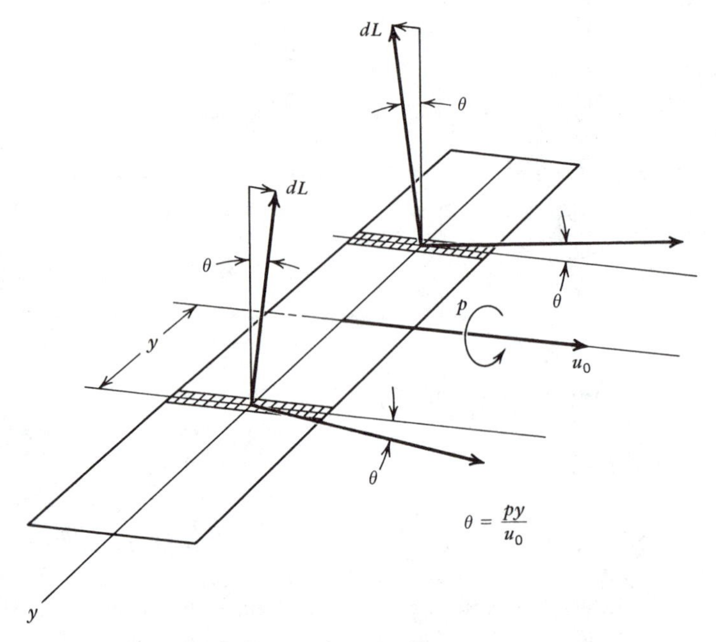

FIG. 5.12 Inclination of C_L vector due to rolling.

elements are shown inclined by the angles $\pm\theta = py/u_0$ from the direction of the vector u_0. Since the local lift is perpendicular to the local relative wind, then the lift vector on the right half of the wing is inclined forward, and that on the left half backward. The result is a *negative* yawing couple, proportional to the product $C_L\hat{p}$. If the wing leading edges are supersonic, then the leading-edge suction is not present, and the local force remains normal to the surface. The increased angle of attack on the right side causes an increase in this normal force there, while the opposite happens on the left side. The result is a positive yawing couple proportional to $\hat{p}$.

The tail contribution to C_{n_p} is easily found from the tail side force given previously (Eq. 5.9,2). The incremental C_n is given by

$$(\Delta C_n)_{\text{tail}} = -\Delta C_{y_F} \frac{S_F}{S} \frac{l_F}{b}$$

where l_F is the distance shown in Fig. 3.13. Therefore

$$(\Delta C_n)_{\text{tail}} = a_F\hat{p} \frac{S_F}{S} \frac{l_F}{b}\left(2\frac{z_F}{b} - \frac{\partial\sigma}{\partial\hat{p}}\right)$$

and

$$(C_{n_p})_{\text{tail}} = a_F V_V\left(2\frac{z_F}{b} - \frac{\partial\sigma}{\partial\hat{p}}\right) \tag{5.9,4}$$

where V_V is the vertical-tail volume ratio.

The Derivative C_{ha_p}

This derivative gives the change of aileron hinge moment due to rolling. It occurs because of the change in wing angle of attack at the ailerons, and because C_{h_α} of the ailerons is usually nonzero. Let y_a be the spanwise coordinate of the right hand mid-aileron section. Then the approximate change in angle of attack at the right-hand aileron is

$$\Delta\alpha = \frac{py_a}{u_0}$$

and

$$\Delta C_{ha} = C_{ha_\alpha}\frac{py_a}{u_0}$$

Therefore

$$C_{ha_p} = \frac{2y_a}{b}C_{ha_\alpha} \tag{5.9,5}$$

The Derivative C_{hr_p}

The change in vertical-tail angle of attack brought about by p produces a change in the rudder hinge moment. This is given by

$$\Delta C_{hr} = -C_{hr_{\alpha_F}}\hat{p}\left(2\frac{z_F}{b} - \frac{\partial\sigma}{\partial\hat{p}}\right)$$

therefore

$$C_{hr_p} = -C_{hr_{\alpha_F}}\left(2\frac{z_F}{b} - \frac{\partial\sigma}{\partial\hat{p}}\right) \tag{5.9,6}$$

When $C_{hr_{\alpha_F}}$ is negative, as for a simple flap control, then a positive roll produces a positive rudder hinge moment.

5.10 THE r DERIVATIVES (C_{y_r}, C_{l_r}, C_{n_r}, C_{ha_r}, C_{hr_r})

When an airplane has a rate of yaw r superimposed on the forward motion u_0, its velocity field is altered significantly. This is illustrated for the wing and vertical tail in Fig. 5.13. The situation on the wing is clearly very complicated when it has much sweepback. The main feature however, is that the velocity of the $\frac{1}{4}$ chord line normal to itself is increased by the yawing on the left-hand side, and decreased on the right side. The aerodynamic forces at each section (lift, drag, moment) are therefore increased on the left-hand side, and decreased on the right-hand side. As in the case of the rolling wing, the unsymmetrical lift distribution leads to an unsymmetrical trailing vortex sheet, and hence a sidewash

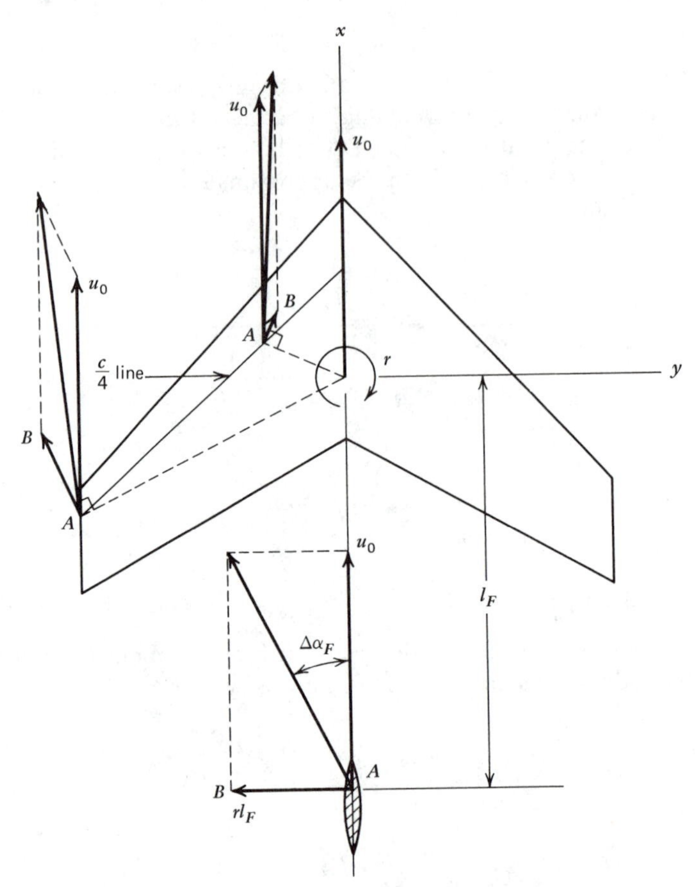

FIG. 5.13 Velocity field due to yawing. $\overrightarrow{AB}$ = velocity vector due to rate of yaw r.

at the tail. The incremental tail angle of attack is then

$$\Delta\alpha_F = \frac{rl_F}{u_0} + r\frac{\partial\sigma}{dr}$$

or

$$\Delta\alpha_F = \hat{r}\left(2\frac{l_F}{b} + \frac{\partial\sigma}{\partial\hat{r}}\right) \tag{5.10,1}$$

The Derivative C_{y_r}

The only contribution to C_{y_r} that is normally important is that of the tail. From the angle-of-attack change we find the incremental C_y to be

$$(\Delta C_y)_{\text{tail}} = a_F\hat{r}\frac{S_F}{S}\left(2\frac{l_F}{b} + \frac{\partial\sigma}{\partial\hat{r}}\right)$$

thus

$$(C_{y_r})_{\text{tail}} = a_F \frac{S_F}{S}\left(2\frac{l_F}{b} + \frac{\partial\sigma}{\partial\hat{r}}\right) \tag{5.10,2}$$

The Derivative C_{l_r}

This is another important cross derivative; the rolling moment due to yawing. The increase in lift on the left wing, and the decrease on the right wing combine to produce a positive rolling moment proportional to the original lift coefficient C_L. Hence this derivative is largest at low speed. Aspect ratio, taper ratio, and sweepback are all important parameters.

When the vertical tail is large, its contribution may be significant. A formula for it can be derived in the same way as for the previous tail contributions, with the result

$$(C_{l_r})_{\text{tail}} = a_F \frac{S_F}{S}\frac{z_F}{b}\left(2 + \frac{l_F}{b} + \frac{\partial\sigma}{\partial\hat{r}}\right) \tag{5.10,3}$$

The Derivative C_{n_r}

C_{n_r} is the *damping-in-yaw* derivative, and is always negative. The body adds a negligible amount to C_{n_r} except when it is very large. The important contributions are those of the wing and tail. The increases in both the profile and induced drag on the left wing and the decreases on the right wing give a negative yawing moment and hence a resistance to the motion. The magnitude of the effect depends on the aspect ratio, taper ratio, and sweepback. For extremely large sweepback, of the order of 60°, the yawing moment associated with the induced drag may be positive; i.e., produce a reduction in the damping.

The side force on the tail also provides a negative yawing moment. The calculation is similar to that for the preceding tail contributions, with the result

$$(C_{n_r})_{\text{tail}} = -a_F V_V\left(2\frac{l_F}{b} + \frac{\partial\sigma}{\partial\hat{r}}\right) \tag{5.10,4}$$

The Derivative C_{ha_r}

The change in aileron hinge moment due to yawing velocity is a consequence of the velocity differential between the right and left ailerons. Let the hinge-moment coefficient of the right-hand aileron, at zero aileron angle, be C_{ha_0}. Then the corresponding hinge moment, with no yawing, is $C_{ha_0}(\rho/2)u_0{}^2S_ac_a$. This hinge moment is normally balanced by that on the left aileron, so that no load is carried to the pilot's control. Now, when yawing is added, the mean forward velocity at the right-hand aileron is changed from u_0 to $(u_0 - ry_a)$, so that the hinge moment is approximately $C_{ha_0}(\rho/2)(u_0 - ry_a)^2S_ac_a$. To the first order in r, the incremental hinge moment is

$$\Delta H_a = -C_{ha_0}\rho r y_a u_0 S_a c_a$$

On the left-hand side, the increment in H is equal to the above but opposite in sign, so that the two are additive with respect to the stick force, just as though the ailerons were deflected through a small positive angle. The coefficient of ΔH_a is

$$\Delta C_{ha} = -2C_{ha_0}\frac{ry_a}{u_0} = -4\hat{r}\frac{y_a}{b}C_{ha_0}$$

Since C_{ha} is defined as the hinge moment on *one* aileron (Chap. 4) then

$$C_{ha_r} = -4\frac{y_a}{b}C_{ha_0} \tag{5.10,5}$$

The Derivative C_{hr_r}

The change in the vertical-tail angle of attack (Eq. 5.10,1) induces a change in the rudder hinge moment. This is given by

$$\Delta C_{hr} = C_{hr_{\alpha_F}}\Delta\alpha_F = C_{hr_{\alpha_F}}\hat{r}\left(2\frac{l_F}{b} + \frac{\partial\sigma}{\partial\hat{r}}\right)$$

where $C_{hr_{\alpha_F}}$ is the derivative, with respect to the vertical-tail angle of attack, of the rudder hinge-moment coefficient. Hence

$$C_{hr_r} = C_{hr_{\alpha_F}}\left(2\frac{l_F}{b} + \frac{\partial\sigma}{\partial\hat{r}}\right) \tag{5.10,6}$$

5.11 THE ξ DERIVATIVES (C_{l_ξ}, C_{n_ξ}, C_{ha_ξ})

The rolling, yawing, and hinge moments due to ailerons are all quantities directly measured in wind-tunnel tests, or estimated from theory (see Sec. 3.10). The derivatives are simply

$$C_{l_\xi} = \frac{\partial C_l}{\partial\delta_a}$$

$$C_{n_\xi} = \frac{\partial C_n}{\partial\delta_a}$$

$$C_{ha_\xi} = \frac{\partial C_{ha}}{\partial\delta_a}$$

5.12 THE ζ DERIVATIVES (C_{y_ζ}, C_{l_ζ}, C_{n_ζ}, C_{hr_ζ})

The incremental side-force coefficient due to deflecting the rudder by the angle ζ is

$$\Delta C_y = \frac{S_F}{S}a_r\zeta$$

where a_r is the vertical-tail lift-curve slope due to rudder deflection, $a_r = \partial C_{L_F}/\partial \delta_r$. It follows that

$$C_{y_\zeta} = a_r \frac{S_F}{S} = \frac{\partial C_y}{\partial \delta_r} \tag{5.12,1}$$

The rolling moment produced by this side force is

$$\Delta C_l = \frac{z_F}{b}\frac{S_F}{S} a_r \zeta$$

thus

$$C_{l_\zeta} = a_r \frac{S_F}{S}\frac{z_F}{b} = \frac{\partial C_l}{\partial \delta_r} \tag{5.12,2}$$

The yawing moment resulting from the tail side force is

$$\Delta C_n = -\frac{l_F}{b}\frac{S_F}{S} a_r \zeta$$

therefore

$$C_{n_\zeta} = -a_r V_V = \frac{\partial C_n}{\partial \delta_r} \tag{5.12,3}$$

(see also Eq. 3.9,9 for the formula when the fin is in the slipstream).

Finally, the hinge-moment derivative is

$$C_{hr_\zeta} = \frac{\partial C_{hr}}{\partial \delta_r} \tag{5.12,4}$$

5.13 THE $\dot{\xi}$ AND $\dot{\zeta}$ DERIVATIVES ($C_{l_{\dot{\xi}}}$, $C_{ha_{\dot{\xi}}}$, $C_{n_{\dot{\zeta}}}$, $C_{hr_{\dot{\zeta}}}$)

The discussion in Sec. 5.7 of the elevator rate derivatives applies to the aileron rate and rudder rate derivatives as well. That is, the required derivatives are obtained from test, or from first-order-in-frequency analysis of oscillating control flaps. $C_{l_{\dot{\xi}}}$ and $C_{n_{\dot{\zeta}}}$, which give the effect of control rate on the airplane moments, may not be large. However, the two hinge-moment derivatives may be quite important, inasmuch as they often supply the main damping of the aileron and rudder control systems.

5.14 SUMMARY OF THE FORMULAS

The formulas that are frequently wanted for reference are collected in Tables 5.1 and 5.2. Where an entry in the table shows only a tail contribution, it is not implied that the wing and body effects are not important, but only that no convenient formula is available.

TABLE 5.1
Summary—Longitudinal Derivatives

	C_x	C_z	C_m	C_{he}
u	$\dfrac{(\partial T/\partial u)_0}{\frac{1}{2}\rho u_0 S} - 2(C_{D_0} + C_{L_0}\tan\theta_0) - \mathbf{M}\dfrac{\partial C_D}{\partial \mathbf{M}}$	$-\mathbf{M}\dfrac{\partial C_L}{\partial \mathbf{M}}$	$\mathbf{M}\dfrac{\partial C_m}{\partial \mathbf{M}} + \rho {u_0}^2 \dfrac{\partial C_m}{\partial p_d}$	$\mathbf{M}\dfrac{\partial C_{he}}{\partial \mathbf{M}} + \rho {u_0}^2 \dfrac{\partial C_{he}}{\partial p_d}$
α	$C_{L_0} - C_{D_\alpha}$	$-(C_{L_\alpha} + C_{D_0})$	$-a(h_n - h)$	$C_{he_{\alpha_t}}\left(1 - \dfrac{\partial \varepsilon}{\partial \alpha}\right)$
$\dot{\alpha}$	Neg.	* $-2a_t V_H \dfrac{\partial \varepsilon}{\partial \alpha}$	* $-2a_t V_H \dfrac{l_t}{\bar{c}} \dfrac{\partial \varepsilon}{\partial \alpha}$	$2C_{he_{\alpha_t}} \dfrac{l_t}{\bar{c}} \dfrac{\partial \varepsilon}{\partial \alpha}$
q	Neg.	* $-2a_t V_H$	* $-2a_t V_H \dfrac{l_t}{\bar{c}}$	$2\dfrac{l_t}{\bar{c}} C_{he_{\alpha_t}}$
η	Neg.	* $-a_e \dfrac{S_t}{S}$	* $-a_e V_H$	$\dfrac{\partial C_{he}}{\partial \delta_e}$
$\dot{\eta}$	Neg.	Neg.	N.A.	N.A.

Neg. means usually negligible.
* means contribution of the *tail only*, formula for wing-body not available.
N.A. means no formula available.

TABLE 5.2
Summary—Lateral Derivatives

	C_y	C_l	C_n	C_{ha}	C_{hr}
β	* $-a_F \frac{S_F}{S}\left(1 - \frac{\partial \sigma}{\partial \beta}\right)$	N.A.	* $a_F V_V \left(1 - \frac{\partial \sigma}{\partial \beta}\right)$	Neg.	$-C_{hr_{\alpha_F}}\left(1 - \frac{\partial \sigma}{\partial \beta}\right)$
p	* $-a_F \frac{S_F}{S}\left(2\frac{z_F}{b} - \frac{\partial \sigma}{\partial \hat{p}}\right)$	N.A.	* $a_F V_V \left(2\frac{z_F}{b} - \frac{\partial \sigma}{\partial \hat{p}}\right)$	$2\frac{y_a}{b}C_{ha_\alpha}$	$-C_{hr_{\alpha_F}}\left(2\frac{z_F}{b} - \frac{\partial \sigma}{\partial \hat{p}}\right)$
r	* $a_F \frac{S_F}{S}\left(2\frac{l_F}{b} + \frac{\partial \sigma}{\partial \hat{r}}\right)$	* $a_F \frac{S_F}{S}\frac{z_F}{b}\left(2\frac{l_F}{b} + \frac{\partial \sigma}{\partial \hat{r}}\right)$	* $-a_F V_V \left(2\frac{l_F}{b} + \frac{\partial \sigma}{\partial \hat{r}}\right)$	$-4\frac{y_a}{b}C_{ha_0}$	$C_{hr_{\alpha_F}}\left(2\frac{l_F}{b} + \frac{\partial \sigma}{\partial \hat{r}}\right)$
ξ	Neg.	$\frac{\partial C_l}{\partial \delta_a}$	$\frac{\partial C_n}{\partial \delta_a}$	$\frac{\partial C_{ha}}{\partial \delta_a}$	Neg.
$\dot{\xi}$	Neg.	N.A.	Neg.	N.A.	Neg.
ζ	* $a_r \frac{S_F}{S}$	* $a_r \frac{S_F}{S}\frac{z_F}{b}$	* $-a_r V_V$	Neg.	$\frac{\partial C_{hr}}{\partial \delta_r}$
$\dot{\zeta}$	Neg.	Neg.	N.A.	Neg.	N.A.

Neg. means usually negligible.
* means contribution of the *tail only*, formula for wing-body not available.
N.A. means no formula available.

5.15 THE DETERMINATION OF AERODYNAMIC TRANSFER FUNCTIONS

It has already been shown (Sec. 4.16) that one approximation to the aerodynamic transfer functions can be obtained directly from the stability derivatives. Let C_a be a typical force or moment coefficient, and v a typical nondimensional disturbance variable; then

$$G_{av} \doteqdot C_{a_v} + C_{a_{\dot{v}}}s + C_{a_{\ddot{v}}}s^2 + \cdots \tag{5.15,1}$$

The use of expressions such as this in Eqs. 4.16,6 and 4.16,7 makes those equations completely equivalent to the conventional equations 4.15,7 and 4.15,8.

It is a significant fact that the true transfer function can be obtained from the same basic information as is required to determine an unsteady derivative such as $C_{a_{\dot{v}}}$. This basic information is either the indicial admittance or the frequency response of C_a. How the transfer function is found from these is shown below.

Determination from the Indicial Admittance

Let the response of C_a to a unit step input of v be known, and designated $A_{av}(\hat{t})$. Since by definition $\bar{C}_a(s) = G_{av}\bar{v}(s)$, and since the transform of the unit step function is $1/s$ (see Table 8.1), then

$$\bar{A}_{av}(s) = G_{av}(s)\frac{1}{s}$$

where $\bar{A}_{av}(s)$ is the Laplace transform of $A_{av}(\hat{t})$. Hence

$$G_{av} = s\bar{A}_{av}(s) \tag{5.15,2}$$

Thus the transfer function is simply s times the transform of the indicial admittance.

Example

R. T. Jones (ref. 5.4) has calculated the indicial lift of elliptic wings in incompressible flow. After converting to the notation used here, and to the semi-mean chord as the reference length; and with the approximation $C_z \doteqdot -C_L$, Jones's (approximate) result for a wing of A.R. 3 becomes

$$A_{z\alpha}(\hat{t}) = -\pi[1.094\,\delta(\hat{t}) + 1.200(1 - 0.283e^{-0.424\hat{t}})] \tag{5.15,3}$$

$\delta(t)$ is Dirac's delta function, defined in Sec. 8.9. It appears in Eq. 5.15,3 because of the additional apparent mass associated with downward acceleration of the wing. For the wing to acquire a downward velocity instantaneously requires the application of an impulse (see Fig. 5.5). The reaction to this impulse is an impulsive lift at $t = 0$. The remaining terms of Eq. 5.15,3 describe the growth of the lift due to circulation.

Following Eq. 5.15,2, we take the Laplace transform of $A_{z\alpha}$ and get

$$\bar{A}_{z\alpha}(s) = -\pi\left[1.094 + 1.200\left(\frac{1}{s} - \frac{0.283}{s + 0.424}\right)\right] \tag{5.15,4}$$

thus

$$G_{z\alpha}(s) = sA_{z\alpha} = -\pi\left(1.200 + 1.094s - \frac{0.340s}{s + 0.424}\right) \tag{5.15,5}$$

For $|s| \ll 0.424$, this expression may be approximated by

$$G_{z\alpha}(s) = -\pi(1.200 + 0.292s)$$

Comparison with Eq. 5.15,1 shows that this corresponds to the stability-derivative representation with

$$C_{z_\alpha} = -1.200\pi$$
$$C_{z_{\dot{\alpha}}} = -0.292\pi$$

These derivatives are exactly the same as are obtained from the given indicial response by applying the method of Sec. 5.5.

Expressions for aerodynamic transfer functions similar to Eq. 5.15,5 can be obtained whenever the indicial admittance can be approximated adequately by a sum of exponential terms. The general form containing an impulse term is

$$A_{av}(\hat{t}) = a_0 + a_1\,\delta(\hat{t}) + a_2e^{b_2\hat{t}} + a_3e^{b_3\hat{t}} + \cdots \tag{5.15,6}$$

The transfer function is found from Eq. 5.15,2 to be

$$G_{av}(s) = a_0 + a_1s + \frac{a_2s}{s - b_2} + \frac{a_3s}{s - b_3} + \cdots. \tag{5.15,7}$$

Determination from the Frequency Response

It is shown in Sec. 8.8 that the frequency response of a linear system is simply obtained from the transfer function. For example, if $G_{av}(s)$ is the transfer function relating C_a and v, and if v is periodic, given by

$$v = v_0e^{i\omega t} = v_0e^{ik\hat{t}} \tag{5.15,8}$$

then the steady-state oscillatory response is given by

$$C_a = G_{av}(ik)v_0e^{ik\hat{t}} \tag{5.15,9}$$

Here $G_{av}(ik)$ is the frequency-response vector, obtained from $G_{av}(s)$ by replacing s with ik. Conversely, if the frequency-response vector for the periodic motion is known, then $G_{av}(s)$ is obtained by replacing k in it with $s/i = -is$.

It is also possible to use the frequency response indirectly. From it the indicial admittance can first be calculated (see ref. 5.5) and then used as shown above to get the transfer function.

Example

The pitching moment of a two-dimensional airfoil in incompressible flow will be used for this example, since the frequency response is known analytically. The Theodorsen solution quoted previously (Eqs. 5.5,6 and 5.5,7) can be used to write an expression in the form of Eq. 5.15,9. This is

$$C_m = \pi[2(h - \tfrac{1}{4})C(k) + ik(h - \tfrac{1}{2})]\alpha_0 e^{ik\hat{t}}$$

It follows that

$$G_{m\alpha} = \pi[2(h - \tfrac{1}{4})C(-is) + s(h - \tfrac{1}{2})] \tag{5.15,10}$$

This expression is useful in its present form for frequency-response calculations, where the analytical form of the Theodorsen function $C(k)$ is not required, only tabulated values being needed. However, for transient-response calculations, the analytical form of $C(k)$ would be needed. Since it is a complicated expression involving Bessel functions, its use in this connection would be inconvenient. A preferable expression for the transfer function can be obtained by using an exponential approximation to the indicial response, such as that given by R. T. Jones (ref. 5.4), and then using the method of the previous example.

5.16 AERORELASTIC DERIVATIVES

In Sec. 4.17 there were introduced aerodynamic derivatives associated with the deformations of the airplane. These are of two kinds: those that appear in the rigid-body equations and those that appear in the added equations of the elastic degrees of freedom. These are illustrated in this section by consideration of the hypothetical vibration mode shown in Fig. 5.14. In this mode it is assumed that the fuselage and tail are rigid, and have a motion of vertical translation only. The flexibility is all in the wing, and it bends without twisting. The functions describing the mode (Eqs. 4.17,1) are therefore:

$$\begin{aligned} x - x_0 &= 0 \\ y - y_0 &= 0 \\ z - z_0 &= h(y)z_T \end{aligned} \tag{5.16,1}$$

For the generalized coordinate, we have used the wing-tip deflection z_T. $h(y)$ is then a normalized function describing the wing bending mode.

Since the elastic degrees of freedom are only important in relation to stability and control when their frequencies are relatively low, approaching those of the rigid-body modes, then it is reasonable to use the same approximation for the aerodynamic forces as is used in calculating stability derivatives. That is, if quasi-

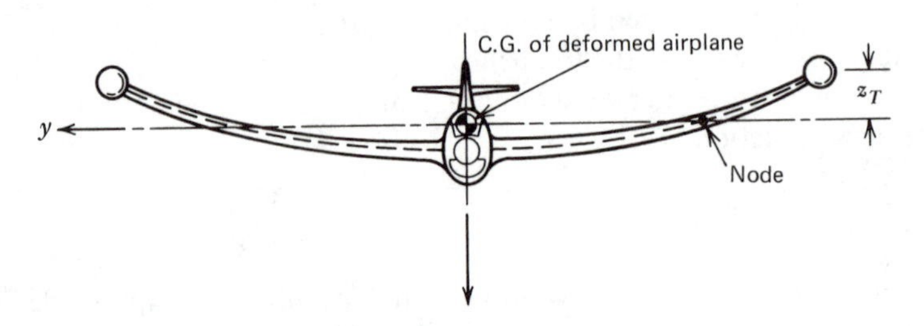

FIG. 5.14 Symmetrical wing bending.

steady flow theory is adequate for the aerodynamic forces associated with the rigid-body motions, then we may use the same theory for the elastic motions.

In the example chosen, we assume that the only significant forces are those on the wing and tail, and that these are to be computed from quasisteady flow theory. In the light of these assumptions, some of the representative derivatives of both types are discussed below. As a preliminary, the forces induced on the wing and tail by the elastic motion are treated first.

Forces on the Wing

The vertical velocity of the wing section distant y from the center line is

$$\dot{z} = h(y)\dot{z}_T \tag{5.16,2}$$

and the corresponding change in wing angle of attack is

$$\Delta\alpha(y) = h(y)\dot{z}_T/u_0 \tag{5.16,3}$$

This angle-of-attack distribution can be used with any applicable steady-flow wing theory to calculate the incremental local section lift. (It will of course be proportional to $\dot{z}_T/u_0$.) Let it be denoted in coefficient form by $C'_l(y)\dot{z}_T/u_0$, and the corresponding increment in wing total lift coefficient by $C'_{L_w}\dot{z}_T/u_0$. $C'_l(y)$ and C'_{L_w} are thus the values corresponding to unit value of the nondimensional quantity $\dot{z}_T/u_0$.

Force on the Tail

The tail experiences a downward velocity $h(0)\dot{z}_T$, and also, because of the altered wing lift distribution, a downwash change $(\partial\varepsilon/\partial\dot{z}_T)\dot{z}_T$. Hence the net change in tail angle of attack is

$$\begin{aligned}\Delta\alpha_t &= h(0)\dot{z}_T/u_0 - \frac{\partial\varepsilon}{\partial\dot{z}_T}\dot{z}_T \\ &= \left[h(0) - \frac{\partial\varepsilon}{\partial(\dot{z}_T/u_0)}\right]\frac{\dot{z}_T}{u_0}\end{aligned}$$

This produces an increment in the tail lift coefficient of amount

$$\Delta C_{L_t} = a_t\left[h(0) - \frac{\partial\varepsilon}{\partial(\dot{z}_T/u_0)}\right]\frac{\dot{z}_T}{u_0} \tag{5.16,5}$$

The Derivative $Z_{\dot{z}_T}$

This derivative describes the contribution of wing bending velocity to the Z force acting on the airplane. A suitable nondimensional form is $\partial C_z/\partial(\dot{z}_T/u_0)$. Since $C_z = -C_L$, we have that

$$\Delta C_z = -C'_{L_w}\frac{\dot{z}_T}{u_0} - a_t\left[h(0) - \frac{\partial\varepsilon}{\partial(\dot{z}_T/u_0)}\right]\frac{\dot{z}_T}{u_0}$$

and hence

$$\frac{\partial C_z}{\partial(\dot{z}_T/u_0)} = -C'_{L_w} - a_t\left[h(0) - \frac{\partial \varepsilon}{\partial(\dot{z}_T/u_0)}\right] \tag{5.16,6}$$

The Derivative A_{nw}

This derivative (see Eq. 4.17,11) represents the contribution to the generalized force in the bending degree of freedom, associated with a change in the w velocity of the airplane. A suitable nondimensional form is obtained by defining

$$C_{\mathscr{F}} = \frac{\mathscr{F}}{\frac{1}{2}\rho u_0^2 S}$$

and using α in place of w ($w = u_0\alpha$). Then the appropriate nondimensional derivative is $C_{\mathscr{F}_\alpha}$.

Let the wing lift distribution due to a perturbation α in the angle of attack (constant across the span) be given by $C_{l_\alpha}(y)\alpha$. Then in a virtual displacement in the wing bending mode δz_T, the work done by this wing loading is

$$\delta W = -\int_{-b/2}^{b/2} \alpha C_{l_\alpha}(y)h(y)\,\delta z_T \frac{1}{2}\rho u_0^2 c(y)\,dy$$

where $c(y)$ is the local wing chord. The corresponding contribution to $\mathscr{F}$ is

$$\frac{\delta W}{\partial z_T} = -\alpha\frac{1}{2}\rho u_0^2 \int_{-b/2}^{b/2} C_{l_\alpha}(y)h(y)c(y)\,dy$$

and to $C_{\mathscr{F}_\alpha}$ is

$$\frac{1}{\frac{1}{2}\rho u_0^2 S}\frac{\partial^2 W}{\partial z_T\,\partial\alpha} = -\frac{1}{S}\int_{-b/2}^{b/2} C_{l_\alpha}(y)h(y)c(y)\,dy \tag{5.16,7}$$

The tail also contributes to this derivative. For the tail lift associated with α is

$$a_t\alpha\left(1 - \frac{\partial\varepsilon}{\partial\alpha}\right)\frac{1}{2}\rho u_0^2 S_t$$

and the work done by this force during the virtual displacement is

$$-a_t\alpha\left(1 - \frac{\partial\varepsilon}{\partial\alpha}\right)\frac{1}{2}\rho u_0^2 S_t h(0)\,\delta z_T$$

Therefore the contribution to $C_{\mathscr{F}}$ is

$$-a_t\alpha\left(1 - \frac{\partial\varepsilon}{\partial\alpha}\right)\frac{S_t}{S}h(0)$$

and to $C_{\mathscr{F}_\alpha}$ is

$$-a_t\frac{S_t}{S}h(0)\left(1 - \frac{\partial\varepsilon}{\partial\alpha}\right) \tag{5.16,8}$$

The total value of $C_{\mathscr{F}_\alpha}$ is then the sum of 5.16,7 and 5.16,8.

The Derivative b_{11} (see Eq. 4.17,11)

This derivative identifies the contribution of $\dot{z}_T$ to the generalized aerodynamic force in the distortion degree of freedom. We have defined the associated wing load distribution above by the local lift coefficient $C_l'(y)\dot{z}_T/u_0$. As in the case of the derivative A_{nw} above, the work done by this loading is calculated, with the result that the wing contributes

$$\frac{\partial C_{\mathscr{F}}}{\partial(\dot{z}_T/u_0)} = \frac{1}{\frac{1}{2}\rho u_0{}^2 S}\,\frac{\partial^2 W}{\partial z_T\,\partial(\dot{z}_T/u_0)} = -\frac{1}{S}\int_{-b/2}^{b/2} C_l'(y)h(y)c(y)\,dy \tag{5.16,9}$$

Likewise, the contribution of the tail is calculated here as for A_{nw}, and is found to be

$$-a_t\frac{S_t}{S}h(0)\left[h(0) - \frac{\partial\varepsilon}{\partial(\dot{z}_T/u_0)}\right] \tag{5.16,10}$$

The total value of $\partial C_{\mathscr{F}}/\partial(\dot{z}_T/u_0)$ is then the sum of 5.16,9 and 5.16,10.

5.17 ADDITIONAL SYMBOLS INTRODUCED IN CHAPTER 5

C_T thrust coefficient, $T/\frac{1}{2}\rho v_c{}^2 S$

k reduced frequency, $\omega\bar{c}/2u_0$

p_d dynamic pressure, $\frac{1}{2}\rho v_c{}^2$

ω circular frequency

See also Secs. 2.11, 3.11, and 4.18.

5.18 BIBLIOGRAPHY

5.1 M. Tobak. On the Use of the Indicial Function Concept in the Analysis of Unsteady Motions of Wings and Wing-Tail Combinations. *NACA Rept. 1188*, 1954.

5.2 J. W. Miles. Unsteady Flow Theory in Dynamic Stability. *J. Aero. Sci.*, vol. 17, no. 1, p. 62, 1950.

5.3 W. H. Michael Jr. Analysis of the Effects of Wing Interference on the Tail Contributions to the Rolling Derivatives. *NACA Rept. 1086*, 1952.

5.4 R. T. Jones. The Unsteady Lift of a Wing of Finite Aspect Ratio. *NACA Rept. 681*, 1940.

5.5 I. E. Garrick. On Some Reciprocal Relations in the Theory of Nonstationary Flows. *NACA Rept. 629*, 1938.

5.6 B. Smilg and L. Wasserman. Application of Three-Dimensional Flutter Theory to Aircraft Structures. *USAAF Tech. Rept. 4798*, 1942.

5.7 K. Orlik-Rückemann and C. O. Olsson. A Method for Determining the Damping-in-Pitch of Semi-Span Models in High-Speed Wind-Tunnels, and Some Results for a Triangular Wing. *Aero. Res. Inst. Sweden FFA Rept. 62*, 1956.

5.8 W. C. Williams, H. M. Drake, and J. Fischel. Comparison of Flight and Wind-Tunnel Measurements of High-Speed-Airplane Stability and Control Characteristics. *NACA TN 3859*, 1956.

5.9 A. F. Donovan, and H. R. Lawrence (editors). *Aerodynamic Components of Aircraft at High Speeds*, vol. VII of Princeton Series. Princeton University Press, Princeton, 1957.

5.10 A. Naysmith. A Collection of Longitudinal Stability Derivatives of Wings at Supersonic Speeds. *Roy. Aircraft Establishment Tech. Note Aero 2423*, 1956.

5.11 J. J. Donegan, S. W. Robinson Jr., and O. B. Gates Jr. Determination of Lateral-Stability Derivatives and Transfer-Function Coefficients from Frequency-Response Data for Lateral Motions, *NACA Rept. 1225*, 1955.

5.12 S. Neumark and A. W. Thorpe. Theoretical Requirements of Tunnel Experiments for Determining Stability Derivatives in Oscillatory Longitudinal Disturbances. *ARC R&M 2903*, 1955.

5.13 C. S. Sinnott. Hinge Moment Derivatives for an Oscillating Control, *ARC R&M 2923*, 1955.

5.14 J. C. Martin, M. S. Diederich, and P. J. Bobbitt. A Theoretical Investigation of the Aerodynamics of Wing-Tail Combinations Performing Time-Dependent Motions at Supersonic Speeds. *NACA TN 3072*, 1954.

5.15 B. H. Beam. A Wind-Tunnel Test Technique for Measuring the Dynamic Rotary Stability Derivatives at Subsonic and Supersonic Speeds. *NACA Rept. 1258*, 1956.

5.16 J. A. Drischler. Calculation and Compilation of the Unsteady-Lift Functions for a Rigid Wing Subjected to Sinusoidal Gusts and to Sinusoidal Sinking Oscillations. *NACA TN 3748*, 1956.

5.17 J. A. Drischler. Approximate Indicial Lift Functions for Several Wings of Finite Span in Incompressible Flow as Obtained from Oscillatory Lift Coefficients. *NACA TN 3639*, 1956.

5.18 W. P. Jones. The Oscillating Airfoil in Subsonic Flow. *ARC R&M 2921*, 1956.

5.19 H. C. Garner. Multhopp's Subsonic Lifting-Surface Theory of Wings in Slow Pitching Oscillations. *ARC R&M 2885*, 1956.

5.20 R. T. Jones. Operational Treatment of the Nonuniform-Lift Theory in Airplane Dynamics. *NACA TN 667*, 1938.

5.21 A. Robinson and J. A. Laurmann. *Wing Theory*. Cambridge University Press, Cambridge, 1956.

5.22 J. P. Campbell and M. O. McKinney. Summary of Methods for Calculating Dynamic Lateral Stability and Response and for Estimating Lateral Stability Derivatives. *NACA Rept. 1098*, 1952.

5.23 H. S. Ribner. The Stability Derivatives of Low-Aspect-Ratio Wings at Subsonic and Supersonic Speeds. *NACA TN 1423*, 1947.

STABILITY OF UNCONTROLLED MOTION (LONGITUDINAL)

CHAPTER 6

6.1 FORM OF SOLUTION OF SMALL-DISTURBANCE EQUATIONS

Experience in the past has shown that Eqs. 4.15,7 and 4.15,8 are adequate to describe the stability of small perturbations of an airplane from a reference state of steady rectilinear flight. They are therefore used as the basis of the discussions of this and the following chapter. These equations suffice because the frequencies and the rates of divergence or convergence in the natural modes are usually small enough to make it unnecessary to use the more precise representation of the unsteady aerodynamic forces which can be accomplished by the use of aerodynamic transfer functions. The equations referred to are of a simple, well-known type: namely ordinary linear differential equations with constant coefficients. Solutions of such equations are always exponential in form. For example, a typical variable such as α is of the form

$$\alpha = a_1 e^{\lambda_1 \hat{t}} + a_2 e^{\lambda_2 \hat{t}} + \cdots \tag{6.1,1}$$

The method of finding the values of λ is illustrated in Sec. 6.2. Both real and complex values may occur, the complex ones always appearing in conjugate pairs, such as $\lambda = n \pm i\omega$. The pair of terms corresponding to such a pair of λ values is

$$a_1 e^{(n+i\omega)\hat{t}} + a_2 e^{(n-i\omega)\hat{t}}$$

By using the fact that $e^{i\theta} = \cos\theta + i\sin\theta$, these terms may be rewritten as

$$e^{n\hat{t}}(A_1 \cos\omega\hat{t} + A_2 \sin\omega\hat{t})$$

where $A_1 = a_1 + a_2$ and $A_2 = i(a_1 - a_2)$. The constants A_1 and A_2 are always real, which implies that a_1 and a_2 are conjugate complex numbers.

There are four possible kinds of term in Eq. 6.1,1 according as λ is real or complex, and has a positive or negative real part. The motion corresponding to each real λ or each complex pair is called a *natural mode* (see Sec. 6.6). The four kinds of mode are illustrated in Fig. 6.1. The disturbances shown in (*a*) and (*c*) increase with time, and hence these are unstable modes. It is conventional to refer to (*a*) as a static instability or *divergence*, since there is no tendency for the disturbance to diminish. By contrast, (*c*) is called dynamic instability or a *divergent oscillation*, since the disturbance quantity alternately increases and diminishes,

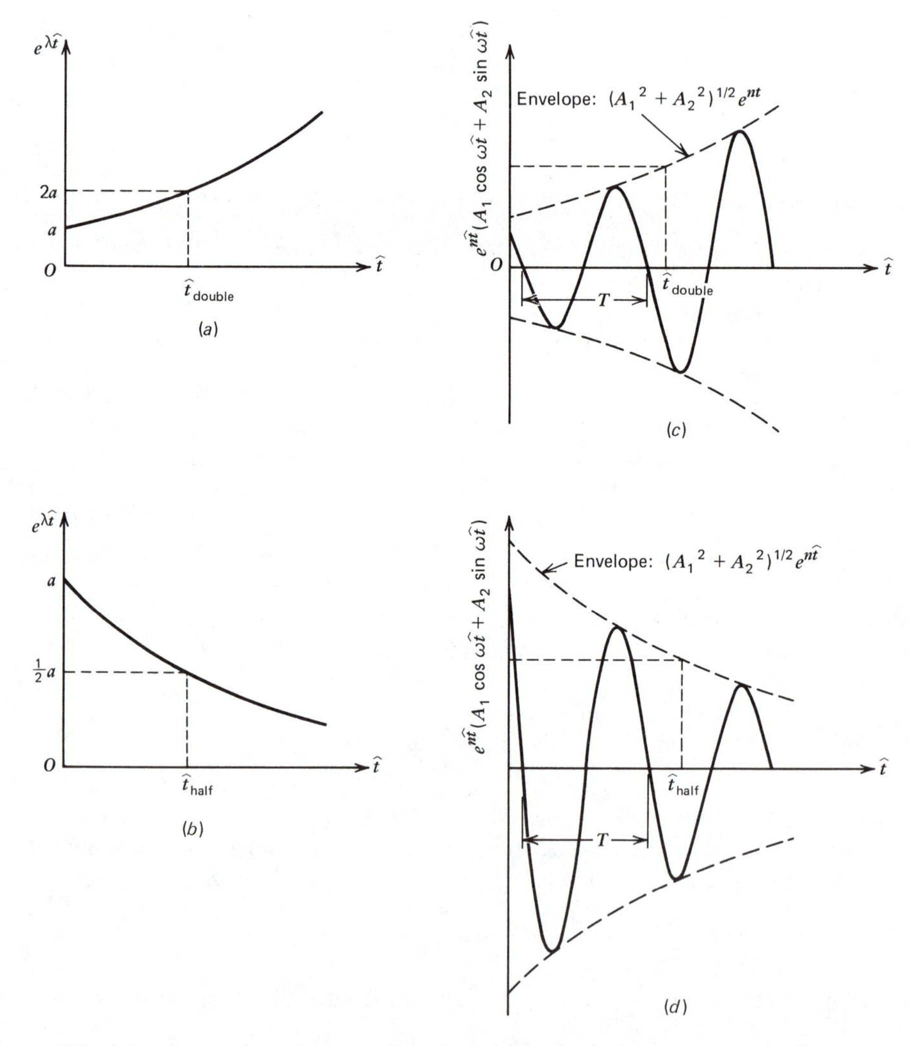

FIG. 6.1 Types of solution. (*a*) λ real, positive. (*b*) λ real, negative, (*c*) λ complex, $n > 0$. (*d*) λ complex, $n < 0$.

the amplitude growing with time. (*b*) illustrates a *subsidence* or *convergence*, and (*d*) a *damped* or *convergent* oscillation. Since in both (*b*) and (*d*) the disturbance quantity ultimately vanishes, they represent stable modes.

It is seen that a "yes" or "no" evaluation of the stability is obtained simply from the signs of the real parts of the λs. If there are no positive real parts, there is no instability. This information is not sufficient, however, to evaluate the flying qualities of an airplane, (see Chap. 1). These are dependent on the quantitative as well as on the qualitative characteristics of the modes. The numerical parameters of primary interest are

1. Period.
2. Time to double or time to half.
3. Cycles to double (N_{double}) or cycles to half (N_{half}).

The first two of these are illustrated in Fig. 6.1. When the roots are real, there is of course no period, and the only parameter is the time to double or half. These are the times that must elapse during which any disturbance quantity will double or halve itself, respectively. When the modes are oscillatory, it is the envelope ordinate that doubles or halves. Since the envelope may be regarded as an amplitude modulation, then we may think of the doubling or halving as applied to the variable amplitude. By noting that $\log_e 2 = -\log_e \frac{1}{2} = 0.69$, the reader will easily verify the following relations:

For real λ*:*

$$\begin{aligned} \hat{t}_{\text{double}} \text{ or } \hat{t}_{\text{half}} &= \frac{0.69}{|\lambda|} \text{ airsec} \\ t_{\text{double}} \text{ or } t_{\text{half}} &= \frac{0.69}{|\lambda|} t^* \text{ sec} \end{aligned} \tag{6.1,2}$$

For complex λ*:*

$$\begin{aligned} \hat{t}_{\text{double}} \text{ or } \hat{t}_{\text{half}} &= \frac{0.69}{|n|} \text{ airsec} \\ t_{\text{double}} \text{ or } t_{\text{half}} &= \frac{0.69}{|n|} t^* \text{ sec} \\ T &= \frac{2\pi}{\omega} \text{ airsec} \\ &= \frac{2\pi}{\omega} t^* \text{ sec} \\ N_{\text{double}} \text{ or } N_{\text{half}} &= \frac{\hat{t}_{\text{double}} \text{ or } \hat{t}_{\text{half}}}{T} = 0.110 \frac{\omega}{|n|} \end{aligned} \tag{6.1,3}$$

6.2 EXAMPLE: LONGITUDINAL STABILITY, CONTROLS FIXED

The relevant equations are obtained from Eqs. 4.15,7. Since the elevator is assumed to be locked in position, then $\eta \equiv 0$, and the elevator equation of motion, Eq. 4.15,7*d*, is dropped.

In the interest of simplicity it is further assumed that the initial flight path is horizontal. We then obtain the following equations:

$$\begin{aligned}
(2\mu D - C_{x_u})\hat{u} - \qquad & C_{x_\alpha}\alpha + \qquad C_{L_0}\theta = 0 \\
(2C_{L_0} - C_{z_u})\hat{u} + (2\mu D - C_{z_{\dot\alpha}}D - C_{z_\alpha})\alpha - & (2\mu + C_{z_q})D\theta = 0 \\
-C_{m_u}\hat{u} - \qquad (C_{m_{\dot\alpha}}D + C_{m_\alpha})\alpha + & (i_B D^2 - C_{m_q}D)\theta = 0
\end{aligned} \tag{6.2,1}$$

$$\mu = \frac{m}{\rho S l}, \qquad t^* = \frac{l}{u_0}, \qquad l = \frac{\bar{c}}{2}$$

We assume that these equations have possible solutions of the form

$$\begin{aligned}
\hat{u} &= \hat{u}_0 e^{\lambda \hat{t}} \\
\alpha &= \alpha_0 e^{\lambda \hat{t}} \\
\theta &= \theta_0 e^{\lambda \hat{t}}
\end{aligned} \tag{6.2,2}$$

This assumption is checked by substituting Eqs. 6.2,2 into Eqs. 6.2,1 to see if they are satisfied. When the substitution is performed, it is found that every term contains the factor $e^{\lambda \hat{t}}$. Since the degenerate case when $e^{\lambda \hat{t}} = 0$ is not of interest, then this factor may be divided out. The result is

$$\begin{aligned}
(2\mu\lambda - C_{x_u})\hat{u}_0 - \qquad & C_{x_\alpha}\alpha_0 + \qquad C_{L_0}\theta_0 = 0 \\
(2C_{L_0} - C_{z_u})\hat{u}_0 + (2\mu\lambda - C_{z_{\dot\alpha}}\lambda - C_{z_\alpha})\alpha_0 - & (2\mu + C_{z_q})\lambda\theta_0 = 0 \\
-C_{m_u}\hat{u}_0 - \qquad (C_{m_{\dot\alpha}}\lambda + C_{m_\alpha})\alpha_0 + & (i_B\lambda^2 - C_{m_q}\lambda)\theta_0 = 0
\end{aligned} \tag{6.2,3}$$

It follows that Eqs. 6.2,2 are a solution of Eqs. 6.2,1 provided that Eqs. 6.2,3 are satisfied. The latter are homogeneous algebraic equations in the unknowns $\hat{u}_0$, α_0, θ_0, and containing the parameter λ. It is a property of this type of equation that there can be nonzero values of the unknowns if and only if the determinant of the coefficients is zero. Setting the determinant equal to zero provides the condition for finding the admissible values of λ.

$$\begin{vmatrix}
(2\mu\lambda - C_{x_u}) & -C_{x_\alpha} & C_{L_0} \\
(2C_{L_0} - C_{z_u}) & (2\mu\lambda - C_{z_{\dot\alpha}}\lambda - C_{z_\alpha}) & -(2\mu + C_{z_q})\lambda \\
-C_{m_u} & -(C_{m_{\dot\alpha}}\lambda + C_{m_\alpha}) & (i_B\lambda^2 - C_{m_q}\lambda)
\end{vmatrix} = 0 \tag{6.2,4}$$

This determinant is known as the *stability determinant*, and Eq. 6.2,4 is called the *characteristic* equation of the dynamic system. Expansion of the determinant leads to a quartic equation for λ.[1]

$$A\lambda^4 + B\lambda^3 + C\lambda^2 + D\lambda + E = 0 \tag{6.2,5}$$

[1] The values of λ that satisfy Eq. 6.2,5 are the *eigenvalues* of the system.

where $A = 2\mu i_B(2\mu - C_{z_{\dot{\alpha}}})$

$$B = -2\mu i_B(C_{z_\alpha} + C_{x_u}) + i_B(C_{x_u}C_{z_{\dot{\alpha}}}) - 2\mu(C_{z_q}C_{m_{\dot{\alpha}}} - C_{m_q}C_{z_{\dot{\alpha}}}) - 4\mu^2(C_{m_{\dot{\alpha}}} + C_{m_q})$$

$$C = i_B(C_{x_u}C_{z_\alpha} - C_{x_\alpha}C_{z_u}) + 2\mu(C_{z_\alpha}C_{m_q} - C_{m_\alpha}C_{z_q} + C_{x_u}C_{m_q} + C_{x_u}C_{m_{\dot{\alpha}}}) - 4\mu^2 C_{m_\alpha} - C_{x_u}(C_{m_q}C_{z_{\dot{\alpha}}} - C_{z_q}C_{m_{\dot{\alpha}}}) + 2C_{L_0}C_{x_\alpha}i_B$$

$$D = -2C_{L_0}{}^2 C_{m_{\dot{\alpha}}} + 2\mu(C_{x_u}C_{m_\alpha} - C_{x_\alpha}C_{m_u} + C_{L_0}C_{m_u}) + C_{x_u}(C_{m_\alpha}C_{z_q} - C_{m_q}C_{z_\alpha}) - C_{x_\alpha}(C_{m_u}C_{z_q} - C_{m_q}C_{z_u}) - C_{L_0}(C_{m_u}C_{z_{\dot{\alpha}}} - C_{z_u}C_{m_{\dot{\alpha}}}) - 2C_{L_0}C_{m_q}C_{x_\alpha}$$

$$E = -C_{L_0}[C_{m_\alpha}(2C_{L_0} - C_{z_u}) + C_{m_u}C_{z_\alpha}]$$

6.3 CRITERIA FOR STABILITY

Before discussing the above example, we shall consider the conditions on the characteristic equation which must be met if there are to be no unstable modes.

Routh's Criteria

These conditions were first stated by Routh (ref. 4.1), who derived them from a theorem of Cauchy. Let the characteristic equation be

$$p_0\lambda^n + p_1\lambda^{n-1} + \cdots + p_n = 0 \qquad (p_0 > 0) \tag{6.3,1}$$

The coefficient p_0 can always be made positive by changing signs throughout, so the requirement $p_0 > 0$ is not restrictive. The necessary and sufficient condition for stability (i.e., that no root of the equation shall be zero or have a positive real part) is that each of a series of test functions shall be positive. The test functions are constructed by the simple scheme shown below. Write the coefficients of Eq. 6.3,1 in two rows as follows:

$$\begin{matrix} p_0 & p_2 & p_4 & \cdots \\ p_1 & p_3 & p_5 & \cdots \end{matrix}$$

Now construct additional rows by cross-multiplication:

$$\begin{matrix} P_{31} & P_{32} & P_{33} & \cdots \\ P_{41} & P_{42} & P_{43} & \cdots \\ P_{51} & \cdots & & \end{matrix}$$

etc.

where

$$P_{31} = p_1p_2 - p_0p_3, \qquad P_{32} = p_1p_4 - p_0p_5, \qquad \text{etc.}$$

and

$$P_{41} = P_{31}p_3 - P_{32}p_1, \qquad P_{42} = P_{31}p_5 - p_1P_{33}, \qquad \text{etc.}$$

$$P_{51} = P_{41}P_{32} - P_{31}P_{42}, \qquad \text{etc.}$$

The required test functions $F_0 \cdots F_n$ are then the elements of the first column, $p_0, p_1, P_{31} \cdots P_{n+1,1}$. If they are all positive, then there are no unstable roots. The number of test functions is $n + 1$, and the last one, F_n, always contains the product $p_n F_{n-1}$. Duncan (ref. 1.10, Sec. 4.10) has shown that the vanishing of p_n and of F_{n-1} represent significant critical cases. If the airplane is stable, and some design parameter is then varied in such a way as to lead to instability, then the following conditions hold:

1. If only p_n changes from $+$ to $-$, then one real root changes from negative to positive; i.e., one divergence appears in the solution (Fig. 6.1*a*).
2. If only F_{n-1} changes from $+$ to $-$, then the real part of one complex pair of roots changes from negative to positive; i.e., one divergent oscillation appears in the solution (Fig. 6.1*c*).

Thus the conditions $p_n = 0$ and $F_{n-1} = 0$ define *boundaries* between stability and instability. The former is the boundary between stability and static instability, and the latter is the boundary between stability and a divergent oscillation.

Test Functions for a Cubic

Let the cubic equation be

$$A\lambda^3 + B\lambda^2 + C\lambda + D = 0 \qquad (A > 0)$$

Then

$$F_0 = A, \qquad F_1 = B, \qquad F_2 = BC - AD, \qquad F_3 = D(BC - AD)$$

The necessary and sufficient conditions for all the test functions to be positive are that A, B, D, and $(BC - AD)$ be positive. It follows that C also must be positive.

Test Functions for a Quartic

Let the quartic equation be

$$A\lambda^4 + B\lambda^3 + C\lambda^2 + D\lambda + E = 0 \qquad (A > 0)$$

Then the test functions are $F_0 = A$, $F_1 = B$, $F_2 = BC - AD$, $F_3 = F_2 D - B^2 E$, $F_4 = F_3 BE$. The necessary and sufficient conditions for these test functions to be positive are

$$A, B, D, E > 0$$

and

$$D(BC - AD) - B^2 E > 0 \tag{6.3,2}$$

It follows that C also must be positive. The quantity on the left-hand side of Eq. 6.3,2 is commonly known as *Routh's discriminant*.

Test Functions for a Quintic

Let the quintic equation be

$$A\lambda^5 + B\lambda^4 + C\lambda^3 + D\lambda^2 + E\lambda + F = 0 \qquad (A > 0)$$

Then the test functions are $F_0 = A$, $F_1 = B$, $F_2 = BC - AD$, $F_3 = F_2D - B(BE - AF)$, $F_4 = F_3(BE - AF) - F_2^2F$, $F_5 = F_4F_2F$. These test functions will all be positive provided that

$$A, B, D, F, F_2, F_4 > 0$$

It follows that C and E also are necessarily positive.

6.4 SOLUTION OF THE CHARACTERISTIC EQUATION

In the early days of aeronautical engineering, the solution of the characteristic equation (the "stability quartic") was a significant task. The ubiquitous computer has now reduced this to a simple computation. Characteristic equations of any order can be solved routinely. Alternatively, the problem can be formulated as the determination of the eigenvalues of the system matrix. For those unusual circumstances in which machine computation is not available, a method that can be carried out by hand is given in the first edition of this book.

6.5 NUMERICAL EXAMPLE

To illustrate the application of the theory, we shall calculate the longitudinal stability of a transport airplane at a high altitude. The numerical data are as follows

$$\begin{array}{ll}
W = 100{,}000 \text{ lb} & S = 1667 \text{ sq ft} \\
A = 7 & \text{Alt.} = 30{,}000 \text{ ft} \\
V = 500 \text{ mph} & \rho = 0.000889 \text{ slug/ft}^3 \\
\quad = 733 \text{ fps} & \\
C_{L_0} = 0.25 & C_{D_0} = 0.0188 \\
C_{x_u} = -0.0376 & C_{x_\alpha} = 0.14 \\
C_{z_u} = 0 & C_{z_\alpha} = -4.90 \\
C_{m_u} = 0 & C_{z_{\dot\alpha}} = 0 \\
C_{m_q} = -22.9 & C_{m_\alpha} = -0.488 \\
C_{z_q} = 0 & C_{m_{\dot\alpha}} = -4.20 \\
l = \dfrac{\bar{c}}{2} = 7.70 \text{ ft} & \mu = 272 \\
t^* = 0.0105 \text{ sec} & i_B = 1900
\end{array}$$

The determinantal equation, Eq. 6.2,4 then becomes, in numerical form,

$$\begin{vmatrix}
(544\lambda + 0.0376) & -0.140 & 0.250 \\
0.500 & (544\lambda + 4.90) & -544\lambda \\
0 & (4.20\lambda + 0.488) & (1900\lambda^2 + 22.92\lambda)
\end{vmatrix} = 0$$

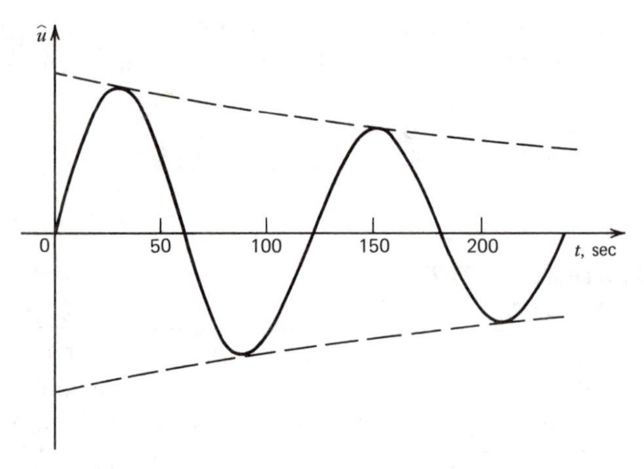

FIG. 6.2 Phugoid mode.

and the characteristic equation is

$$5.64(100\lambda)^4 + 13.14(100\lambda)^3 + 20.65(100\lambda)^2 + 0.1630(100\lambda) + 0.0610 = 0$$

All the constants of the equation are positive, and Routh's discriminant (see Sec. 6.3) is

$$R = 0.1630(13.14 \times 20.65 - 5.64 \times 0.1630) - (13.14)^2 \times 0.061 > 0$$

Hence there are no unstable roots of the equation.

The characteristic equation has been solved, and the roots were found to be

$$\text{Mode 1:} \quad 100\lambda_{1,2} = -0.00302 \pm 0.0545i$$
$$\text{Mode 2:} \quad 100\lambda_{3,4} = -1.162 \pm 1.515i$$

Mode 1

The significant characteristics are

Period, $$T = \frac{2\pi}{0.0545} \times 0.0105 \times 100 = 121 \text{ sec}$$

Damping, $$t_{\text{half}} = \frac{0.69}{0.00302} \times 0.0105 \times 100 = 240 \text{ sec}$$

$$N_{\text{half}} = \frac{240}{121} = 1.98 \text{ cycles}$$

This long-period lightly damped mode is called the *phugoid*. It was first described by Lanchester, who also named it.[2] A typical variation with time of one of the variables in this mode is illustrated in Fig. 6.2 (the time zero is arbitrary). For a further discussion of the phugoid see Secs. 6.6 and 6.8.

[2] *Phugoid* is from the Greek root for *flee*, as in *fugitive*. Lanchester actually wanted *fly*.

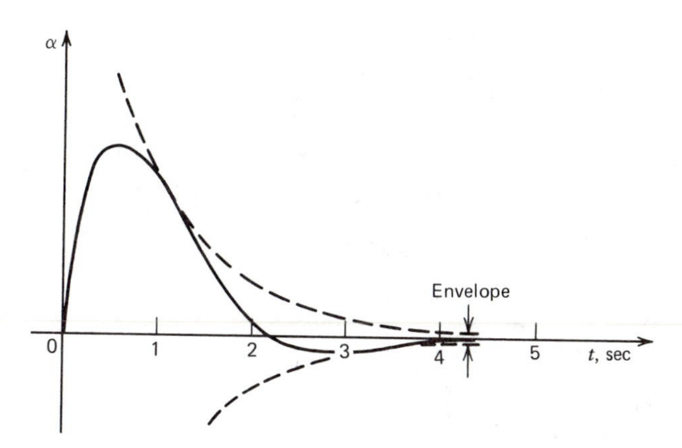

FIG. 6.3 Short-period mode.

Mode 2

The significant characteristics of the second mode are

Period, $$T = \frac{2\pi}{1.515} \times 0.0105 \times 100 = 4.37 \text{ sec}$$

Damping, $$t_{\text{half}} = \frac{0.69}{1.162} \times 0.0105 \times 100 = 0.624 \text{ sec}$$

$$N_{\text{half}} = \frac{0.624}{4.37} = 0.143 \text{ cycles}$$

This mode is seen to be of much shorter period than the phugoid, and very heavily damped. It is called the *short-period mode.* Figure 6.3 illustrates the manner in which a typical flight variable changes with time in this mode (the time zero is again arbitrary). For a further discussion, see Secs. 6.6 and 6.8.

The two modes described above are typical of the control-fixed longitudinal characteristics of stable airplanes over a wide range of conditions. That is, such airplanes usually have one long-period slightly damped mode, and one with short period and heavy damping.

6.6 THE "SHAPES" OF THE NATURAL MODES[3]

There is more to the natural modes than the characteristics discussed in the foregoing. In addition to the period and the rate of growth or decay, each mode has what may be called a *shape*. To clarify this concept, consider a simple system of two degrees of freedom (Fig. 6.4). It consists of two massive beads on a light

[3] The author is indebted to Professor E. E. Larrabee of MIT for drawing attention to the technique used in this section to picture the modes.

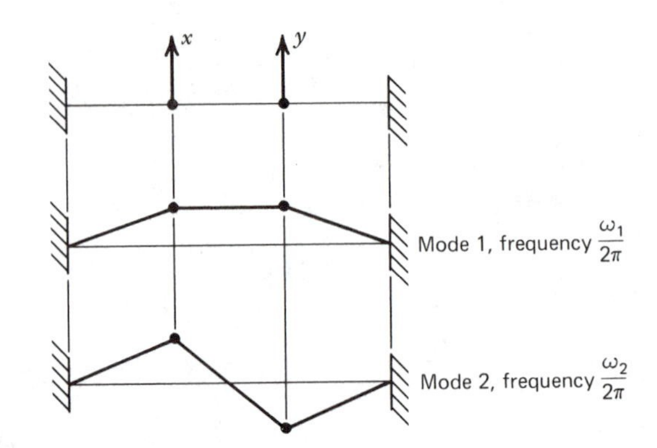

FIG. 6.4 Natural modes.

stretched string without damping. It is intuitively obvious, as well as demonstrable from the equations of motion, that the system has two possible modes of simple harmonic motion, as illustrated. (In this case, the modes are *orthogonal*, or *normal*). Each of the two normal modes is characterized by two features: its frequency and its shape. The shape can be described quantitatively by the ratio of the amplitudes of x and y, and the phase angle between them. This information is most conveniently represented on a vector diagram. Each periodic variable, e.g., $x = x_1 \cos(\omega t + \varphi)$, is represented as a rotating vector[4] whose projection on the real axis is the value of the variable. That is, $x = R[x_1 e^{i(\omega t + \varphi)}] = R[x_1 e^{i\varphi} e^{i\omega t}]$, where R indicates the real part. The two modes of Fig. 6.4 are then completely described by the vector diagrams of Fig. 6.5. Figure 6.5*a* shows that in mode 1 the amplitudes of the two degrees of freedom are the same, that they are in phase, and that the circular frequency is ω_1. Likewise Fig. 6.5*b* gives all the characteristics of mode 2.

Damping

When the system is undamped, the rotating vectors are of constant length. When the system is damped, a typical variable is of the form $x = x_1 e^{nt} \cos(\omega t + \varphi)$. The corresponding complex variable is $x_1 e^{nt} e^{i\varphi} e^{i\omega t}$. The modulus of this variable, which equals the amplitude and the length of the vector, is $x_1 e^{nt}$. This varies exponentially with time, so that the tip of the vector traces out a spiral, instead of a circle.

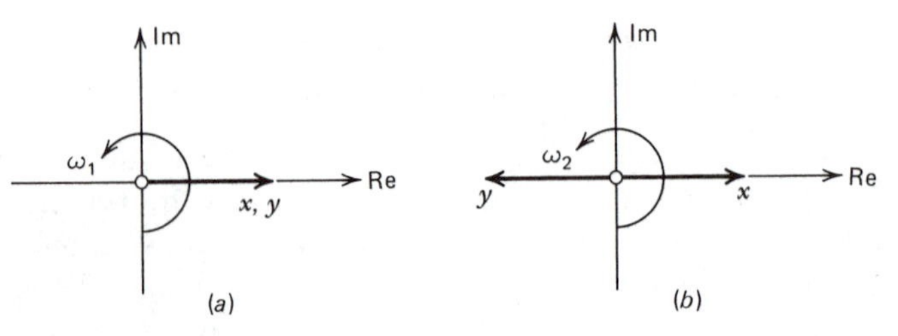

FIG. 6.5 Vector representation of the modes of Fig. 6.4.

[4] The set of these rotating vectors comprises the *eigenvector* of the mode. (See Ref 1.10, Sec. 3.3)

Real Roots

When the mode corresponds to a real root of the characteristic equation, as in Figs. 6.1*a* and *b*, the situation is simpler. In that case, the motion being aperiodic, no phase angle is involved, and the vector representation is not applicable. The "shape" of the mode is then defined simply by the *ratios* of the aperiodic variables, which remain fixed as the variables themselves change with time.

In the simple example dealt with above, we were able to find the shapes of the modes intuitively. We must now consider how to find them for the more complicated cases that arise in airplane dynamics.

The Longitudinal Modes

We shall illustrate the determination of natural modes with the example of longitudinal motion calculated in Sec. 6.5. Let us suppose that the disturbed motion of the airplane is initiated in such a way that only one of the modes is excited. (This is always possible theoretically. See also Sec. 6.7.) Let the roots for this mode be $\lambda = n \pm i\omega$. With a suitable choice of the time origin, the solution may be written

$$\begin{aligned} \hat{u} &= \hat{u}_1 e^{n\hat{t}} \cos \omega\hat{t} \\ \alpha &= \alpha_1 e^{n\hat{t}} \cos (\omega\hat{t} + \varepsilon) \\ \theta &= \theta_1 e^{n\hat{t}} \cos (\omega\hat{t} + \delta) \end{aligned} \tag{6.6,1}$$

In keeping with our complex representation, let the right-hand side of Eqs. 6.6,1 be the real parts of [5]

$$\begin{aligned} &\hat{u}_0 e^{n\hat{t}} e^{i\omega\hat{t}}, &\quad \hat{u}_0 &= \hat{u}_1 \\ &\alpha_0 e^{n\hat{t}} e^{i\omega\hat{t}}, &\quad \alpha_0 &= \alpha_1 e^{i\varepsilon} \\ &\theta_0 e^{n\hat{t}} e^{i\omega\hat{t}}, &\quad \theta_0 &= \theta_1 e^{i\delta} \end{aligned} \tag{6.6,2}$$

$\hat{u}_0$, α_0, and θ_0 are the *complex amplitudes* of $\hat{u}$, α, and θ, and ε and δ are phase angles by which α and θ lead $\hat{u}$. As written above, $\hat{u}_0$ is arbitrarily chosen to be real. Since the governing differential equations (6.2,1) are linear, then, if they are satisfied by the complex variables (6.6,2), they are also satisfied by the real variables (6.6,1). Therefore we substitute the complex variables into Eqs. 6.2,1. When this is done, we find that the factor $e^{n\hat{t}}e^{i\omega\hat{t}}$ is contained in every term, and may be divided out. The equations that result are like Eqs. 6.2,3 except that λ is replaced by $n + i\omega$. If we divide these equations through by θ_0, we get three equations for the two unknowns $\hat{u}_0/\theta_0$ and α_0/θ_0. Any two of the equations may be used to solve for these rations, which are the required quantities as is demonstrated below.[6]

[5] The set $\{\hat{u}_0 \alpha_0 \theta_0 \hat{q}_0\}$ constitutes the *eigenvector* corresponding to the eigenvalue λ.

[6] That the same answers will be obtained for $\hat{u}_0/\theta_0$ and α_0/θ_0, no matter which pair of equations is used, is ensured by the vanishing of the determinant. In fact it is just this requirement for homogeneous equations that leads to the determinantal condition.

Using the first and third equations, we obtain

$$[2\mu(n+i\omega) - C_{x_u}]\frac{\hat{u}_0}{\theta_0} - C_{x_\alpha}\frac{\alpha_0}{\theta_0} = -C_{L_0}$$

$$-C_{m_u}\frac{\hat{u}_0}{\theta_0} - [C_{m_{\dot{\alpha}}}(n+i\omega) + C_{m_\alpha}]\frac{\alpha_0}{\theta_0} = -[i_B(n+i\omega)^2 - C_{m_q}(n+i\omega)] \tag{6.6,3}$$

These equations are best handled in numerical form for each case. They are solved for the ratios $\hat{u}_0/\theta_0$, and α_0/θ_0. In general, these ratios are complex numbers, e.g., $\hat{u}_0/\theta_0 = a + ib$. However, Eqs. 6.6,2 show that

$$\frac{\hat{u}_0}{\theta_0} = \frac{\hat{u}_1 e^{-i\delta}}{\theta_1}$$

Therefore the required amplitude and phase relations for θ are

$$\frac{\hat{u}_1}{\theta_1} = [a^2 + b^2]^{1/2}$$

$$\delta = -\tan^{-1}\frac{b}{a}$$

Similarly, if

$$\frac{\alpha_0}{\theta_0} = c + id = \frac{\alpha_1}{\theta_1}e^{i(\varepsilon-\delta)}$$

then the amplitude and phase relations for α are

$$\frac{\alpha_1}{\theta_1} = [c^2 + d^2]^{1/2}$$

$$\varepsilon - \delta = \tan^{-1}\frac{d}{c}$$

or

$$\varepsilon = \delta + \tan^{-1}\frac{d}{c}$$

The theory given above is now applied to the numerical example of Sec. 6.5. The numerical equations corresponding to Eqs. 6.6,3 are

$$[544(n+i\omega) + 0.0376]\frac{\hat{u}_0}{\theta_0} - 0.140\frac{\alpha_0}{\theta_0} = -0.250$$

$$0 + [4.20(n+i\omega) + 0.488]\frac{\alpha_0}{\theta_0} = -[1900(n+i\omega)^2 + 22.9(n+i\omega)]$$

When the values of $(n + i\omega)$ for the phugoid and short-period modes are substituted, we obtain the following results.

Phugoid Mode

$$n + i\omega = (-0.00302 + 0.0545i)10^{-2}$$

$$\hat{u}_1/\theta_1 = 0.841, \qquad \delta = -94.9^\circ$$

$$\alpha_1/\theta_1 = 0.0256, \qquad \varepsilon - \delta = -84.5^\circ, \qquad \varepsilon = -179.4^\circ$$

Short-Period Mode

$$n + i\omega = (-1.162 + 1.515i)10^{-2}$$

$$\hat{u}_1/\theta_1 = 0.012, \qquad \delta = -12.8^\circ$$

$$\alpha_1/\theta_1 = 1.24, \qquad \varepsilon - \delta = 27.6^\circ, \qquad \varepsilon = 14.8^\circ$$

The vector diagrams for the modes are drawn from the above data, and are shown in Fig. 6.6. Two highly significant characteristics are clearly evident from

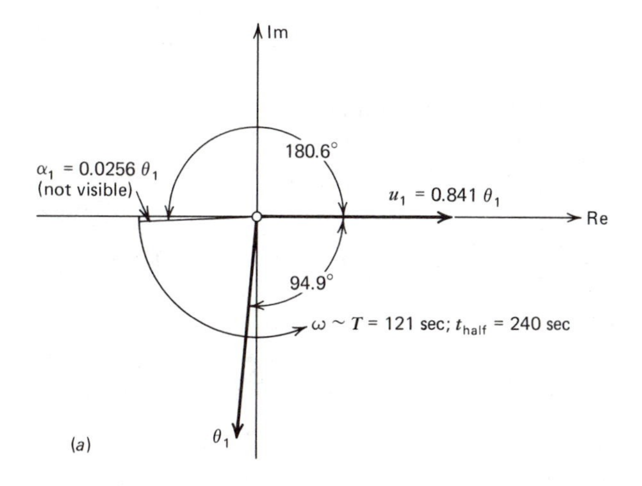

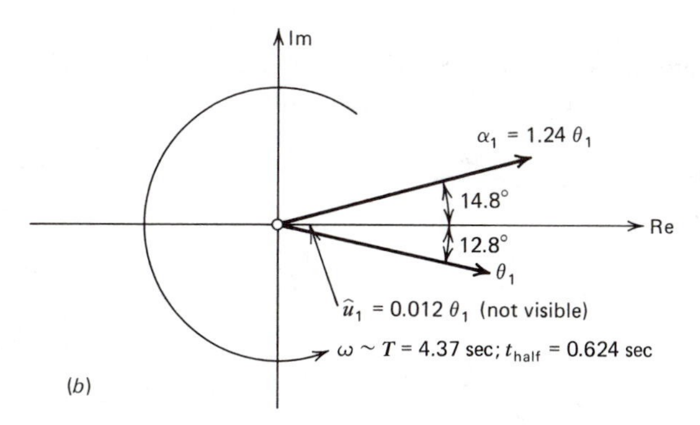

FIG. 6.6 Vector diagrams of longitudinal modes. (*a*) Phugoid mode. (*b*) Short-period mode.

the diagrams. These are that the angle-of-attack disturbance is negligibly small in the phugoid mode, and the speed disturbance is negligible in the short-period mode. *They are essentially motions in two degrees of freedom* (see also Sec. 6.7).

Flight Paths in the Natural Modes

Additional insight into the modes is gained by studying the flight path of the airplane. This may be computed from Eqs. 4.14,6. For initially horizontal flight, $\theta_0 = 0$, and they become, for symmetric motion:

$$\frac{dx'}{dt} = u_0 + u = u_0(1 + \hat{u})$$

$$\frac{dy'}{dt} = 0$$

$$\frac{dz'}{dt} = -u_0\theta + w = u_0(-\theta + \alpha)$$

Using Eqs. 6.6,1 for flight in a natural mode, we get

$$\frac{1}{t^*}\frac{dx'}{d\hat{t}} = u_0(1 + \hat{u}_1 e^{n\hat{t}} \cos \omega\hat{t})$$

$$\frac{1}{t^*}\frac{dz'}{d\hat{t}} = u_0 e^{n\hat{t}}[\alpha_1 \cos(\omega\hat{t} + \varepsilon) - \theta_1 \cos(\omega\hat{t} + \delta)]$$

Integration yields

$$x' = u_0 t^*\left[\hat{t} + \frac{\hat{u}_1}{n^2 + \omega^2} e^{n\hat{t}}(n \cos \omega\hat{t} + \omega \sin \omega\hat{t})\right]$$

$$z' = u_0 t^* \frac{e^{n\hat{t}}}{n^2 + \omega^2}\{\alpha_1[n \cos(\omega\hat{t} + \varepsilon) + \omega \sin(\omega\hat{t} + \varepsilon)$$
$$- \theta_1[n \cos(\omega\hat{t} + \delta) + \omega \sin(\omega\hat{t} + \delta)\}$$

By way of example, the amplitude of θ is assumed to be $\theta_1 = 0.2$ radian $= 11.46°$ for both modes. Using the results for the amplitude ratios and phase angles shown

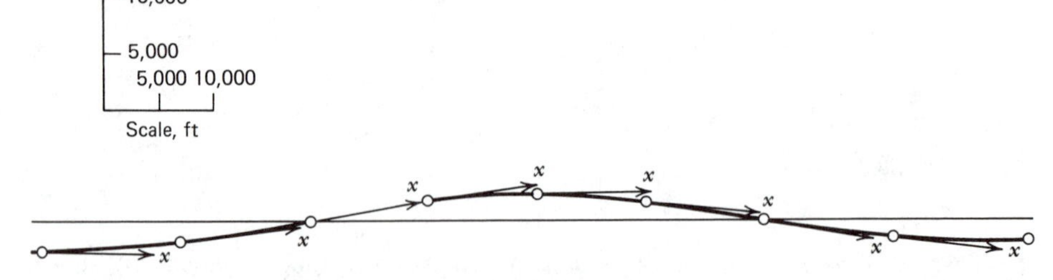

FIG. 6.7*a* Phugoid flight path (fixed reference frame).

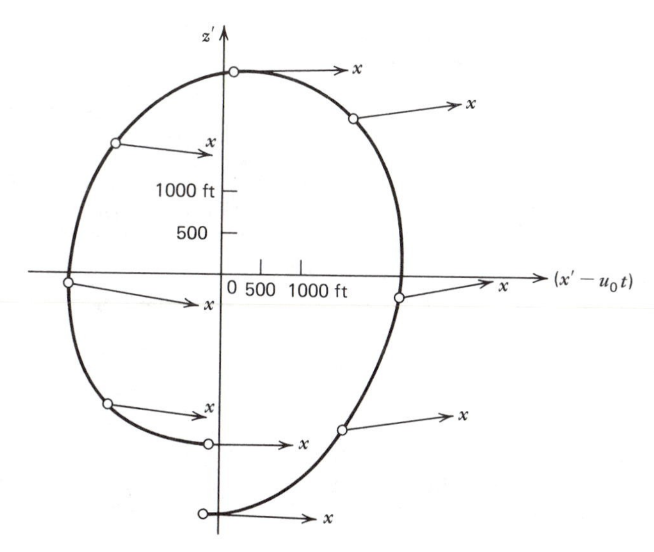

FIG. 6.7*b* Phugoid flight path (moving reference frame).

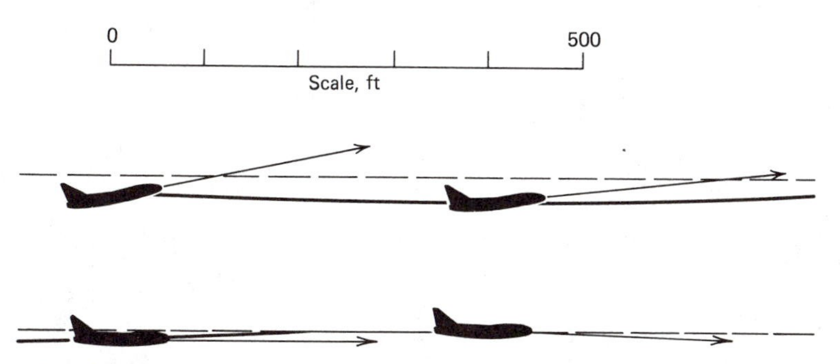

FIG. 6.7*c* Short-period flight path.

in Fig. 6.6, we have all the information required to compute the flight paths in the two modes. These computations have been carried out, and the results are displayed in Fig. 6.7. The physical characteristics of the modes are clearly evident in this figure. In the phugoid mode, the x axis of the airplane remains essentially tangent to the flight path, the principal feature of the motion being the slow rising and falling of the airplane, accompanied by change in speed. The speed is greatest when the height is least; there is a continual interchange between kinetic and potential energy. Figure 6.7*a* shows the path relative to fixed axes, and Fig. 6.7*b* shows the path as it would appear to an observer flying alongside at the steady speed u_0. At the scales of these figures, the airplane would be seen only as a point.

Figure 6.7*c* shows the initial part of the short-period motion. It is so rapidly damped out that the transient has virtually disappeared within 1,000 ft of flight. The deviation of the flight path from a straight line is small, the principal feature of the motion being the rapid rotation of the airplane in pitch.

6.7 APPROXIMATE EQUATIONS FOR THE LONGITUDINAL MODES

The calculations of Sec. 6.6 show that, in the example treated, the two natural modes are virtually motions with two degrees of freedom—in the phugoid $\alpha \doteqdot 0$, and in the short-period motion $\hat{u} \doteqdot 0$. This suggests that by dropping these particular variables from the original equations of motion, simpler approximate equations could be obtained for these modes. This is indeed the case for a wide range of configurations and flight conditions.

Phugoid Mode

To obtain the simplified equations for the phugoid, we set $\alpha = 0$, and drop the pitching-moment equation from Eqs. 6.2,1. (Since there are only two variables left, $\hat{u}$ and θ, then one of the three equations must be discarded. The choice here is governed by the fact that, with $\alpha \doteqdot 0$, the pitching-moment changes will be small.) The resulting equations are

$$\begin{aligned} (2\mu D - C_{x_u})\hat{u} + C_{L_0}\theta = 0 \\ (2C_{L_0} - C_{z_u})\hat{u} - (2\mu + C_{z_q}) D\theta = 0 \end{aligned} \tag{6.7,1}$$

The characteristic equation for this system is found by the method of Sec. 6.2, and turns out to be

$$2\mu(2\mu + C_{z_q})\lambda^2 - C_{x_u}(2\mu + C_{z_q})\lambda + C_{L_0}(2C_{L_0} - C_{z_u}) = 0 \tag{6.7,2}$$

To verify that this equation does give an approximation to the phugoid mode, the numerical data of Sec. 6.5 is substituted into it. The result is

$$29.60(100\lambda)^2 + 0.2045(100\lambda) + 0.125 = 0$$

so that

$$100\lambda = -0.00346 \pm 0.0649i$$

The roots previously obtained for the phugoid mode from the quartic characteristic equation are

$$100\lambda = -0.00302 \pm 0.0545i$$

The approximate roots are seen to give the damping and period with errors of 19.1% and 14.6%, respectively. The precision of the approximate result is low. It is nevertheless useful, particularly for assessing the influence of those parameters that were retained in the approximate equations. To this end we write Eq. 6.7,2 in a form that is very convenient for second-order systems, namely,

$$\lambda^2 + 2\zeta\omega_n\lambda + \omega_n{}^2 = 0 \tag{6.7,3}$$

where

$$\omega_n = \left[\frac{C_{L_0}(2C_{L_0} - C_{z_u})}{2\mu(2\mu + C_{z_q})}\right]^{1/2} \tag{6.7,4a}$$

and

$$\zeta = -\frac{C_{x_u}(2\mu + C_{z_q})}{2[C_{L_0}(2C_{L_0} - C_{z_u})2\mu(2\mu + C_{z_q})]^{1/2}} \tag{6.7,4b}$$

It is a convenience to identify the coefficients of Eq. 6.7,3 with the physical properties of a linear spring-mass-damper system (which is governed by the same equation). When the mass is unity, then ${\omega_n}^2$ is the spring stiffness, and $2\zeta\omega_n$ is the viscous damping constant. In terms of ω_n and ζ, the roots are

$$\lambda = -\zeta\omega_n \pm i\omega_n\sqrt{1 - \zeta^2} \tag{6.7,5}$$

ω_n is called the *undamped circular frequency*, since, when the damping is zero ($\zeta = 0$), then $\lambda = \pm i\omega_n$, and the motion is simple harmonic with frequency $\omega_n/2\pi$. ζ is called the *damping ratio*. As noted above, $\zeta = 0$ corresponds to zero damping, and, when $\zeta = 1$, the imaginary part of Eq. 6.7,5 vanishes. Thus $\zeta = 1$ represents the boundary between oscillatory and aperiodic motion. This is the condition of *critical damping*.

In the phugoid oscillation being discussed, C_{z_u} and C_{z_q} are frequently negligible, and in any case are not dominant parameters. Let us assume them to be zero. Then ω_n and ζ take the simpler forms:

$$\omega_n = \frac{C_{L_0}}{\sqrt{2\mu}} \qquad \zeta = -\frac{C_{x_u}}{2\sqrt{2}C_{L_0}}$$

The undamped natural frequency is seen to increase with C_{L_0}; i.e., as the reference flight speed decreases. As the altitude increases, so does μ, and therefore the phugoid frequency diminishes with height at constant C_{L_0}. The damping ratio ζ depends solely on C_{x_u}/C_{L_0}. In level flight at constant thrust, and no compressibility effects, $C_{x_u} = -2C_{D_0}$. Then $\zeta = (1/\sqrt{2})(C_{D_0}/C_{L_0})$. The damping of the phugoid in these circumstances is seen to be inversely proportional to the lift/drag ratio of the airplane. When compressibility effects are present, with substantial values of $\partial C_{D_0}/\partial \mathbf{M}$, then the value of C_{x_u} may be materially changed, and hence the damping as well. In particular, in the low supersonic range, where the wave drag coefficient is decreasing rapidly with $\mathbf{M}$, C_{x_u} may even become positive. If it does, then $\zeta < 0$, and the oscillation will become divergent (see Eq. 6.7,5). The motion will remain oscillatory unless $\zeta^2 > 1$.

The above conclusions relative to the phugoid oscillation have been derived from the simplified approximate equations. The effect of C.G. position, C_{m_q} and $C_{m_{\dot\alpha}}$, are not brought out by these, since they enter only into the pitching-moment equation which was neglected. The influence of C.G. position is shown in Sec. 6.9.

Short-Period Mode

To obtain the simplified equations for the short-period mode, we set $\hat{u} = 0$, and discard the X force equation. From Eqs. 6.2,1, upon replacing $D\theta$ by $\hat{q}$, we obtain

$$\begin{aligned} (2\mu D - C_{z_{\dot\alpha}}D - C_{z_\alpha})\alpha - (2\mu + C_{z_q})\hat{q} &= 0 \\ -(C_{m_{\dot\alpha}}D + C_{m_\alpha})\alpha + (i_B D - C_{m_q})\hat{q} &= 0 \end{aligned} \tag{6.7,6}$$

The characteristic equation for this system is

$$(\lambda^2 + 2\zeta\omega_n\lambda + \omega_n^2) = 0$$

where

$$\omega_n^2 = \frac{C_{z_\alpha}C_{m_q} - C_{z_q}C_{m_\alpha} - 2\mu C_{m_\alpha}}{i_B(2\mu - C_{z_{\dot\alpha}})}$$

$$\zeta = -\frac{(2\mu - C_{z_{\dot\alpha}})C_{m_q} + i_B C_{z_\alpha} + C_{m_{\dot\alpha}}(2\mu + C_{z_q})}{2[i_B(2\mu - C_{z_{\dot\alpha}})(C_{z_\alpha}C_{m_q} - C_{z_q}C_{m_\alpha} - 2\mu C_{m_\alpha})]^{1/2}} \tag{6.7,7}$$

Somewhat simpler expressions are obtained for ω_n and ζ if we neglect C_{z_q} and $C_{z_{\dot\alpha}}$, which are frequently small. We then get

$$\omega_n^2 = \frac{C_{z_\alpha}C_{m_q} - 2\mu C_{m_\alpha}}{2\mu i_B}$$

$$\zeta = -\frac{2\mu C_{m_q} + i_B C_{z_\alpha} + 2\mu C_{m_{\dot\alpha}}}{2[2\mu i_B(C_{z_\alpha}C_{m_q} - 2\mu C_{m_\alpha})]^{1/2}}$$

As long as ω_n and $\zeta\omega_n$ are real and positive, the motion is a damped oscillation. Although several of the airplane parameters exert significant influences on these quantities, perhaps the most important is the C.G. position, which controls the magnitude and sign of C_{m_α}. We note that a critical condition occurs when the C.G. is in that position for which

$$C_{z_\alpha}C_{m_q} = 2\mu C_{m_\alpha} \tag{6.7,8}$$

Then

$$\omega_n = 0, \quad \text{and} \quad \zeta\omega_n = -\frac{2\mu(C_{m_q} + C_{m_{\dot\alpha}}) + i_B C_{z_\alpha}}{4\mu i_B} > 0$$

For these values of the variables, the characteristic equation is $\lambda(\lambda + 2\zeta\omega_n) = 0$, with the two roots $\lambda = 0$ and $\lambda = -2\zeta\omega_n$. The mode corresponding to $\lambda = 0$ is ae^0, a constant. Thus the variations of α and $\hat{q}$ with time are of the form $a_1 + a_2e^{-2\zeta\omega_n\hat{t}}$. Since $\zeta\omega_n$ is positive, then the second term dies out, and only the constant a_1 is left. A longitudinal motion at constant speed with constant values of α and $\hat{q}$ will be recognized by the reader as a steady pull-up (see Chap. 3). Thus, when the C.G. is in the critical position defined by Eq. 6.7,8 the airplane can respond to a transient disturbance by developing a steady load factor, with θ increasing linearly with time.

When the C.G. is moved aft of the critical position noted, then C_{m_α} becomes more positive, and ω_n^2 becomes negative. $\zeta\omega_n$ is not affected by the change in C_{m_α}, and so remains positive. The system is then equivalent to a spring-mass-damper with a *negative* spring; that is, a spring that drives the mass away from equilibrium instead of pulling it back. In this condition the roots of the characteristic equation are

$$\lambda_{1,2} = -(\zeta\omega_n) \pm \sqrt{(\zeta\omega_n)^2 - \omega_n^2}$$

Since $(\zeta\omega_n)$ is positive, and ω_n^2 is negative, then the roots are real, and the motion is aperiodic. The root corresponding to the + sign is positive, and hence the motion is statically unstable.

The critical C.G. position is readily found. From Eq. 2.3,2, $C_{m_\alpha} = a(h - h_n)$, and $C_{z_\alpha} \doteqdot -C_{L_\alpha} = -a$. When these are substituted into Eq. 6.7,8, the result is

$$h = h_n - \frac{1}{2\mu} C_{m_q}$$

This C.G. position is exactly the same as the stick-fixed maneuver point h_m (see Eq. 3.1,14). In this light the maneuver point is seen to be a criterion for divergence of the short-period longitudinal motion. It is an approximate criterion in that C_{z_q} and $C_{z_{\dot{\alpha}}}$ were neglected in the derivation above.

The accuracy of Eqs. 6.7,6 as a representation of the short-period mode is checked by using the numerical values of Sec. 6.5. When this is done, we get, from Eqs. 6.7,7,

$$\omega_n^2 = 3.65 \times 10^{-4}$$
$$\zeta = 0.609$$

The approximate roots are therefore

$$\begin{aligned} \lambda &= -\zeta\omega_n \pm i\omega_n\sqrt{1-\zeta^2} \\ &= (-1.162 \pm 1.515i)10^{-2} \end{aligned}$$

The value obtained from the quartic equation (Sec. 6.5) was

$$\lambda = (-1.162 + 1.515i)10^{-2}$$

The exact and approximate roots are seen to be the same within the accuracy of the calculation. The approximate equations for the short-period mode give acceptable results over a wide range of configuration and flight variables. They are also useful in obtaining better values for the phugoid mode than were found from Eq. 6.7,2. The approximate short-period roots give a quadratic factor of the quartic characteristic equation. By synthetic division the phugoid roots can then be extracted.

6.8 THE ACTUAL TRANSIENT MOTION

The calculation of transient responses to specific disturbances will be treated in detail in Chaps. 8 and 9. However, it is worth while at this point to draw attention to the relation between the natural modes and actual transients. In the foregoing sections, we have, in effect, found that it is possible for the airplane to have certain simple special motions: the natural modes. The most general motion it can have is a linear superposition of these simpler motions (cf. Eq. 6.1,1).

Now each aperiodic or nonoscillatory natural mode has associated with it one arbitrary constant (the initial value of any one of the variables); and each periodic or oscillatory normal mode has associated with it two arbitrary constants (the amplitude and phase angle of any one of the variables). The total number of arbitrary constants is then equal to the number of aperiodic modes plus twice the number of periodic modes: i.e., to the degree of the characteristic equation, or the order of the system. Specification of the same number of initial conditions will serve to determine all the arbitrary constants, and hence the constitution of the transient in terms of the normal modes.

In the longitudinal motion treated in the preceding sections, there are two periodic modes, the degree of the characteristic equation being four. The four initial conditions might be given on $\hat{u}$, θ, α, and $D\theta = \hat{q}$. As mentioned previously, it is always theoretically possible to choose initial conditions such that only one mode is excited.

6.9 VARIATION OF THE ROOTS WITH C.G. POSITION

The example of Sec. 6.5 is now extended to include other positions of the C.G. It is assumed that C_{m_α} is the only parameter affected by the C.G. location, and that it varies thus

$$\begin{aligned} C_{m_\alpha} &= a(h - h_n) \\ &= 4.88(h - h_n) \end{aligned} \tag{6.9,1}$$

Values of $(h - h_n)$ from $-0.20 + 0.10$ were used, and the characteristic equations were formed and solved.

The results of the calculations are shown in the form of root locus plots in Fig. 6.8. Figure 6.8*a* shows the variation of the real and imaginary parts of the short-period roots as the C.G. is moved backward through the neutral point. It is seen that the damping n remains essentially constant, while the frequency ω decreases to zero (point A) at a C.G. position between the neutral point and 2% of $\bar{c}$ forward of it. The root locus then splits off into a pair of real roots, branches AB and AC of the locus. These represent damped aperiodic modes, or subsidences. Figure 6.8*b* shows that the phugoid mode has a similar behavior as the C.G. is moved back. The damping remains almost constant, while the frequency decreases. At point D, when the C.G. is just forward of the neutral point, the oscillatory mode degenerates into a pair of aperiodic modes, as shown by the branches DF and DE of the locus. DF is a subsidence, and that portion of DE to the right of the origin represents a divergence—i.e., the airplane is statically unstable when the C.G. is aft of the neutral point.

The behavior of the solution is very interesting for the range of C.G. positions $h > h_n + 0.02$. It is seen that the branch AB of the short-period mode, Fig. 6.8*a*, and the branch DF of the phugoid meet at the point F when the C.G. is somewhere between 2 and 3% of $\bar{c}$ aft of the neutral point. *A new oscillatory mode then arises*, corresponding to the branches FG of the locus. This is a stable oscillation whose damping and period both lie between those of the short-period mode and the phugoid. For example, when $h_n - h = -0.03$, the roots are

$$100\lambda = -0.1124 \pm 0.08627i$$

The corresponding period and damping are

$$T = 76.5 \text{ sec}$$
$$N_{\text{half}} = 0.0845 \text{ cycle}$$

The method of Sec. 6.6 for calculating the mode shape has been applied to this case with the result shown in Fig. 6.9. The mode is seen to be entirely different from the short-period and phugoid modes previously described (cf. Fig. 6.6). The most important difference is that in this mode all three degrees of freedom are significantly excited. For this reason there is no simple approximation to it based on a two-degrees-of-freedom approach.

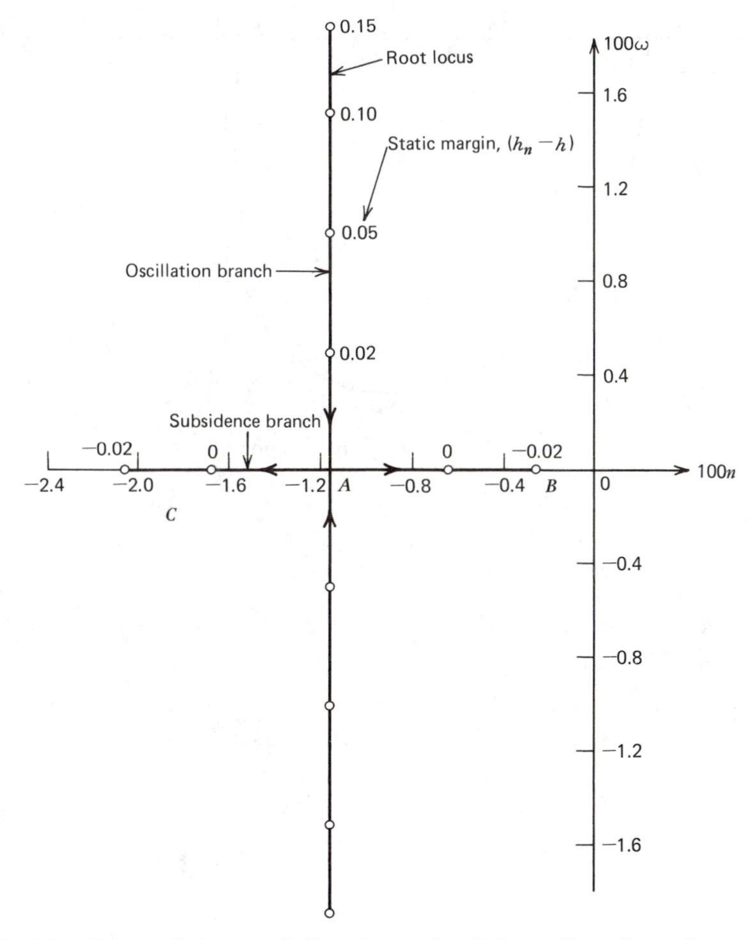

FIG. 6.8a Locus of short-period roots, varying C.G. position. $C_{m_u} = 0$.

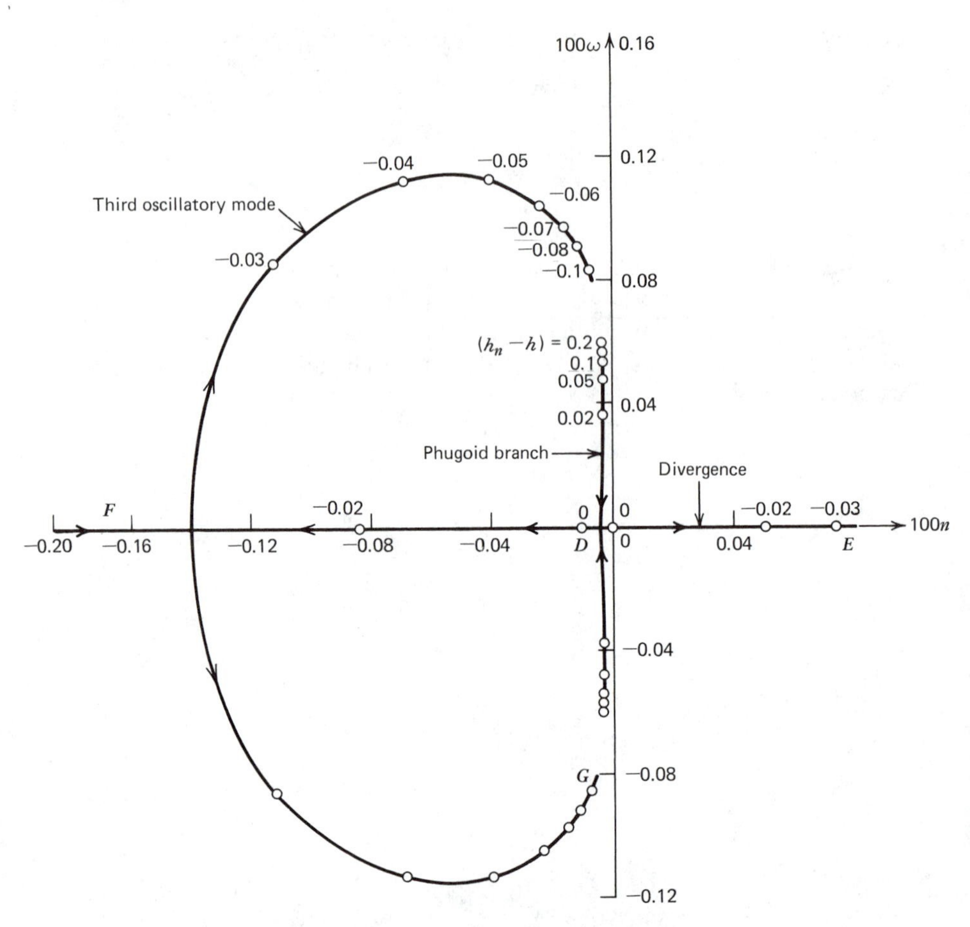

FIG. 6.8*b* Locus of phugoid roots, varying C.G. position. $C_{m_u} = 0$.

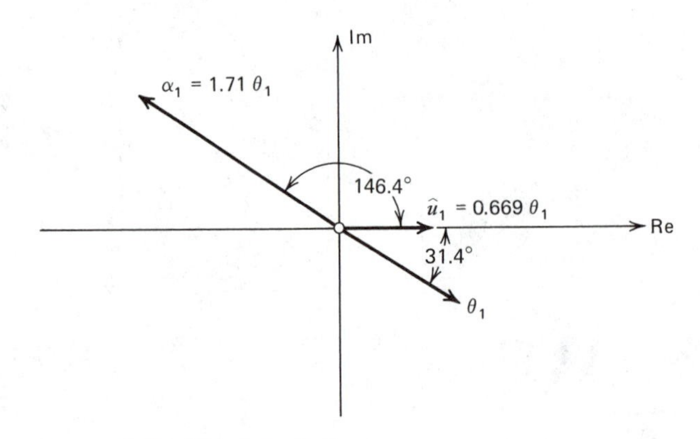

FIG. 6.9 Shape of the "third mode."

6.10 STATIC STABILITY

It was pointed out in Chap. 2 that the concept of static stability introduced there was limited in scope. The limitation was imposed by considering the pitching moment induced by a disturbance to α only, while the speed remains constant. We are now in a position to consider a more general criterion for static stability. In Sec. 6.3 we discussed the conditions on the coefficients of the characteristic equation that must be met if the airplane is to be stable. It was noted there that, when the last term of the equation (e.g., E of Eq. 6.2,5) becomes negative, then a divergence occurs in the solution. That is, the motion is statically unstable.

The more general criterion for static longitudinal stability is then, from Eq. 6.2,5 (which applies to elevator-fixed level flight),

$$E = -C_{L0}[C_{m_\alpha}(2C_{L_0} - C_{z_u}) + C_{m_u}C_{z_\alpha}] > 0 \tag{6.10,1}$$

When there are no speed effects due to compressibility, airframe distortion, or the propulsion system, then $C_{m_u} = C_{z_u} = 0$, and the criterion is $-2C_{L_0}{}^2C_{m_\alpha} > 0$, or simply $C_{m_\alpha} < 0$. This is identical with the condition used in Chap. 2, and is by definition satisfied so long as the C.G. is forward of the neutral point, stick-fixed.

When C_{z_u} and C_{m_u} are not zero, as in flight near $\mathbf{M} = 1$, then they may exert a large influence on E. It may be found, for example, that, in order for E to be positive, the C.G. must lie forward of the leading edge of the mean chord. It is pertinent to ask the following question: "Does it matter whether a negative value of E is caused by the C_{m_α} term or by the C_{m_u} term? That is, is the numerical value of E, however brought about, an index of the static stability?" The answer to this question is best given by a numerical example. We shall therefore consider once more the example of Secs. 6.5, and 6.9, but this time add a destabilizing C_{m_u}:

$$C_{m_u} = -0.10$$

With C_{m_α} given by Eq. (6.9,1), and the other numerical values from Sec. 6.5, the expression for E is

$$\begin{aligned} E &= -0.25[4.88(h - h_n)(0.5) + 0.49] \\ &= -0.1227 - 0.610(h - h_n) \end{aligned} \tag{6.10,2}$$

Thus E is negative, and the airplane is statically unstable, for all values of $h > h_n - 0.201$.

The Shape of the Divergent Mode

In order to understand fully the behavior of the unstable airplane, we must know the shape of the unstable mode. By way of illustration, we shall calculate it for the example airplane with the C.G. location 10% forward of the neutral point: i.e., $h - h_n = -0.10$. The unstable root for this C.G. position has been found by solving the appropriate quartic, and is $\lambda = 0.000500$. The calculation of the shape is made with Eqs. 6.6,3 by setting $\omega = 0$, and n equal to the above value of λ. When the numerical values have been substituted, the following real equations are

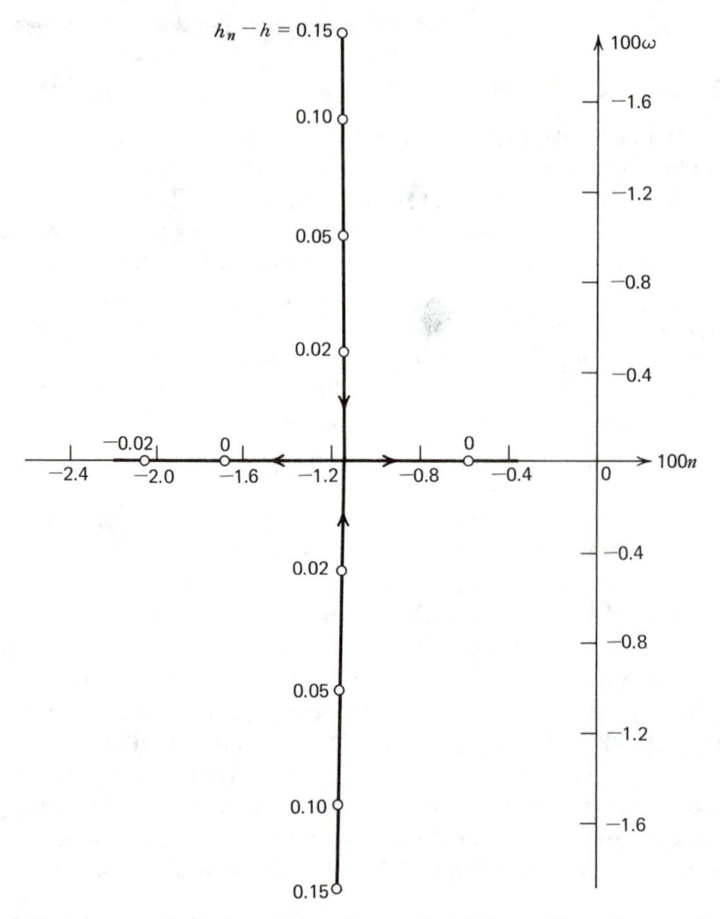

FIG. 6.10*a* Locus of short-period roots, varying C.G. position. $C_{m_u} = -0.10$.

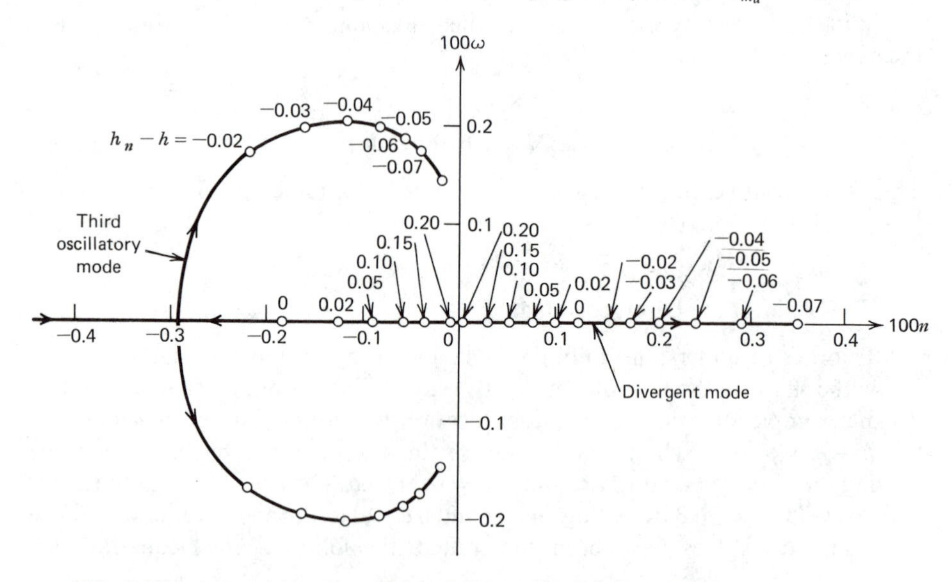

FIG. 6.10*b* Locus of roots, varying C.G. position. $C_{m_u} = -0.10$.

obtained:

$$0.3096\frac{\hat{u}_0}{\theta_0} - 0.140\frac{\alpha_0}{\theta_0} = -0.250$$

$$0.100\frac{\hat{u}_0}{\theta_0} + 0.490\frac{\alpha_0}{\theta_0} = -0.01192$$

The solutions are

$$\frac{\hat{u}_0}{\theta_0} = -0.749, \qquad \frac{\alpha_0}{\theta_0} = 0.129$$

or

$$\alpha_0:\hat{u}_0:\theta_0 = 0.129:-0.749:1.00$$

If an initial disturbance were to excite only this mode, then at the time when the angle of pitch had changed by say $-10°$, the speed would have increased by $0.749 \times (10/57.3) \times 733 = 95.8$ fps, and the angle of attack would have decreased by 1.29°. The airplane is seen to be in a dive of continuously increasing speed and angle. As in all cases of unstable modes, however, it must be remembered that the linear solution may not be valid for the large values of the disturbances that ensue as time passes. This is especially so when compressibility is an important factor. The change in speed noted above corresponds to a change in **M** of the order of 0.1. In the transonic flight range, this might be enough to alter appreciably the values of aerodynamic coefficients such as C_{L_α} and C_{m_α}.

The Loci of the Roots

Calculations were carried out for the same range of C.G. positions as was used in the example of Sec. 6.9. The coefficients of the quartics obtained were the same as in Sec. 6.9 except for E, which was computed from Eq. 6.10,2. The root loci are shown in Figs. 6.10*a* and *b*. Figure 6.10*a* shows the short-period roots, which are almost the same as when $C_{m_u} = 0$ (cf. Fig. 6.8*a*). This is to be expected; it was shown in Sec. 6.6 that the short-period mode proceeds with nearly no variation in speed, so that the introduction of a value of C_{m_u} different from zero has little influence on the motion.

Figure 6.10*b* shows the loci of the remaining roots. For all the C.G. positions shown, E is negative, and hence there is a positive real root, giving a divergent mode. This root, together with the small negative one, replaces the phugoid roots of the stable airplane.

The third oscillatory mode, which appeared previously, is found once more, although the roots are somewhat changed from the values on Fig. 6.8*b*.

Rate of Divergence

We return now to the question posed at the beginning of this section. The answer is found on Figs. 6.11 and 6.12. These show how the time to double of the divergent mode varies with the static margin, and with the value of E for the aircraft with and without a speed instability (i.e., negative C_{m_u}). Figure 6.12 shows

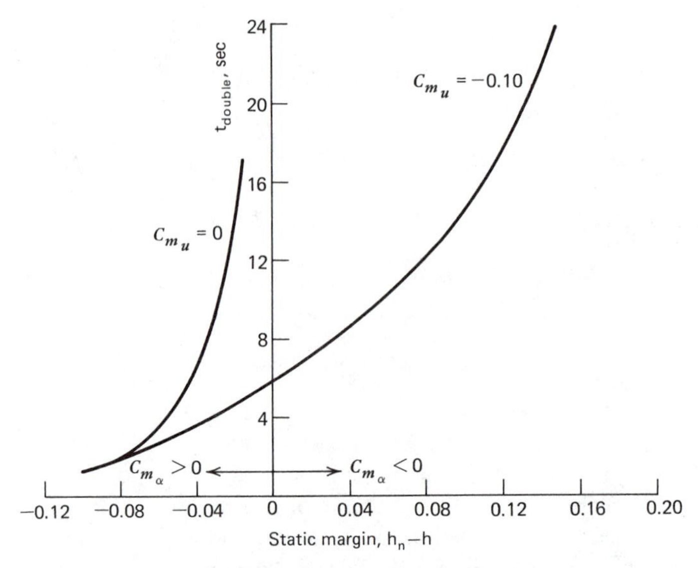

FIG. 6.11 Time to double for the divergent mode.

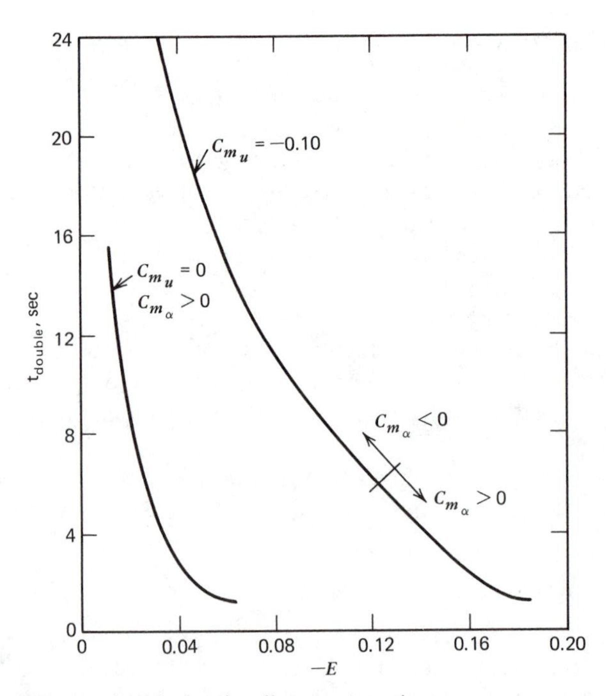

FIG. 6.12 Time to double for the divergent mode.

that the rate of divergence is not determined by E alone, but depends strongly on whether the negative value of E results from a positive C_{m_α} or a negative C_{m_u}. For the airplane and flight condition considered, the presence of the negative C_{m_u} would not lead to dangerous handling qualities of the airplane for normal C.G. positions ($h_n - h > 0.03$), the time to double being then in excess of 8 sec.

6.11 STABILITY BOUNDARIES

It was pointed out in Sec. 6.3 that the vanishing of certain test functions of the characteristic equation define boundaries between stability and instability. This idea was applied to the discussion of static stability in Sec. 6.10. It is sometimes convenient to plot what are known as *stability boundaries*. These are obtained by setting one of the critical test functions equal to zero, and allowing two of the airplane configuration or flight variables to change. For each value of one of these variables, the vanishing of the test function fixes the value of the other. Thus a curve is defined in the plane of the two variables which separates a region of stability from one of instability. The type of instability (static or dynamic) is known from the test function used. For example, the vanishing of Routh's discriminant for the longitudinal motion has been found to correspond to a divergence of the phugoid oscillation. Significant governing parameters are C_{m_α} and C_{m_q}. (The importance of C_{m_α} was shown in Sec. 6.9.) Zimmerman (ref. 6.1) has calculated a large number of the stability boundaries for this case and plotted them as curves in what is essentially the C_{m_α}, C_{m_q} plane. In the notation of this book, the coordinates are as shown in the sample of Fig. 6.13. The region on the shaded side of the curve defines stable combinations of the variables, and the region on the other side defines combinations for which the oscillation is divergent.

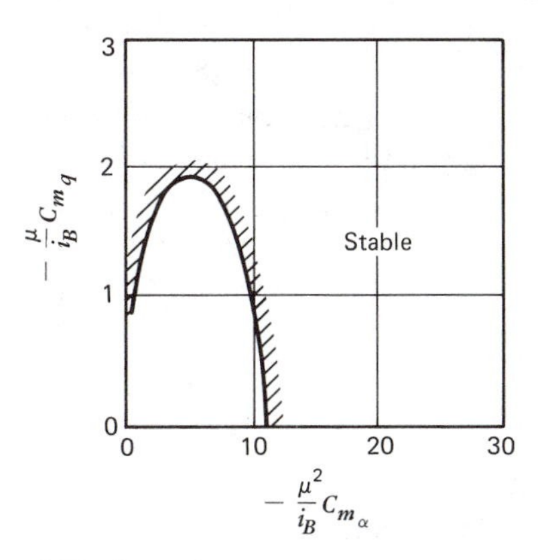

FIG. 6.13 Phugoid stability boundary.

$C_L = 1.90, \qquad C_D = 0.30$

$C_{L_\alpha} = 4.00, \qquad dC_D/d\alpha = 1.50$

(Reproduced from *NACA Rept. 521*, 1935, by C. H. Zimmerman.)

6.12 LONGITUDINAL STABILITY—ELEVATOR FREE

When the elevator is free to move, the mathematical system is obtained by adding the elevator-angle perturbation η and the elevator equation, Eq. 4.15,7*d*, to the system of equations used in the foregoing sections. If all the variables and equations are retained, a sixth-order system results, with a sextic characteristic equation.

Let us consider what may be expected of such a system by a physical argument. Imagine first that the airplane is fully constrained, as in a wind tunnel, and that the elevator is free to flap. In these circumstances all the variables except η are zero, and its variation is governed by a second-order equation obtained from Eq. 4.15,7*d*:

$$(i_e D^2 - C_{he_{\dot{\eta}}} D - C_{he_\eta})\eta = 0 \tag{6.12,1}$$

This system may be identified with the simple mass-spring-damper used before. As long as the elevator is not overbalanced aerodynamically, $C_{he_\eta} < 0$, and the effective spring is positive. $C_{he_{\dot{\eta}}}$ is also negative; so the system is damped. The usually large hinge moment, and the usually small inertia may be expected to combine to give a high natural frequency for the system (a representative value might be of the order of 10 cps).

Effect on the Phugoid

Imagine now that the elevator is locked, and the airplane subject to an initial disturbance such that a pure phugoid oscillation is excited. Now let the elevator be freed. Its motion will be governed by Eq. 4.15,7*d*, which we may write in the form

$$(i_e D^2 - C_{he_{\dot{\eta}}} D - C_{he_\eta})\eta = f(\hat{t}) \tag{6.12,2}$$

where

$$f(\hat{t}) = (2C_{he_0} + C_{he_u})\hat{u} + (C_{he_{\dot{\alpha}}} D + C_{he_\alpha})\alpha + C_{he_q} D\theta$$

The quantities $\hat{u}$, α, and θ on the right-hand side vary periodically with the stick-free phugoid period, so that $f(\hat{t})$ is a periodic function having the frequency of the phugoid oscillation. This is usually of the order of 0.01 cps. Thus we find that the elevator system is being forced at a frequency of the order of 1/1000 of its natural frequency. Now we know that a second-order system being driven at such a low frequency will respond in a quasistatic fashion. That is, at each moment the value of η will be that which it would take up if the aerodynamic forces were applied to it statically,

$$\eta = -\frac{f(\hat{t})}{C_{he_\eta}}$$

Now consider the terms that make up $f(\hat{t})$. We have already seen (Sec. 6.6) that α is negligible in the phugoid mode, so that the α term in $f(\hat{t})$ may be neglected. In setting up the motion described, we have tacitly assumed that freeing the elevator did not change the trim speed of the airplane. The implication is that the elevator is trimmed by a tab so that the initial hinge moment C_{he_0} is zero. Let us further assume that the effect of speed on the hinge-moment coefficient is negligible, so that $C_{he_u} = 0$. The $\hat{u}$ term in $f(\hat{t})$ then disappears as well. Now the amplitude of θ is a first-order quantity, but, owing to the low frequency of the phugoid, the amplitude of $D\theta$ is very small. For the example of Sec. 6.6 we have

Initial amplitude of θ, $\theta_1 = 0.2$ radian

Initial amplitude of $D\theta$, $\hat{q}_1 = 0.00011$ radian/airsec

Thus we see that the θ term of $f(\hat{t})$ is also negligible, so that finally $f(\hat{t}) \doteqdot 0$ and $\eta \doteqdot 0$ in the phugoid mode. As a first approximation then, we may neglect altogether the influence of the free elevator on the phugoid oscillation, and assume that one of the modes of the complete elevator-free system will be identical with the stick-fixed phugoid.

Effect on the Short-Period Oscillation

Let us now imagine that the conditions of the preceding paragraphs are altered by substituting the short-period motion for the phugoid. The separation in frequency between the basic aircraft motion and that of the elevator as a single-degree-of-freedom system is now greatly reduced, so that important couplings can occur. To find these effects we must turn to the equations of motion. We should expect that the result previously obtained for the short-period oscillation (i.e., variation in $\hat{u}$ is negligible) should hold as well with the free control. With the assumption of zero $\hat{u}$ then, the reduced equations of motion for level flight are obtained from Eqs. 4.15,7 as

$$\begin{aligned}(2\mu D - C_{z_{\dot{\alpha}}}D - C_{z_\alpha})\alpha - (2\mu + C_{z_q})\hat{q} - C_{z_\eta}\eta &= 0\\ -(C_{m_{\dot{\alpha}}}D + C_{m_\alpha})\alpha + (i_B D - C_{m_q})\hat{q} - (C_{m_{\dot{\eta}}}D + C_{m_\eta})\eta &= 0\\ -(C_{he_{\dot{\alpha}}}D + C_{he_\alpha})\alpha - C_{he_q}\hat{q} - (i_e D^2 - C_{he_{\dot{\eta}}}D - C_{he_\eta})\eta &= 0\end{aligned} \tag{6.12,3}$$

This is a system of the fourth order. It has been studied theoretically by Jones and Cohen (ref. 6.2), and by Greenberg and Sternfield (ref. 6.3) when there is friction in the control system. This conclusion of these studies, which has been verified in flight, is that practically important couplings can occur between the elevator motion and that of the aircraft. The coupling is strongest when the control is closely balanced aerodynamically and when the elevator is not mass-balanced. The short-period oscillations may then become poorly damped or even divergent. For further details, the reader is referred to the original literature.

6.13 ADDITIONAL SYMBOLS INTRODUCED IN CHAPTER 6

$A, B, \ldots, E$	coefficients of characteristic equation
n	real part of λ, damping parameter
N_{half}	number of cycles to half amplitude
T	period of oscillation

t_{half}	time to half amplitude
t_{double}	time to double amplitude
$(\hat{u}_0, \alpha_0, \theta_0)$	complex amplitudes of $(\hat{u}, \alpha, \theta)$
$(\hat{u}_1, \alpha_1, \theta_1)$	real amplitudes of $(\hat{u}, \alpha, \theta)$
ε, δ	phase angles of α and θ
λ	root of characteristic equation
ω	imaginary part of λ, circular frequency
ω_n	undamped natural frequency
ζ	damping ratio

See also Secs. 2.11, 3.11, 4.18, and 5.17.

6.14 BIBLIOGRAPHY

6.1 C. H. Zimmerman. An Analysis of Longitudinal Stability in Power-off Flight with Charts for Use in Design. *NACA Rept. 521*, 1935.

6.2 R. T. Jones and D. Cohen. An Analysis of the Stability of an Airplane with Free Controls. *NACA Rept. 709*, 1941.

6.3 H. Greenberg and L. Sternfield. A Theoretical Investigation of Longitudinal Stability of Airplanes with Free Controls including the Effect of Friction in the Control System. *NACA Rept. 791*, 1944.

6.4 H. S. Sharp. A Comparison of Methods for Evaluating the Complex Roots of Quartic Equations. *J. Math. Phys.*, vol. 20, no. 3, 1941.

6.5 S. Lim. A Method of Successive Approximation of Evaluating the Real and Complex Roots of Cubic and Higher-Order Equations. *J. Math. Phys.*, vol. 20, no. 3, 1941.

6.6 R. A. Frazer and W. J. Duncan. On the Numerical Solution of Equations with Complex Roots. *Proc. Roy. Soc. A*, vol. 125, p. 68, 1929.

6.7 B. B. Klawans and H. I. Johnson. Some Effects of Fuselage Flexibility on Longitudinal Stability and Control. *NACA TN 3543*, 1956.

6.8 M. D. McLaughlin. A Theoretical Investigation of the Short-Period Dynamic Longitudinal Stability of Airplanes Having Elastic Wings of 0° to 60° Sweepback. *NACA TN 3251*, 1954.

6.9 H. A. Cole, S. C. Brown, and E. C. Holleman. The Effects of Flexibility on the Longitudinal Dynamic Response of the B-47 Airplane. *IAS Preprint 678*, 1957.

6.10 S. Neumark. Analysis of Short-Period Longitudinal Oscillations of an Aircraft—Interpretation of Flight Tests. *ARC R&M 2940*, 1956.

6.11 B. Etkin. On a Third Mode of Longitudinal Control-Fixed Oscillation. *J. Aero. Sci.* (Readers Forum), vol. 24, no. 1, 1957.

STABILITY OF UNCONTROLLED MOTION (LATERAL)

CHAPTER 7

The methods given in Chap. 6 for the analysis of longitudinal stability are applicable to the lateral motions of aircraft as well. We shall apply them in this chapter to the study of representative lateral modes.

7.1 LATERAL STABILITY, CONTROLS FIXED

The relevant equations are obtained from Eqs. 4.15,8 by setting $\xi = \zeta = 0$, and dropping the two control-system equations. We shall consider here only level flight, so that $\theta_0 = 0$. We then have, from Eq. 4.15,8*f*, that $\hat{p} = D\phi$, which enables $\hat{p}$ to be eliminated. The resulting equations are

$$\begin{aligned} (2\mu D - C_{y_\beta})\beta - (C_{y_p}D + C_{L_0})\phi + (2\mu - C_{y_r})\hat{r} &= 0 \\ -C_{l_\beta}\beta + (i_A D^2 - C_{l_p}D)\phi - (i_E D + C_{l_r})\hat{r} &= 0 \\ -C_{n_\beta}\beta - (i_E D^2 + C_{n_p}D)\phi + (i_c D - C_{n_r})\hat{r} &= 0 \end{aligned} \tag{7.1,1}$$

$$\mu = \frac{m}{\rho S l}, \qquad t^* = \frac{l}{u_0}, \qquad l = \frac{b}{2}$$

By proceeding as in Sec. 6.2, we find the stability determinant to be

$$\Delta = \begin{vmatrix} (2\mu\lambda - C_{y_\beta}) & -(C_{y_p}\lambda + C_{L_0}) & (2\mu - C_{y_r}) \\ -C_{l_\beta} & (i_A\lambda^2 - C_{l_p}\lambda) & -(i_E\lambda + C_{l_r}) \\ -C_{n_\beta} & -(i_E\lambda^2 + C_{n_p}\lambda) & (i_C\lambda - C_{n_r}) \end{vmatrix} \tag{7.1,2}$$

This expands to give a fourth-order characteristic equation.

$$A\lambda^4 + B\lambda^3 + C\lambda^2 + D\lambda + E = 0 \tag{7.1,3}$$

$$
\begin{aligned}
\text{where } A &= 2\mu(i_A i_C - i_E{}^2) \\
B &= C_{y_\beta}(i_E{}^2 - i_A i_C) - 2\mu[i_C C_{l_p} + i_A C_{n_r} + i_E(C_{l_r} + C_{n_p})] \\
C &= 2\mu(C_{n_r}C_{l_p} - C_{n_p}C_{l_r} + i_A C_{n_\beta} + i_E C_{l_\beta}) \\
&\quad + i_A(C_{y_\beta}C_{n_r} - C_{n_\beta}C_{y_r}) + i_C(C_{y_\beta}C_{l_p} - C_{l_\beta}C_{y_p}) \\
&\quad + i_E(C_{y_\beta}C_{n_p} - C_{n_\beta}C_{y_p} + C_{l_r}C_{y_\beta} - C_{y_r}C_{l_\beta}) \\
D &= C_{y_\beta}(C_{l_r}C_{n_p} - C_{n_r}C_{l_p}) + C_{y_p}(C_{l_\beta}C_{n_r} - C_{n_\beta}C_{l_r}) \\
&\quad + (2\mu - C_{y_r})(C_{l_\beta}C_{n_p} - C_{n_\beta}C_{l_p}) - C_{L_0}(i_C C_{l_\beta} + i_E C_{n_\beta}) \\
E &= C_{L_0}(C_{l_\beta}C_{n_r} - C_{n_\beta}C_{l_r})
\end{aligned}
$$

7.2 NUMERICAL EXAMPLE

To provide a representative illustration of lateral characteristics, we shall use once more the airplane and flight conditions employed in the examples of Chap. 6. The data are as follows:

$$
\begin{array}{lll}
W = 100{,}000 \text{ lb} & S = 1667 \text{ ft}^2 & \\
A = 7, \quad \Lambda = 30^\circ & \text{Alt.} = 30{,}000 \text{ ft} & \\
u_0 = 500 \text{ mph} & l = \dfrac{b}{2} = 54.0 \text{ ft} & \\
\quad = 733 \text{ fps} & & \\
\mu = 38.8 & C_{L_0} = 0.25 & t^* = 0.0737 \\
i_A = 3.69 & i_C = 9.22 & i_E = -0.39 \\
C_{y_\beta} = -0.168 & C_{y_p} = 0 & C_{y_r} = 0.192 \\
C_{l_\beta} = -0.022 - 0.10C_{L_0} & C_{l_p} = -0.43 & C_{l_r} = 0.0077 + 0.25C_{L_0} \\
C_{n_\beta} = 0.036 + 0.04C_{L_0}{}^2 & C_{n_p} = 0.008 - 0.10C_{L_0} & C_{n_r} = -0.116 - 0.020C_{L_0}{}^2
\end{array}
$$

A number of the stability derivatives are dependent on the initial lift coefficient C_{L_0}, and this dependence is shown explicitly for completeness. The value $C_{L_0} = 0.25$ is substituted into these expressions to give the numerical values for this example. Using these, the stability determinant takes the following numerical form:

$$
\Delta = \begin{vmatrix}
(77.6\lambda + 0.168) & -0.25 & 77.4 \\
0.047 & (3.69\lambda^2 + 0.43\lambda) & (0.39\lambda - 0.0702) \\
-0.0385 & (0.39\lambda^2 + 0.017\lambda) & (9.22\lambda + 0.117)
\end{vmatrix}
$$

Expansion of the determinant leads to the following characteristic equation:

$$
0.2625(10\lambda)^4 + 0.348(10\lambda)^3 + 0.1700(10\lambda)^2 + 0.1465(10\lambda) + 0.000697 = 0
$$

The characteristic equation has been solved, yielding the following four roots:

$$
\begin{aligned}
&\text{Mode 1:} \quad \lambda_1 = -0.000479 \\
&\text{Mode 2:} \quad \lambda_2 = -0.1178 \\
&\text{Mode 3:} \quad \lambda_{3,4} = -0.00723 \pm 0.0682i
\end{aligned}
$$

The characteristics of the three normal modes are:

Mode 1 (Convergence)

$$t_{\text{half}} = \frac{0.69}{0.000479} \times 0.0737 = 106 \text{ sec}$$

Mode 2 (Convergence)

$$t_{\text{half}} = \frac{0.69}{0.1178} \times 0.0737 = 0.43 \text{ sec}$$

Mode 3 (Damped Oscillation)

$$T = \frac{2\pi}{0.0682} \times 0.0737 = 6.79 \text{ sec}$$

$$t_{\text{half}} = \frac{0.69}{0.00723} \times 0.0737 = 7.03 \text{ sec}$$

$$N_{\text{half}} = 1.04 \text{ cycles}$$

All three modes are seen to be stable. Two are aperiodic convergences, and the third is a damped oscillation. The convergence of mode 1 is very slow, while that of mode 2 is extremely rapid. The oscillatory mode is of a period and damping such that it would easily be controlled by the pilot.

7.3 THE SHAPES OF THE MODES

The method introduced in Sec. 6.6 for finding the shapes of the longitudinal normal modes is used here to study the above lateral modes. If β_0, ϕ_0, and $\hat{r}_0$ are the amplitudes of the variables (real for aperiodic modes, and complex for oscillatory modes), then from the last two of Eqs. 7.1,1, the relations governing the mode shapes are

$$\begin{aligned} -C_{l_\beta}\left(\frac{\beta_0}{\phi_0}\right) - [i_E(n+i\omega) + C_{l_r}]\left(\frac{\hat{r}_0}{\phi_0}\right) &= -[i_A(n+i\omega)^2 - C_{l_p}(n+i\omega)] \\ -C_{n_\beta}\left(\frac{\beta_0}{\phi_0}\right) + [i_C(n+i\omega) - C_{n_r}]\left(\frac{\hat{r}_0}{\phi_0}\right) &= [i_E(n+i\omega)^2 + C_{n_p}(n+i\omega)] \end{aligned} \tag{7.3,1}$$

As before, the values of n and ω correspond to the roots $\lambda = n + i\omega$. The shapes of the modes found in Sec. 7.2 are now obtained by substituting the roots into Eqs. 7.3,1.

Mode 1 (Spiral Mode)

$$n = -0.000479, \qquad \omega = 0$$

Hence

$$0.047\left(\frac{\beta_0}{\phi_0}\right) - 0.0704\left(\frac{\hat{r}_0}{\phi_0}\right) = 0.000205$$

$$-0.0385\left(\frac{\beta_0}{\phi_0}\right) + 0.113\left(\frac{\hat{r}_0}{\phi_0}\right) = 0.00000805$$

These have the solution

$$\frac{\beta_0}{\phi_0} = 0.00911, \qquad \frac{\hat{r}_0}{\phi_0} = 0.00317$$

Now $\hat{r}_0$ is a nondimensional rate of rotation, while β_0 and ϕ_0 are angles. It is preferable to deal only with angles when considering the lateral modes. The conversion from $\hat{r}$ to angle of yaw ψ is readily made by using Eq. 4.15,8g; i.e., for level flight, $\theta_0 = 0$ and

$$\hat{r} = D\psi$$

In the pure mode, $\hat{r} = R[\hat{r}_0 e^{(n+i\omega)\hat{t}}]$, and $\psi = R[\psi_0 e^{(n+i\omega)\hat{t}}]$, therefore

$$\hat{r}_0 = (n + i\omega)\psi_0$$

or

$$\psi_0 = \frac{\hat{r}_0}{n + i\omega} \tag{7.3,2}$$

Thus, in mode 1,

$$\psi_0 = \frac{\hat{r}_0}{-0.000479}$$

and

$$\frac{\psi_0}{\phi_0} = -6.63$$

so that

$$\beta_0 : \psi_0 : \phi_0 = 0.00911 : -6.63 : 1$$

The flight path of the airplane after time $t = 0$ is obtained from Eqs. 4.14,6. When applied to lateral disturbances from level flight, they give the following relations:

$$\frac{dx'}{dt} = u_0$$

$$\frac{dy'}{dt} = u_0\psi + v = u_0(\psi + \beta) = u_0 R[(\psi_0 + \beta_0)e^{(n+i\omega)\hat{t}}] \tag{7.3,3}$$

$$\frac{dz'}{dt} = 0$$

As an example of the spiral mode, let $\psi_0 = -10° = -0.1745$ radian. Then $\beta_0 \doteqdot 0$, and $\phi_0 = 1.51°$. Then integration of Eqs. 7.3,3 yields

$$x' = 733t \text{ ft}$$
$$y' = +19{,}700(e^{n\hat{t}} - 1)$$
$$z' = 0$$

The flight path is shown in Fig. 7.1. The motion is seen to be a flat curve in a horizontal plane, in which the aircraft slowly approaches the reference heading (for which $\psi = 0$). Since $\beta \doteqdot 0$, the motion may be thought of as a truly banked turn of gradually increasing radius.

When this mode is unstable (which happens when E of Eqs. 7.1,3 is negative), then ψ increases with time instead of decreasing. The flight of the airplane is then as in a truly banked turn of ever-decreasing radius, i.e., a tightening spiral. This is indicated by the dotted line in Fig. 7.1. It is this motion that gives the mode its name.

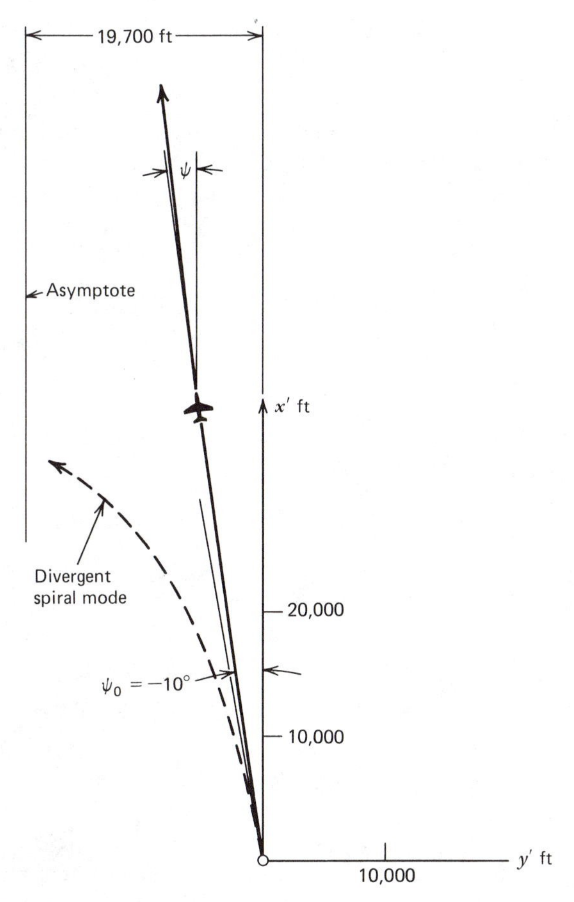

FIG. 7.1 Spiral mode.

Because β tends to be very small in this mode compared to ϕ and ψ, one might at first expect that a simplified analysis, in which β is set equal to zero, would give a good approximation to the mode. In fact this is not so. The reason is that the *aerodynamic forces* are dependent on β, $\hat{p}$, and $\hat{r}$, not on β, ϕ, and ψ, and the former three quantities are of comparable magnitudes. In the example $\beta:\hat{p}:\hat{r} = 1:-0.052:0.35$. The external forces acting on the airplane in the spiral mode are inherently small, and so it may be termed a *weak* mode.

Mode 2 (Rolling Convergence)

$$n = -0.1178, \qquad \omega = 0$$

With these values of n and ω, Eqs. 7.3,1 are solved just as for the spiral mode to give

$$\frac{\beta_0}{\phi_0} = -0.0017, \qquad \frac{\hat{r}_0}{\phi_0} = 0.0036$$

On conversion from $\hat{r}$ to ψ by the use of Eq. 7.3,2 we get

$$\beta_0:\psi_0:\phi_0 = -0.0017:-0.0305:1$$

The reason for the name of this mode is now clear. The motion is practically a single-degree-of-freedom rotation about the x axis. We would therefore expect to obtain an approximation to the rolling mode by suppressing β and $\hat{r}$ in the equations of motion and retaining only the rolling equation. As a check on the quantities relevant to the aerodynamic forces (see Spiral Mode above), we note that

$$\beta:\hat{p}:\hat{r} = 0.0144:1:-0.0305$$

and conclude that the rolling moments due to β and $\hat{r}$ should be negligible compared to that due to $\hat{p}$.

Approximation to the Rolling Mode

When we retain only the rolling-moment equation, and neglect the β and $\hat{r}$ terms, we get, from Eq. 7.1,1,

$$(i_A D^2 - C_{l_p} D)\phi = 0$$

or

$$(i_A D - C_{l_p})\hat{p} = 0 \tag{7.3,4}$$

Thus we have a very simple first-order equation for the rolling mode. Its characteristic root is

$$\lambda = \frac{C_{l_p}}{i_A}$$

which for our numerical example is $\lambda = -0.1165$. This differs by only 1% from the exact value -0.1178

Mode 3 (Lateral Oscillation, or "Dutch Roll")

$$n = -0.00723, \qquad \omega = 0.0682$$

The values of n and ω are used in Eq. 7.3,1 to obtain the following results:

$$\frac{\beta_0}{\phi_0} = 0.681e^{i(-58.5°)}$$

$$\frac{\hat{r}_0}{\phi_0} = 0.0479e^{i(-135.3°)}$$

Equation 7.3,2 is used to find ψ_0, with the result

$$\frac{\psi_0}{\phi_0} = 0.699e^{i(128.6°)}$$

The vector diagram for the mode is given in Fig. 7.2.

The flight path in the oscillatory mode is obtained from Eqs. 7.3,3. That is

$$x' = u_0 t$$
$$z' = 0$$

and

$$y' = u_0 R \int_0^t (\psi_0 + \beta_0) e^{(n+i\omega)\hat{t}}\, dt$$

When the numerical values are substituted, and the integration carried out, the result is

$$x' = 733t \text{ ft}$$
$$z' = 0$$
$$y' = 4.00 - e^{n\hat{t}}(4.00 \cos \omega\hat{t} + 13.1 \sin \omega\hat{t}) \text{ ft}$$

The amplitude of y' is seen to be of the order of 13 ft, while the wavelength of the path is of the order $733 \times 7 \doteqdot 5000$ ft. That is, the deviation of the flight path

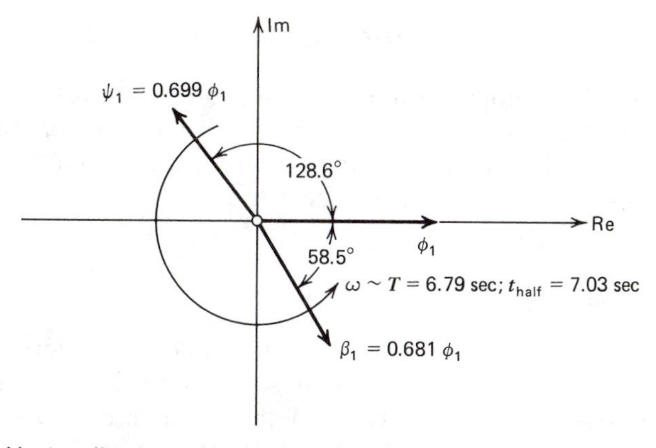

FIG. 7.2 Vector diagram of lateral oscillation.

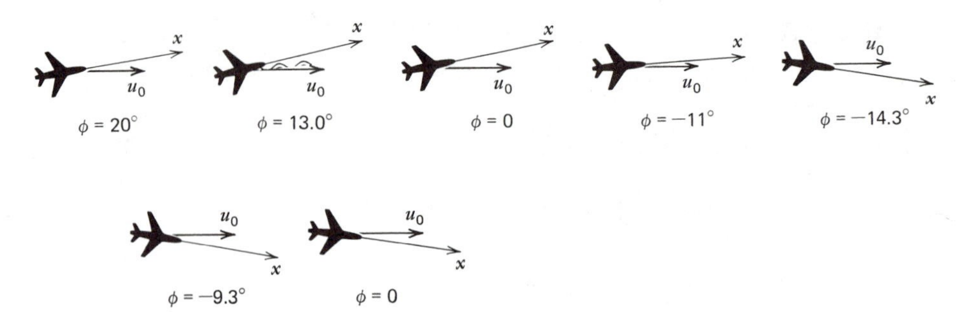

FIG. 7.3 Lateral oscillatory mode.

from a straight line is negligible for this example. A plan view of the motion is shown in Fig. 7.3.

Approximation to the Lateral Oscillation

The vector diagram of Fig. 7.2 shows that β and ψ are nearly equal and opposite. This indicates that an approximation to the mode can be obtained by setting $\psi = -\beta$ (or $\hat{r} = -D\beta$) in Eqs. 7.1,1, and neglecting the side-force equation. When this is done, the equations obtained are

$$\begin{aligned}(i_E D^2 + C_{l_r} D - C_{l_\beta})\beta + (i_A D - C_{l_p})\hat{p} = 0\\ (i_c D^2 - C_{n_r} D + C_{n_\beta})\beta + (i_E D + C_{n_p})\hat{p} = 0\end{aligned} \tag{7.3,5}$$

These equations have a cubic characteristic. When the numerical values for the example airplane are used, the roots of the cubic are found to be

$$\lambda = -0.115$$

and

$$\lambda = -0.00765 \pm 0.0664i$$

The first of these is an approximation to the root for the rolling convergence ($\lambda = -0.1178$), and the complex pair do indeed give a rough approximation to the oscillatory mode. They differ from the exact roots by about 6%. The damping is too high by this amount, and the frequency too low.

Heading Stability

It is intuitively evident, although not demonstrated in the foregoing analysis and examples, that a stable airplane will not seek any particular heading. That is, if it is traveling due North when disturbed, then after recovery of straight level flight it will in general have some heading other than North. The formal proof of this follows from considering the equations of motion in the variables β, ϕ, ψ, instead of β, ϕ, r. Since, in level flight, $\hat{r} = D\psi$, the change in Eqs. 7.1,1 is easily made. The stability determinant, Eq. 7.1,2, will then be altered by multiplying the last column by λ; and the characteristic equation becomes of the fifth order, with one root zero. The analysis of the normal mode corresponding to the zero root, by a method similar to that already given, but using the first two of Eqs. 7.1,1, leads to the result

$$\frac{\beta_0}{\psi_0} = \frac{\phi_0}{\psi_0} = 0$$

This means that there is a mode which we may call "heading displacement," in which only ψ differs from zero, and moreover, is constant with time.

The effect of the zero root on the transient corresponding to given initial conditions is to introduce an extra constant term into the solution for ψ. This is in general not zero, indicating that the headings at the beginning and end of the transient are different.

7.4 LATERAL-STABILITY BOUNDARIES

When the planform of an airplane is fixed, there still remain two important design parameters that exert a decisive influence on the lateral stability. These are the wing dihedral angle and the vertical-tail size.

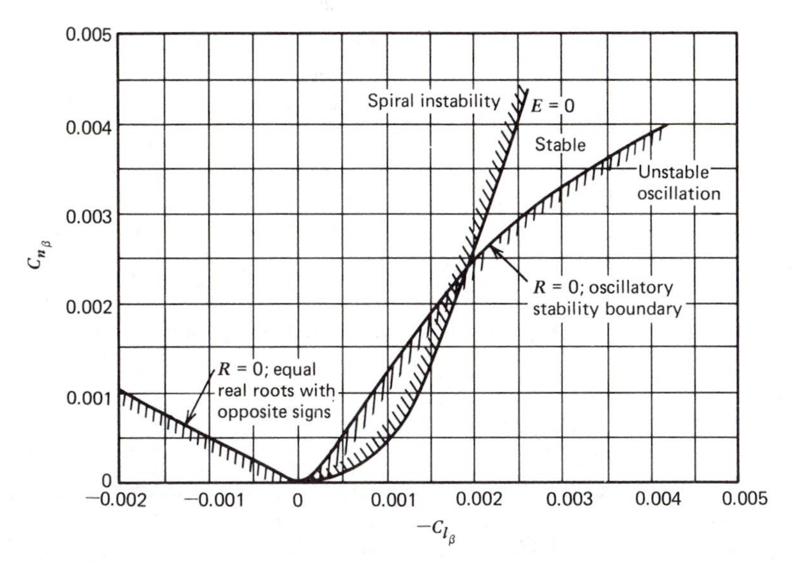

FIG. 7.4 Lateral stability boundaries. (R=Routh's discriminant)
(Reproduced from *NACA Rept. 1098*, 1952, by J. P. Campbell and M. O. McKinney.)

The dihedral angle influences primarily the derivative C_{l_β}, and the tail makes major contributions to C_{n_β}, C_{y_β}, C_{n_r}, and C_{y_r}. The values of C_{l_β} and C_{n_β} are convenient parameters to use in calculating the stability boundaries for which $E = 0$ and Routh's discriminant $= 0$ (see ref. 7.1). Figure 7.4 is an example of the lateral stability boundaries. It is seen that, for the particular airplane and flight condition to which the figure applies, there is a limited range of values of C_{n_β} and C_{l_β} for which the airplane is stable. Generally speaking, increasing $|C_{l_\beta}|$ (increasing dihedral effect) tends to produce oscillatory instability and to avoid spiral instability, whereas increasing C_{n_β} (increasing tail size) tends to do the opposite. The effects of the two parameters on spiral instability is clearly evident from the expression for E given in Eq. 7.1,3.

$$E = C_{L_0}(C_{l_\beta}C_{n_r} - C_{n_\beta}C_{l_r})$$

The first term of E is positive, and the second term is negative. Hence increasing $|C_{l_\beta}|$ makes E larger, and increasing C_{n_β} reduces E.

7.5 VARIATION OF THE LATERAL MODES WITH C_{L_0}

Changes in flight condition can alter the lateral modes in several ways. Variation of the speed of flight may bring about changes in the stability derivatives because of aeroelastic, compressibility, and propulsive effects. Changes in flight speed also introduce changes in several of the stability derivatives because of their dependence on C_{L_0}. Also, changes in altitude increase the relative density μ and the inertia parameters i_A, i_C, and i_E. These effects are such that the periods tend to increase at high altitude, and the damping to decrease. The necessity of designing for satisfactory flying qualities over a wide range of speeds and heights makes the problem a difficult one in practice.

An illustration of these effects is given in this section. The example airplane has swept-back wings, for which several of the stability derivatives vary quite strongly with C_{L_0}. For example, at low speeds, or high C_{L_0}, the dihedral effect is much larger than at high speeds. To show the influence of these variations, the roots of the characteristic equation are found for a range of speeds at the given height (30,000 ft). The quantities that vary with speed, or C_{L_0}, are t^* and the five stability derivatives C_{l_β}, C_{l_r}, C_{n_β}, C_{n_p}, and C_{n_r}. The dependence of these on C_{L_0} is as shown in Sec. 7.2. The varying angle of attack also introduces some variation in the inertia coefficients i_A, i_C, and i_E (see Eqs. 4.5,7). This variation is included in the example.

The results are plotted in Fig. 7.5, which shows the variation of the principal characteristics of the three modes with speed. The stability of both the spiral and oscillatory modes is seen to deteriorate as the speed is reduced, until approximately 250 mph, at which point the spiral mode goes unstable. Below this speed the damping of the oscillatory mode improves again.

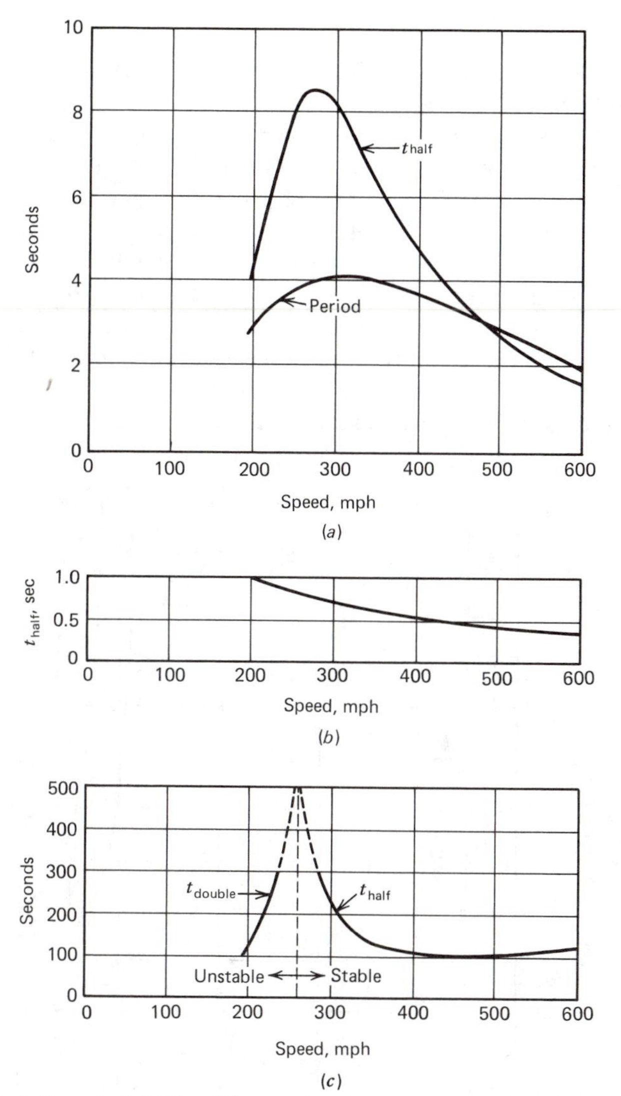

FIG. 7.5 Variation of stability with speed. Altitude 30,000 ft. (*a*) Oscillatory mode. (*b*) Rolling mode. (*c*) Spiral mode.

7.6 LATERAL STABILITY WITH FREE CONTROLS

The equations that govern the lateral stability of an airplane with ailerons and rudders free are Eqs. 4.15,8 with the control-force terms ΔC_{fa} and ΔC_{fr} set equal to zero. The technical importance of the control-free modes of motion is very much dependent on the airplane configuration being studied. It is obvious, of course, that, if the machine is fitted with irreversible power controls, the free-control cases are of no interest at all. In general, the freeing of one control will modify the control-fixed modes and introduce new ones. Freeing both controls simultaneously will lead to a still more complicated situation, with coupling of the two control degrees of freedom.

Analysis of typical modes that occur when the ailerons and rudder are separately free have been made at the NACA (refs. 7.2,3,4). These have shown in general that close aerodynamic balance of the controls, and mass unbalance may lead to lightly damped or unstable oscillatory modes, or to rapid divergence.

It was found possible in the studies noted to simplify the analysis by suppressing certain of the degrees of freedom. In *NACA Rept.* 787, Cohen showed for a particular (representative) airplane, that the important conclusions could be drawn from a two-degree-of-freedom analysis, in which only the aileron angle and the rolling motion were retained. In *NACA Rept.* 762, Greenberg and Sternfield found that the important modes related to the free rudder could be reproduced

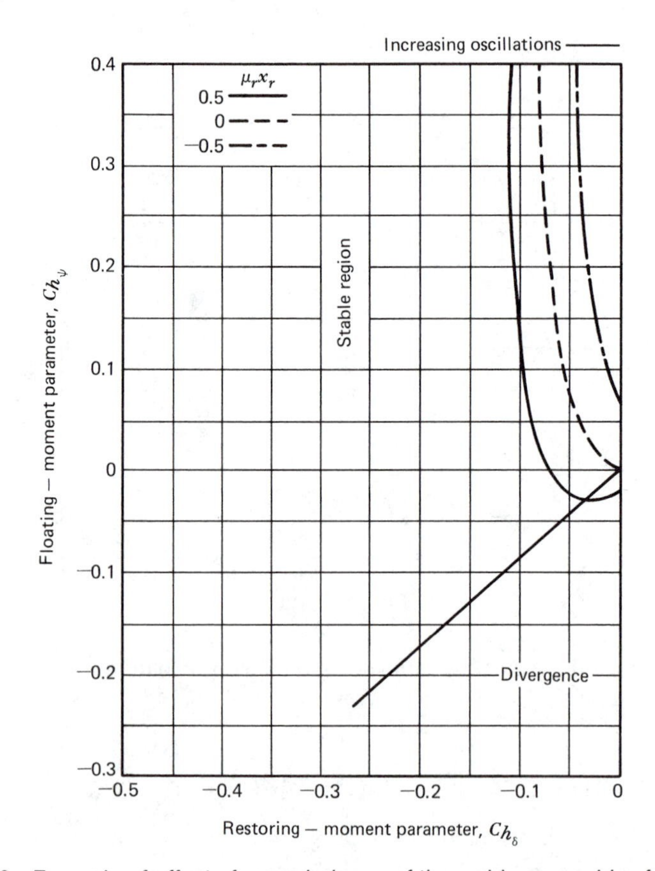

FIG. 7.6 Example of effect of mass balance of the rudder on rudder-free stability.

$$\mu_r = m_r/\rho s c_r, \qquad x_r = \frac{\text{dist. of rudder C.G. from hinge line}}{b/2}$$

(Reproduced from *NACA Rept. 762*, 1943, by H. Greenberg and L. Sternfield.)

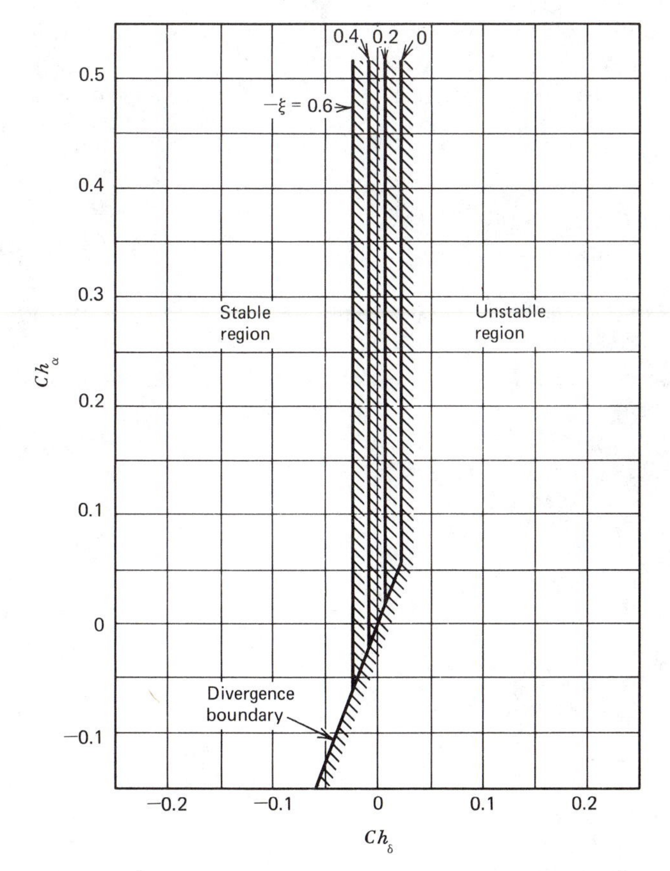

FIG. 7.7 Example of stability boundaries for aileron-free case. ξ = mass-unbalance parameter.
(Reproduced from *NACA Rept. 787*, 1944 by D. Cohen.)

by making the assumption $\psi = -\beta$ (see Sec. 7.3) and suppressing the rolling freedom ($\phi \equiv 0$).

For details of these analyses, the reader is referred to the original reports. However, to show some typical results, Figs. 7.6 and 7.7 are reproduced from the NACA reports.

7.7 BIBLIOGRAPHY

7.1 J. P. Campbell and M. O. McKinney. Summary of Methods for Calculating Dynamic Lateral Stability and Response and for Estimating Lateral Stability Derivatives. *NACA Rept. 1098*, 1952.

7.2 R. T. Jones and D. Cohen. An Analysis of the Stability of an Airplane with free Controls. *NACA Rept. 709*, 1941.

7.3 D. Cohen. A Theoretical Investigation of the Rolling Oscillations of an Airplane with Ailerons Free. *NACA Rept. 787*, 1944.

7.4 H. Greenberg and L. Sternfield. A Theoretical Investigation of the Lateral Oscillations of an Airplane with Free Rudder, with Special Reference to the Effect of Friction. *NACA Rept. 762*, 1943.

7.5 C. H. Zimmerman. An Analysis of Lateral Stability in Power-off Flight with Charts for Use in Design. *NACA Rept. 589*, 1937.

7.6 L. Sternfield. Effect of Product of Inertia on Lateral Stability. *NACA TN 1193*, 1947.

7.7 R. O. Schade and J. L. Hassell. The Effect on Dynamic Lateral Stability and Control of Large Variations in the Rotary Stability Derivatives. *NACA Rept. 1151*, 1953.

7.8 B. Klawans. A Simple Method of Calculating the Characteristics of the Dutch Roll Motion of an Airplane. *NACA TN 3754*, 1956.

7.9 J. M. Hedgepeth, P. G. Waner and R. J. Kell. A Simplified Method for Calculating Aeroelastic Effects on the Roll of Aircraft. *NACA TN 3370*, 1955.

SOME MATHEMATICAL AIDS

CHAPTER 8

8.1 INTRODUCTION

This chapter contains a review of some mathematical concepts and formulas that are used in the following two chapters on airplane control. It is presented in order to provide a convenient reference in consistent notation. The topics covered are Laplace transforms, indicial and impulsive admittances, convolution integrals, transfer functions, and frequency response. The treatment is intended to be explanatory rather than rigorous, although some essential derivations are included.

8.2 THE LAPLACE TRANSFORM

Let $x(t)$ be a known function of t for values of $t > 0$. Then the Laplace transform of $x(t)$ is defined by the integral relation

$$\bar{x}(s) = \mathscr{L}[x(t)] = \int_0^\infty e^{-st}x(t)\,dt \tag{8.2,1}$$

The integral is convergent only for certain functions $x(t)$ and for certain values of s. The Laplace transform is defined only when the integral converges. This restriction is weak and excludes few cases of interest to engineers. It should be noted that the original function $x(t)$ is converted into a new function of the transform variable s by the transformation. The two notations for the transform shown on the left-hand side of Eq. 8.2,1 will be used interchangeably. By way of example, the transforms of some functions that commonly occur in problems of linear systems are listed in Table 8.1. The student should verify this table by direct application of Eq. 8.2,1 noting at the same time the admissible values of s.

TABLE 8.1

	$x(t)$	$\bar{x}(s)$
1	1	$\frac{1}{s}$
2	t	$\frac{1}{s^2}$
3	$\frac{t^{n-1}}{(n-1)!}$	$\frac{1}{s^n}$
4	e^{at}	$\frac{1}{s-a}$
5	$\sin at$	$\frac{a}{s^2+a^2}$
6	$\cos at$	$\frac{s}{s^2+a^2}$
7	te^{at}	$\frac{1}{(s-a)^2}$
8	$\frac{t^{n-1}}{(n-1)!}e^{at}$	$\frac{1}{(s-a)^n}$
9	$e^{at}\sin bt$	$\frac{b}{(s-a)^2+b^2}$
10	$e^{at}\cos bt$	$\frac{s-a}{(s-a)^2+b^2}$
11	$\sinh at$	$\frac{a}{s^2-a^2}$
12	$\cosh at$	$\frac{s}{s^2-a^2}$
13	$e^{at}\sinh bt$	$\frac{b}{(s-a)^2-b^2}$
14	$e^{at}\cosh bt$	$\frac{s-a}{(s-a)^2-b^2}$
15	$\delta(t)$	1
16	$\delta(t-T)$	e^{-sT}
17	$\mathbf{1}(t-T)$	$\frac{e^{-sT}}{s}$
18	$f(t-T)\mathbf{1}(t-T)$	$e^{-sT}\mathscr{L}[f(t)]$
19	$\dot{x}(t)$	$s\bar{x}(s)-x(0)$
20	$e^{at}x(t)$	$\bar{x}(s-a)$

Transforms of Derivatives

Given the function $x(t)$, the transforms of its derivatives can be found from Eq. 8.2,1.

$$\mathscr{L}\left[\frac{dx}{dt}\right] = \int_0^\infty e^{-st}\frac{dx}{dt}\,dt$$

$$= \int_{t=0}^\infty e^{-st}\,dx = xe^{-st}\Big]_{t=0}^\infty + s\int_0^\infty xe^{-st}\,dt$$

When $xe^{-st} \to 0$ as $t \to \infty$ (only this case is considered), then

$$\mathscr{L}\left[\frac{dx}{dt}\right] = -x(0) + s\bar{x}(s) \tag{8.2,2}$$

where $x(0)$ is the value of $x(t)$ when $t = 0$.[1] The process may be repeated to find the higher derivatives by replacing $x(t)$ in Eq. 8.2,2 by $\dot{x}(t)$, and so on. The result is

$$\mathscr{L}\left[\frac{d^n x}{dt^n}\right] = -\frac{d^{n-1}x}{dt^{n-1}}(0) - s\frac{d^{n-2}x}{dt^{n-2}}(0) - \cdots + s^n\bar{x} \tag{8.2,3}$$

Transform of an Integral

The transform of an integral can readily be found from that derived above for a derivative. Let the integral be

$$y = \int x(t)\,dt$$

and let it be required to find $\bar{y}(s)$. By differentiating with respect to t, we get

$$\frac{dy}{dt} = x(t)$$

thus

$$\bar{x}(s) = \mathscr{L}\left[\frac{dy}{dt}\right] = s\bar{y}(s) - y(0)$$

and

$$\bar{y}(s) = \frac{1}{s}\bar{x}(s) + \frac{1}{s}y(0) \tag{8.2,4}$$

8.3 APPLICATION TO DIFFERENTIAL EQUATIONS

The Laplace transform finds one of its most important uses in the solution of linear differential equations. The commonest application in airplane dynamics is to ordinary equations with constant coefficients. The technique for such equations is illustrated by applying it to the equation of a simple spring-mass-damper

[1] To avoid ambiguity when dealing with step functions, $t = 0$ should always be interpreted as $t = 0^+$.

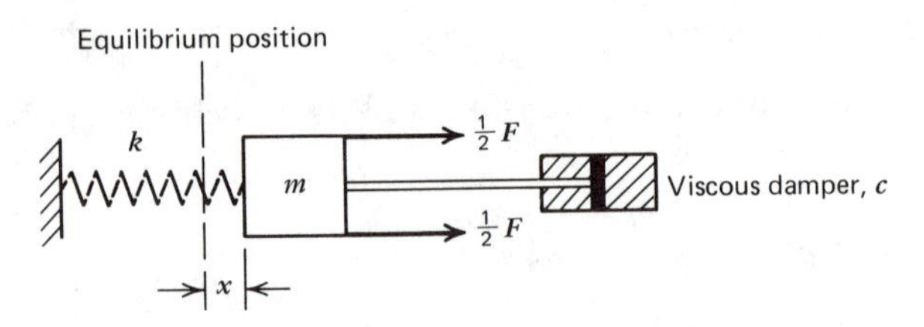

FIG. 8.1 Linear second-order system. $m\ddot{x} = F - kx - c\dot{x}$.

system acted on by an external force (Fig. 8.1). The differential equation of the system is

$$\ddot{x} + 2\zeta\omega_n\dot{x} + \omega_n^2 x = f(t) \tag{8.3,1}$$

$2\zeta\omega_n$ is the viscous resistance per unit mass, c/m, ω_n^2 is the spring rate per unit mass, k/m, and $f(t)$ is the external force per unit mass. The Laplace transform of Eq. 8.3,1 is formed by multiplying through by e^{-st} and integrating term by term from zero to infinity. This gives

$$\mathscr{L}[\ddot{x}] + 2\zeta\omega_n\mathscr{L}[\dot{x}] + \omega_n^2\mathscr{L}[x] = \mathscr{L}[f(t)] \tag{8.3,2}$$

Upon using the results of Sec. 8.2, this equation may be written

$$s^2\bar{x} + 2\zeta\omega_n s\bar{x} + \omega_n^2\bar{x} = \bar{f} + \dot{x}(0) + sx(0) + 2\zeta\omega_n x(0) \tag{8.3,3}$$

or

$$\bar{x}(s) = \frac{\bar{f} + \dot{x}(0) + (s + 2\zeta\omega_n)x(0)}{s^2 + 2\zeta\omega_n s + \omega_n^2} \tag{8.3,4}$$

The original differential equation 8.3,1 has been converted by the transformation into the algebraic equation 8.3,3 which is easily solved (Eq. 8.3,4) to find the transform of the unknown function. In the numerator of the right-hand side of Eq. 8.3,4 we find a term dependent on the excitation ($\bar{f}$), and terms dependent on the initial conditions [$\dot{x}(0)$ and $x(0)$]. The denominator is recognized as the left-hand side of the characteristic equation of the system (see Eq. 6.7,3). We shall call it the *characteristic polynomial.* As exemplified here, finding the Laplace transform of the desired solution $x(t)$ is usually a very simple process. The heart of the problem is the passage from the transform $\bar{x}(s)$ to the function $x(t)$. Methods for carrying out the inverse transformation are described in Sec. 8.4. Before proceeding to these, however, some general comments on the method are in order.

One of the most important advantages of solving differential equations by the Laplace transform is that the initial conditions are automatically taken into account. When the inverse transformation of Eq. 8.3,4 is carried out, the solution applies for the given forcing function $f(t)$ and the given initial conditions. This is preferable to finding a general solution that has in it a number of arbitrary constants which must subsequently be fitted to the initial conditions. This process, although simple in principle, becomes extremely tedious for systems of order higher than the third. A second convenience made possible by the transform method is that in systems of many degrees of freedom, represented by simultaneous differential equations, the solution for any one variable may be found independently of the others.

8.4 METHODS FOR THE INVERSE TRANSFORMATION

The Use of Tables of Transforms

Extensive tables of transforms (like Table 8.1) have been published (see bibliography) that are useful in carrying out the inverse process. When the transform involved can be found in the tables, the function $x(t)$ is obtained directly.

The Method of Partial Fractions

In some cases it is convenient to expand the transform $\bar{x}(s)$ in partial fractions, so that the elements are all simple ones like those in Table 8.1. The function $x(t)$ can then be obtained simply from the table. We shall demonstrate this procedure with an example. Let the second-order system of Sec. 8.3 be initially quiescent, i.e., $x(0) = 0$, and $\dot{x}(0) = 0$, and let it be acted upon by a constant unit force applied at time $t = 0$. Then $f(t) = 1$, and $\bar{f}(s) = 1/s$ (see Table 8.1). From Eq. 8.3,4, we find that

$$\bar{x}(s) = \frac{1}{s(s^2 + 2\zeta\omega_n s + \omega_n^2)} \tag{8.4,1}$$

Let us assume that the system is aperiodic; i.e., that $\zeta > 1$. Then the roots of the characteristic equation are real and equal to

$$\lambda_{1,2} = n \pm \omega' \tag{8.4,1a}$$

where

$$n = -\zeta\omega_n$$
$$\omega' = \omega_n(\zeta^2 - 1)^{1/2}$$

The denominator of Eq. 8.4,1 can be written in factored form so that

$$\bar{x}(s) = \frac{1}{s(s - \lambda_1)(s - \lambda_2)} \tag{8.4,2}$$

Now let Eq. 8.4,2 be expanded in partial fractions,

$$\bar{x}(s) = \frac{A}{s} + \frac{B}{(s - \lambda_1)} + \frac{C}{(s - \lambda_2)} \tag{8.4,3}$$

By the usual method of equating Eqs. 8.4,2 and 8.4,3, we find

$$A = \frac{1}{\lambda_1\lambda_2}$$

$$B = \frac{1}{\lambda_1(\lambda_1 - \lambda_2)}$$

$$C = \frac{1}{\lambda_2(\lambda_2 - \lambda_1)}$$

Therefore

$$\bar{x}(s) = \frac{1/\lambda_1\lambda_2}{s} + \frac{1/\lambda_1(\lambda_1 - \lambda_2)}{s - \lambda_1} + \frac{1/\lambda_2(\lambda_2 - \lambda_1)}{s - \lambda_2}$$

By comparing these three terms with the first and fourth entries in Table 8.1, we may write down the solution immediately as

$$x(t) = \frac{1}{\lambda_1\lambda_2} + \frac{1}{\lambda_1(\lambda_1 - \lambda_2)} e^{\lambda_1 t} - \frac{1}{\lambda_2(\lambda_1 - \lambda_2)} e^{\lambda_2 t}$$

$$= \frac{1}{{\omega_n}^2}\left[1 + \frac{n - \omega'}{2\omega'} e^{(n+\omega')t} - \frac{n + \omega'}{2\omega'} e^{(n-\omega')t}\right] \tag{8.4,4}$$

Heaviside Expansion Theorem

When the transform is a ratio of two polynomials in s, the method of partial fractions can be generalized. Let

$$\bar{x}(s) = \frac{N(s)}{D(s)}$$

where $N(s)$ and $D(s)$ are polynomials and the degree of $D(s)$ is higher than that of $N(s)$. Let the roots of $D(s) = 0$ be a_r, so that

$$D(s) = (s - a_1)(s - a_2)\cdots(s - a_n)$$

Then the inverse of the transform is

$$x(t) = \sum_{r=1}^{n} \left\{\frac{(s - a_r)N(s)}{D(s)}\right\}_{s=a_r} e^{a_r t} \tag{8.4,5}$$

The effect of the factor $(s - a_r)$ in the numerator is to cancel out the same factor of the denominator. The substitution $s = a_r$ is then made in the reduced expression.

In applying this theorem to Eq. 8.4,2, we have the three roots $a_1 = 0$, $a_2 = \lambda_1$, $a_3 = \lambda_2$, and $N(s) = 1$. With these roots, Eq. 8.4,4 follows immediately from Eq. 8.4,5.

Repeated Roots

When two or more of the roots are the same, then the expansion theorem given above fails. For then, after canceling one of the repeated factors from $D(s)$ by the factor $(s - a_r)$ of the numerator, still another remains and becomes zero when s is set equal to a_r. Some particular cases of equal roots are shown in Table 8.1, items 2, 3, 7, and 8. The method of partial fractions, coupled with these entries in the table, suffices to deal conveniently with most cases encountered in stability and control work. However, for cases not conveniently handled in this way, a general formula is available for dealing with repeated roots (p. 12, ref. 8.2). Equation 8.4,5 is used to find that part of the solution which corresponds to single roots. To this is added the solution corresponding to each multiple factor $(p - a_r)^m$ of $D(s)$. This is given by

$$\left[\frac{(s-a_r)^2N(s)}{D(s)}\right]_{s=a_r} te^{a_rt} + \left\{\frac{d}{ds}\left[\frac{(s-a_r)^2N(s)}{D(s)}\right]\right\}_{s=a_r} e^{a_rt} \quad \text{for } m=2 \tag{8.4,6}$$

and by

$$\sum_{n=0}^{m-1}\left\{\frac{d^n}{ds^n}\left[\frac{(s-a_r)^mN(s)}{D(s)}\right]\right\}_{s=a_r}\frac{t^{m-n-1}}{n!(m-n-1)!}e^{a_rt} \quad \text{for } m>2$$

The Inversion Theorem

The function $x(t)$ can be found formally from its transform $\bar{x}(s)$ by the application of the inversion theorem (refs. 8.1,2). It is given by the line integral

$$x(t) = \frac{1}{2\pi i}\lim_{\omega\to\infty}\int_{\gamma-i\omega}^{\gamma+i\omega} e^{st}\bar{x}(s)\,ds \tag{8.4,7}$$

where γ is a real number greater than the real part of all values of s for which $\bar{x}(s)$ diverges. That is, $s = \gamma$ is a straight line on the s plane lying parallel to the imaginary axis, and to the right of all the poles of $\bar{x}(s)$. This theorem can be used, employing the methods of contour integrals in the complex plane, to evaluate the inverse of the transform.

8.5 INDICIAL AND IMPULSIVE ADMITTANCE

Consider a physical system governed by linear differential equations, and subject to an external excitation; for example, the second-order system of Fig. 8.1. Let the system be initially quiescent; i.e., in an equilibrium configuration and with all time derivatives zero. Let it then be acted upon by the *unit step forcing function* $\mathbf{1}(t)$, which is as shown in Fig. 8.2. For the system of Fig. 8.1, the forcing function

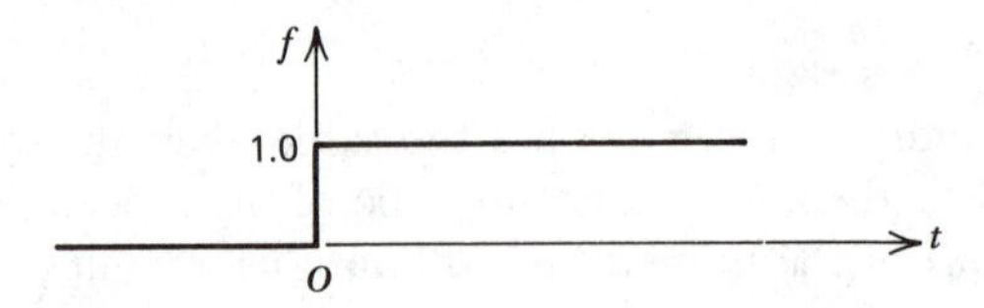

FIG. 8.2 Unit-step function. $f = 0$ for $t < 0$, $f = 1$ for $t \geq 0$.

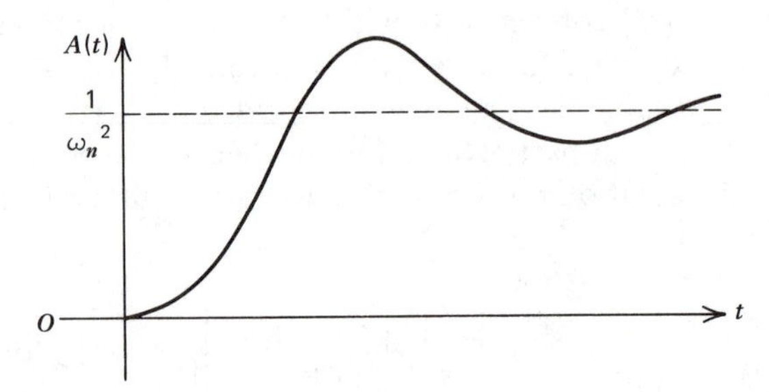

FIG. 8.3 Indicial admittance—second-order system.

is the external force per unit mass (see Eq. 8.3,1). The response of the system is then the *indicial admittance* $A(t)$. It is illustrated in Fig. 8.3 for an underdamped second-order system.

Consider now the same system acted upon by a positive step function at time $t = 0$, followed by a negative step function at time $t = \Delta t$, as illustrated in Fig. 8.4. The magnitude of the step is such that the area under the f–t diagram (i.e., the

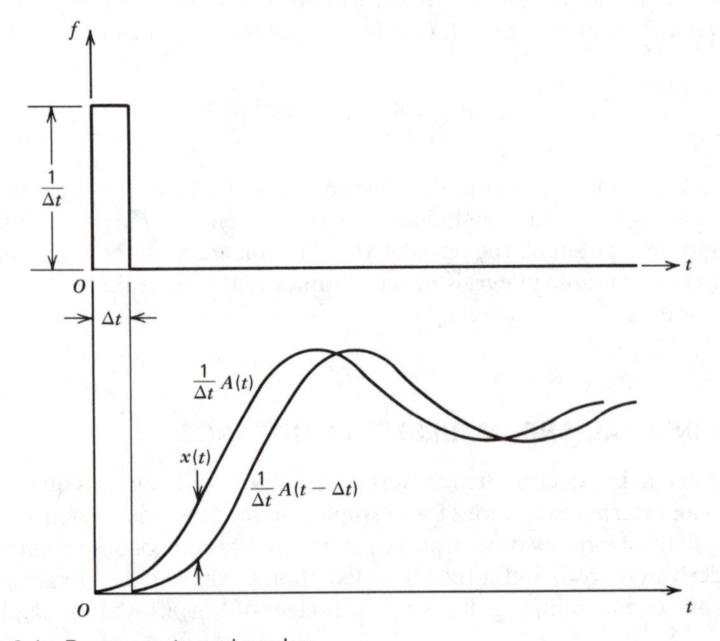

FIG. 8.4 Response to an impulse.

impulse) is unity. When the forcing function is actually a force, then the area truly represents a mechanical impulse. When it is a physical quantity of a different kind (e.g., voltage) then we use the term impulse by analogy. Since the differential equation has been presumed to be linear, we may obtain the response to this forcing function from the indicial admittance as follows:

$$x(t) = \left(\frac{1}{\Delta t}\right)A(t) - \left(\frac{1}{\Delta t}\right)A(t - \Delta t)$$
$$= \frac{A(t) - A(t - \Delta t)}{\Delta t}$$

The function $A(t - \Delta t)$ is simply $A(t)$ displaced to the right by Δt, as shown on Fig. 8.4. The *impulsive admittance* $h(t)$ is defined as the limit of $x(t)$ as $\Delta t \to 0$. That is,

$$h(t) = \lim_{\Delta t \to 0} \frac{A(t) - A(t - \Delta t)}{\Delta t}$$
$$= \frac{dA(t)}{dt} \tag{8.5,1}$$

The impulsive admittance (the response to a unit impulse) is thus shown to be the derivative of the indicial admittance (the response to a unit step). The *unit impulse* is the delta function $\delta(t)$, discussed more fully in Sec. 8.9.

Since first- and second-order systems are encountered frequently in airplane stability and control, it is useful to have the admittances of such systems in a ready reference form. The appropriate derivations follow.

Admittances of a First-Order System

An example of a first-order system with zero forcing function was met in Sec. 7.3 (see Eq. 7.3,4). When the excitation is not zero, the equation is most conveniently written

$$\dot{x} + \frac{1}{T}x = f(t) \tag{8.5,2}$$

The quantity T is the only parameter of the equation, and is known as the *time constant* of the system. To compute the indicial admittance we set $f(t) = 1$, $t \geq 0$. Since the initial conditions are zero, the Laplace transform of the equation is

$$\left(s + \frac{1}{T}\right)\bar{x} = \bar{f} = \frac{1}{s}$$

or

$$\bar{x} = \frac{1}{s(s + 1/T)} \tag{8.5,3}$$

The roots of the denominator are 0 and $-1/T$, so that application of the expansion theorem, Eq. 8.4,5 gives directly

$$A(t) = Te^{0} - Te^{-t/T}$$

or

$$A(t) = T(1 - e^{-t/T}) \tag{8.5,4}$$

Differentiation of $A(t)$ gives the impulsive admittance

$$h(t) = e^{-t/T} \tag{8.5,5}$$

These admittances are shown in Fig. 8.5. The significance of the time constant T is also shown in the figure.

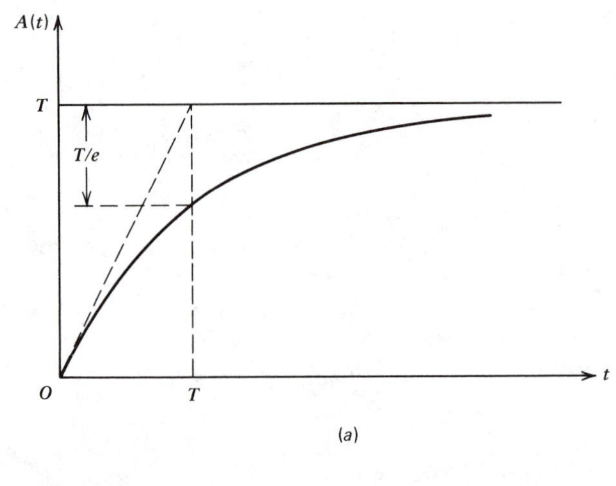

(a)

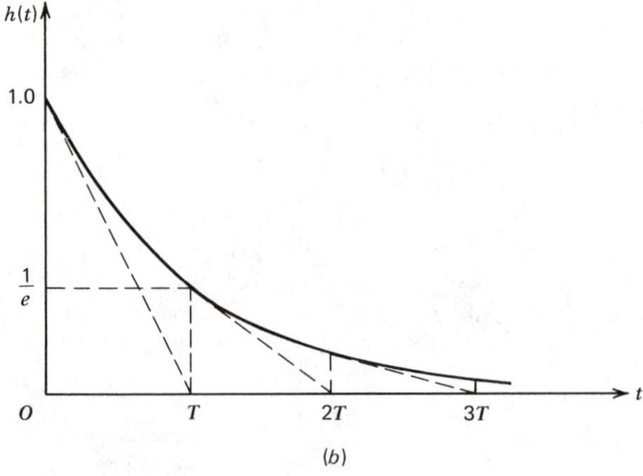

(b)

FIG. 8.5 (a) Indicial admittance of first-order system. (b) Impulsive admittance of first-order system.

Admittances of a Second-Order System

The differential equation of a second-order system is

$$\ddot{x} + 2\zeta\omega_n\dot{x} + \omega_n^2 x = f(t)$$

The Laplace transform of x has already been obtained in Sec. 8.4 for unit forcing function.

$$\bar{x}(s) = \frac{1}{s(s^2 + 2\zeta\omega_n s + \omega_n^2)} \tag{8.5,6}$$

The solution when the roots are real, and the motion is aperiodic, is given by Eqs. 8.4,4. When $\zeta < 1$, the roots are complex and the solution is different. In that case, let the roots be

$$\begin{aligned} \lambda_{1,2} &= n \pm i\omega \\ n &= -\zeta\omega_n \\ \omega &= \omega_n(1 - \zeta^2)^{1/2} \end{aligned} \tag{8.5,7}$$

The denominator of $\bar{x}(s)$ is expressed in factored form, so that

$$\bar{x}(s) = \frac{1}{s(s - \lambda_1)(s - \lambda_2)} \tag{8.5,8}$$

Using the given roots, Eq. 8.5,8 can be rewritten as

$$\bar{x}(s) = \frac{1}{s[(s - n)^2 + \omega^2]} \tag{8.5,9}$$

Equation 8.5,9 can be expanded by partial fractions into

$$\begin{aligned} \bar{x}(s) &= \frac{1}{\omega_n^2}\left[\frac{1}{s} + \frac{2n - s}{(s - n)^2 + \omega^2}\right] \\ &= \frac{1}{\omega_n^2}\left[\frac{1}{s} + \frac{n}{(s - n)^2 + \omega^2} - \frac{(s - n)}{(s - n)^2 + \omega^2}\right] \end{aligned} \tag{8.5,10}$$

From Table 8.1, we find the inverse of Eq. 8.5,10 to be

$$x(t) = \frac{1}{\omega_n^2}\left(1 + \frac{n}{\omega}e^{nt}\sin\omega t - e^{nt}\cos\omega t\right)$$

Therefore

$$A(t) = \frac{1}{\omega_n^2}\left[1 - e^{nt}\left(\cos\omega t - \frac{n}{\omega}\sin\omega t\right)\right] \tag{8.5,11}$$

The impulsive admittance when $\zeta < 1$ is the derivative of Eq. 8.5,11, that is,

$$h(t) = \frac{1}{\omega}e^{nt}\sin\omega t \tag{8.5,12}$$

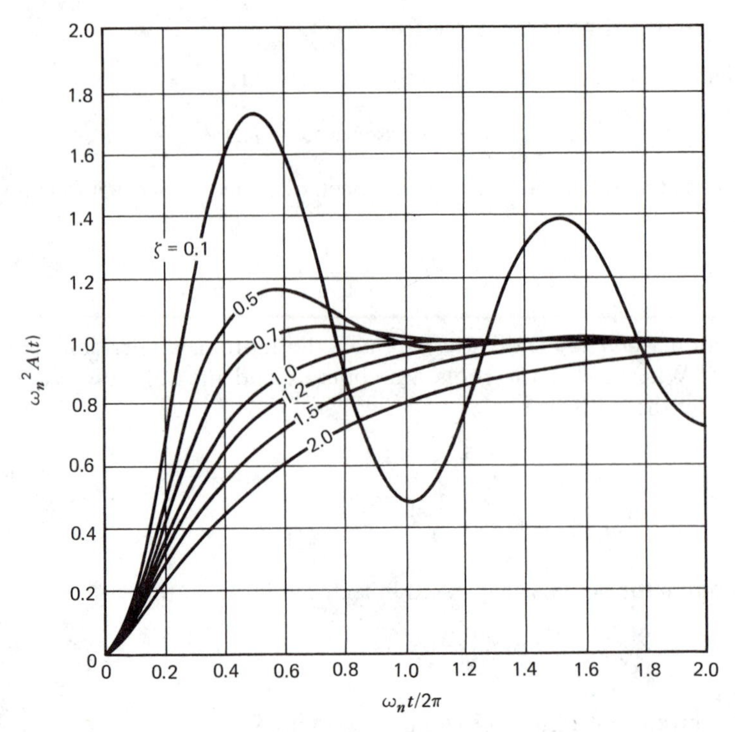

FIG. 8.6 Indicial admittance of second-order system.
(By permission from *Instrument Engineering*, vol. III, by Draper, McKay, and Lees, copyright 1955, McGraw-Hill Book Co.)

Likewise, for the nonoscillatory case ($\zeta > 1$) the impulsive admittance is the derivative of Eq. 8.4,4[2]; that is,

$$
\begin{aligned}
h(t) &= \frac{1}{2\omega'}\left[e^{(n+\omega')t} - e^{(n-\omega')t}\right] \\
&= \frac{1}{\omega'} e^{nt} \frac{e^{\omega' t} - e^{-\omega' t}}{2} \\
&= \frac{1}{\omega'} e^{nt} \sinh \omega' t
\end{aligned}
\tag{8.5,13}
$$

Charts of the indicial and impulsive admittances are given in Figs. 8.6 and 8.7. The coordinates of these figures are so chosen that the only parameter of the curves is ζ, the damping ratio. Both periodic ($\zeta < 1$) and aperiodic ($\zeta \geq 1$) cases are included. When $\zeta = 1$, the system is said to be critically damped. This is the minimum damping for which there is no overshoot.

[2] The equations for the nonoscillatory case can be obtained formally from those of the oscillatory case by noting that $\omega' = i\omega$.

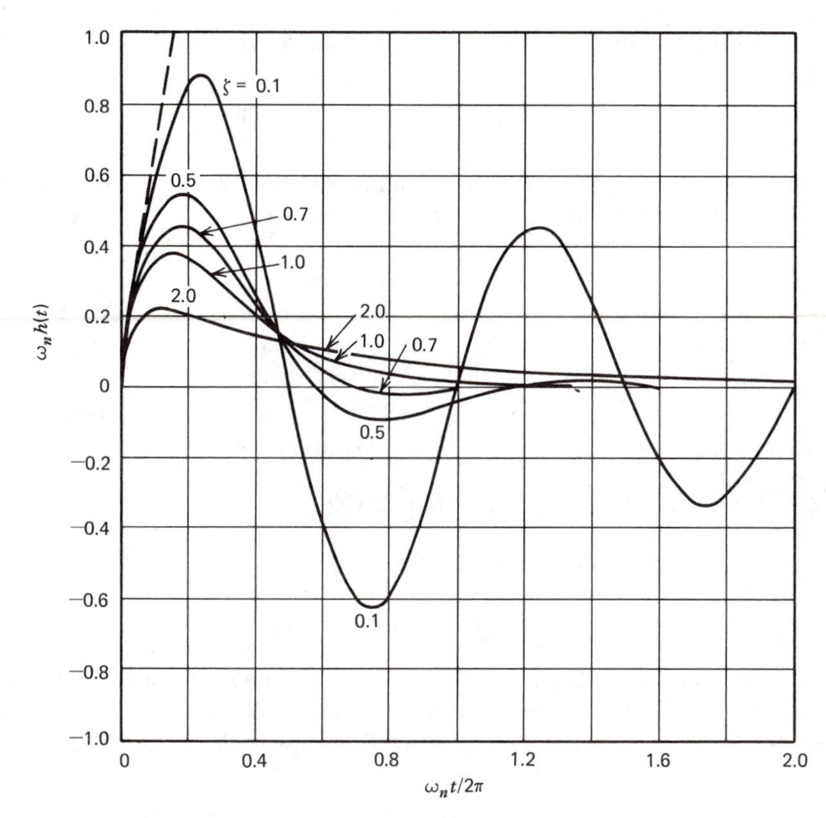

FIG. 8.7 Impulsive admittance of second-order system.

8.6 SUPERPOSITION THEOREM (CONVOLUTION INTEGRAL, DUHAMEL'S INTEGRAL)

The theorems of this section facilitate the calculation of transient responses of linear systems to complicated forcing functions. The general response appears as the superposition of responses to a sequence of steps or impulses that simulate the actual forcing function. Because of the profound importance of these theorems, a relatively complete derivation is presented.

Let

$$\bar{x}_1(s) \quad \text{be the transform of } x_1(t)$$

and

$$\bar{x}_2(s) \quad \text{be the transform of } x_2(t)$$

Then the function $x_3(t)$ whose transform is the product $\bar{x}_3 = \bar{x}_1\bar{x}_2$ is

$$x_3(t) = \int_{\tau=0}^{t} x_1(\tau)x_2(t-\tau)\,d\tau \tag{8.6,1}$$

Proof:

$$\bar{x}_3 = \int_0^\infty e^{-su}x_1(u)\,du \times \int_0^\infty e^{-sv}x_2(v)\,dv$$

where u and v are dummy variables of integration. This is equivalent to the double integral

$$\bar{x}_3(s) = \iint_S e^{-s(u+v)} x_1(u) x_2(v)\, du\, dv$$

where S is the area of integration shown in Fig. 8.8. Now let the region of integration be transformed into the t,τ plane by the substitution

$$u + v = t$$
$$v = \tau$$

Then

$$\bar{x}_3(s) = \iint_{S'} e^{-st} x_1(t-\tau) x_2(\tau)\, dS'$$

where S' is the region shown in Fig. 8.9. Integration first with respect to τ gives

$$\bar{x}_3(s) = \int_{t=0}^{\infty} e^{-st}\, dt \int_{\tau=0}^{t} x_1(t-\tau) x_2(\tau)\, d\tau$$

Therefore, by definition (Eq. 8.2,1)

$$x_3(t) = \int_{t=0}^{t} x_1(t-\tau) x_2(\tau)\, d\tau \qquad \text{Q.E.D.}$$

Consider a linear system acted upon by a forcing function $f(t)$. Let the characteristic polynomial of the system be $C(s)$ (see Sec. 8.3). When the system is initially quiescent (initial conditions all zero), then its response to the excitation is given by

$$\bar{x}(s) = \frac{\bar{f}(s)}{C(s)} \tag{8.6,2}$$

The indicial admittance is the response when $f(t)$ is the unit step function, with transform $1/s$. Hence

$$\bar{A}(s) = \frac{1}{sC(s)}$$

or

$$\frac{1}{C(s)} = s\bar{A}$$

Therefore

$$\bar{x} = s\bar{A}\bar{f} \tag{8.6,3}$$

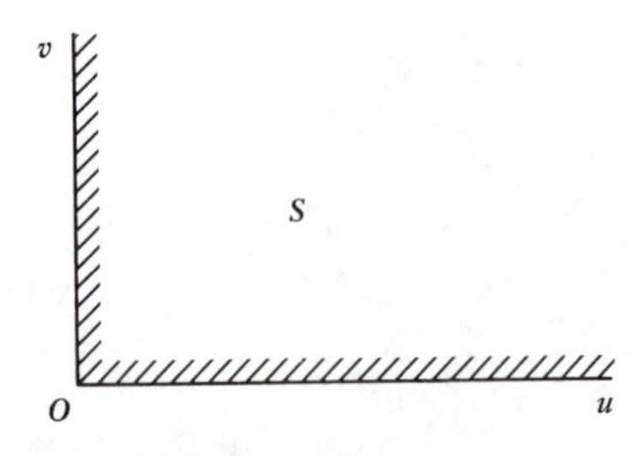

FIG. 8.8 The u, v plane.

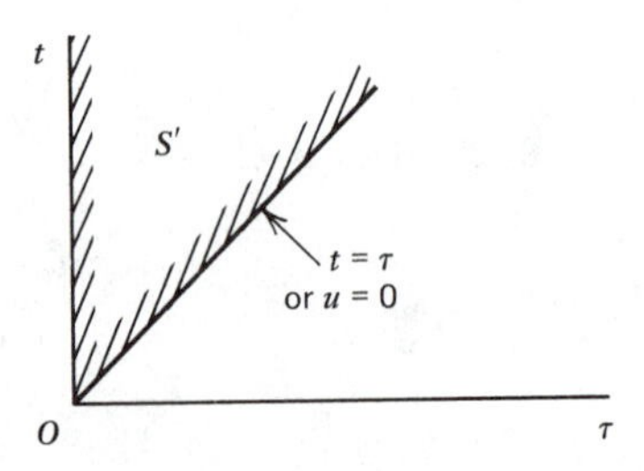

FIG. 8.9 The t, τ plane.

But

$$s\bar{A} = \mathscr{L}[A'(t)] = \bar{h}(s)$$

where $h(t)$ is the impulsive admittance.

Also, when $f(0) = 0$, $\quad s\bar{f} = \mathscr{L}[f'(t)]$

Therefore, by combining s first with $\bar{A}$, and then with $\bar{f}$, Eq. 8.6,3 may be written in the alternative forms:

$$\bar{x} = \bar{h}\bar{f} \qquad (a)$$

or

$$\bar{x} = \bar{A}\mathscr{L}[f'(t)] \qquad (b) \qquad (8.6,4)$$

By applying the superposition theorem to Eq. 8.6,4 we get the two forms of the convolution integral, or Duhamel's integral

$$x(t) = \int_{\tau=0}^{t} h(t-\tau)f(\tau)\,d\tau \qquad (a)$$

$$x(t) = \int_{\tau=0}^{t} A(t-\tau)f'(\tau)\,d\tau \qquad (f(0)=0) \qquad (b) \qquad (8.6,5)$$

When $f(0)$ is not zero, then there must be added to Eq. 8.6,5b a term to allow for the initial step in $f(t)$; i.e.,

$$x(t) = f(0)A(t) + \int_{\tau=0}^{t} A(t-\tau)f'(\tau)\,d\tau \qquad (8.6,5c)$$

The physical significance of these integrals is brought out by considering them as the limits of the following sums

$$x(t) = \Sigma h(t-\tau)f(\tau)\Delta\tau \qquad (a)$$

$$x(t) = A(t)f(0) + \Sigma A(t-\tau)f'(\tau)\Delta\tau \qquad (b) \qquad (8.6,6)$$

Typical terms of the summations are illustrated in Figs. 8.10 and 8.11. The summation forms are quite convenient for computation, especially when the interval $\Delta\tau$ is kept constant.

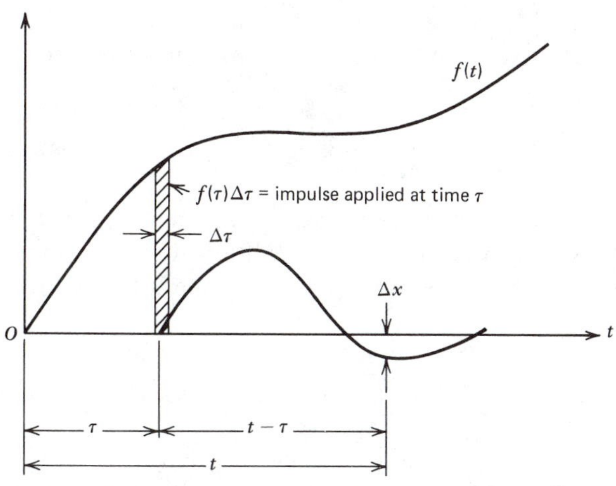

FIG. 8.10 Duhamel's integral, impulsive form. $\Delta x = h(t-\tau)f(\tau)\,\Delta\tau$ = response at time t to impulse at time τ.

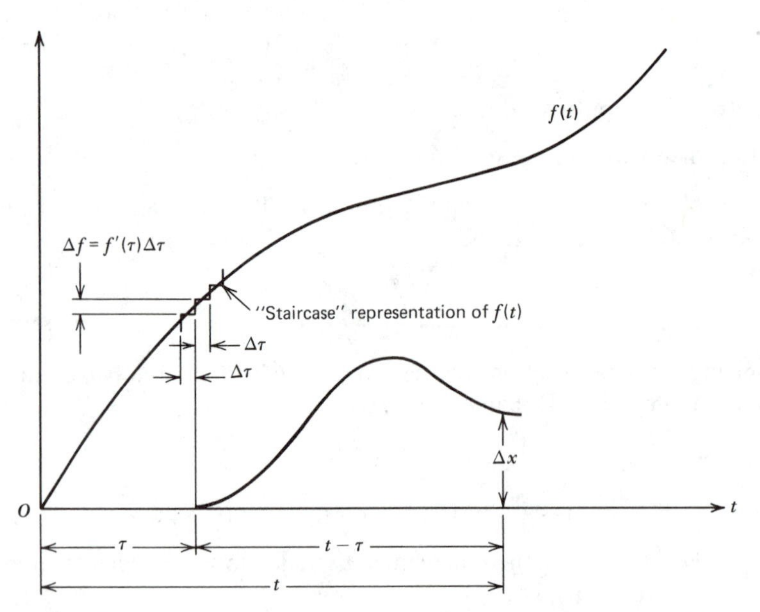

FIG. 8.11 Duhamel's integral, indicial form. Δf = step input applied at time τ, $\Delta x = A(t-\tau)\,\Delta f$ = response at time t to step input Δf.

8.7 TRANSFER FUNCTIONS

Consider a system or component, the dynamics of which is governed by linear equations, and which has two variables associated with it. Let these variables be interpreted as *input* and *output* or *excitation* and *response*. For example, an electronic amplifier might be such a system, with an input voltage and an output voltage. The spring-mass-damper system of Fig. 8.1 is also a system of this kind. Here the *input* or *excitation* is the force F acting on the mass, and the *output* or *response* is the displacement x. A convenient diagrammatic representation of this point of view is the *block diagram*, as shown in Fig. 8.12.

Definition of the Transfer Function

Imagine that the system is initially quiescent; that is, that the input is zero, and the system is at rest in an equilibrium condition. After time zero, an input $x_i(t)$ is applied, and an output $x_o(t)$ ensues. We define the transfer function $G(s)$ to be the

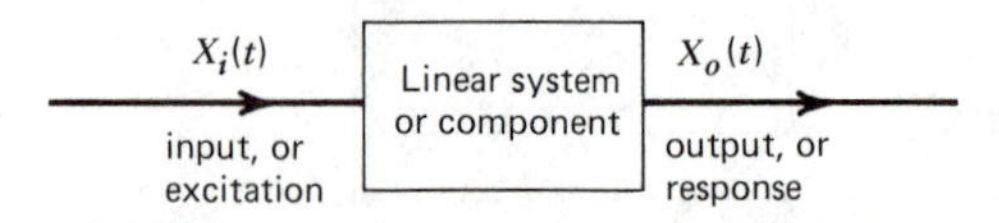

FIG. 8.12 Block diagram.

ratio of the Laplace transforms of the output and the input. That is

$$G(s) = \frac{\bar{x}_o(s)}{\bar{x}_i(s)} \tag{8.7,1}$$

We shall define as the *static gain* the limiting value of $G(s)$ as $s \to 0$; i.e.,

$$\lim_{s \to 0} G(s) = K \tag{8.7,2}$$

(A physical interpretation of the static gain is given in the example below.)

From Eq. 8.7,1 we see that the output of the system is given in terms of the input and the transfer function thus

$$\bar{x}_o(s) = G(s)\bar{x}_i(s)$$

In this form the elements that determine the output are clearly distinguished. The transfer function describes the *system characteristics*, and $\bar{x}_i(s)$ describes the input. Given these two, the Laplace transform of the output is known at once, and application of the methods for finding the inverse of the transform leads directly to the desired response.

Example

Let us find the transfer function of the second-order system of Fig. 8.1. The governing differential equation is Eq. 8.3,1, in which $f(t)$ is the input and $x(t)$ is the output. The Laplace transform is Eq. 8.3,3. Since the initial conditions $x(0)$ and $\dot{x}(0)$ are specified to be zero, then

$$\bar{x}(s)(s^2 + 2\zeta\omega_n s + \omega_n^{\ 2}) = \bar{f}(s)$$

or from Eq. 8.7,1

$$G(s) = \frac{\bar{x}(s)}{\bar{f}(s)} = \frac{1}{s^2 + 2\zeta\omega_n s + \omega_n^{\ 2}} \tag{8.7,3}$$

The transfer function is seen to be just the inverse of the characteristic polynomial of the system. The static gain K is found to be

$$K = \lim_{s \to 0} G(s) = \frac{1}{\omega_n^{\ 2}} \tag{8.7,4}$$

It was found in Sec. 8.5 (Eq. 8.5,11) that $1/\omega_n^{\ 2}$ is the asymptotic response of the system to a unit step input. Thus we have an interpretation of the static gain (valid for systems of any order); namely that K is the ultimate response of the system to a constant unit input.[4] If the constant input is I, then the final steady-state output is KI. The term static gain is seen to be very descriptive, in that it is the multiplicative factor by which a steady output is obtained from a steady input.

[3] The specification of an initially quiescent system eliminates any transient associated with nonzero initial conditions, and thereby makes the transfer function unique.

[4] It is implied that the system is stable, so that the transients die out with time.

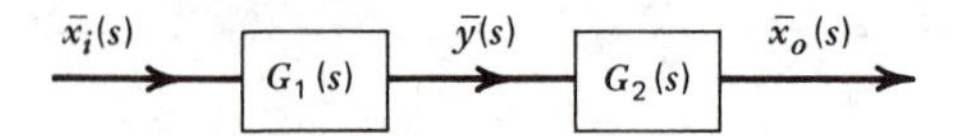

FIG. 8.13 Components in series.

Components in Series

Suppose we have two components in series, so that the output $y(t)$ of the first is the input to the second. The situation is illustrated in Fig. 8.13. From the definition of the transfer function we have

$$\bar{y}(s) = G_1(s)\bar{x}_i(s)$$

and

$$\bar{x}_o(s) = G_2(s)\bar{y}(s)$$

thus

$$\bar{x}_o(s) = G_1(s)G_2(s)\bar{x}_i(s)$$

If we denote the transfer function of the whole system by $G(s)$, i.e., $G(s) = \bar{x}_0(s)/\bar{x}_i(s)$, then

$$G(s) = G_1(s)G_2(s) \tag{8.7,5}$$

Thus the overall transfer function of components in series is obtained simply by multiplying together the individual transfer functions of the components. (The result is proved here for only two components. The extension to any number is trivial.)

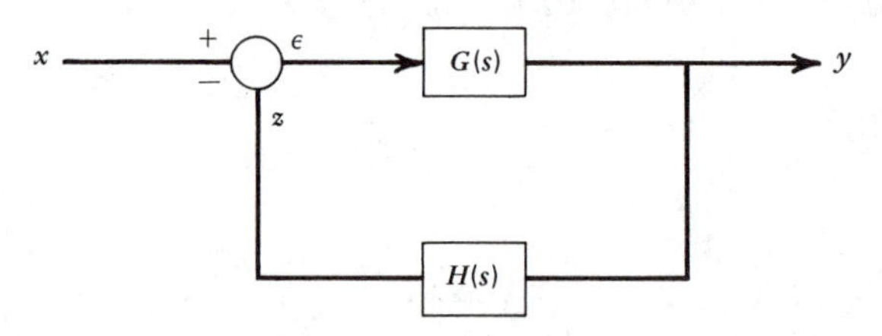

FIG. 8.14 General feedback system.

System with Feedback

Figure 8.14 shows a general *feedback* arrangement, containing two subsystems. When used as a feedback controller, ε is called the *actuating signal*,[5] $G(s)$ the forward-path transfer function and $H(s)$ the feedback transfer function. As indicated ε is the difference between x and z, so

$$\bar{\varepsilon} = \bar{x} - \bar{z}$$

$$\bar{y} = G(\bar{x} - \bar{z})$$

$$\bar{z} = H\bar{y}$$

[5] The designation *error* is reserved for the difference $x - y$, the aim of such a control system being to force y to be equal to x.

from this it follows easily that the overall transfer function is

$$\frac{\bar{y}}{\bar{x}} = \frac{G}{1 + GH} \tag{8.7,6}$$

and the actuating-signal transfer function is

$$\frac{\bar{\varepsilon}}{\bar{x}} = \frac{1}{1 + GH} \tag{8.7,7}$$

8.8 FREQUENCY RESPONSE

A very useful index of the response characteristics of a system is the frequency response. It is the variation with frequency of the steady-state amplitude and phase of the output when the input is a sinusoidal function of fixed amplitude. The frequency response is found to be very simply related to the transfer function.

Let the input be $x_i(t) = A_i \cos \omega t$, which describes an input of amplitude A_i and frequency $\omega/2\pi$. Let it be the real part of $X_i(t) = A_i e^{i\omega t}$. If we use the complex number $X_i(t)$ as the input in Eq. 8.7,1, then the steady-state output $X_o(t)$ is also complex. The physical output corresponding to $x_i(t)$ is then $x_o(t) =$ real part of $X_o(t)$. As usual, $X_i(t)$ and $X_o(t)$ are interpreted as rotating vectors whose projections on the real axis give the relevant physical variables (see Fig. 8.15).

From Table 8.1, item 4, the transform of X_i is

$$\bar{X}_i = \frac{A_i}{s - i\omega} \tag{8.8,1}$$

Therefore

$$\bar{X}_o = G(s)\bar{X}_i = A_i \frac{G(s)}{s - i\omega}$$

Let the function $G(s)$ be the ratio of two polynomials

$$G(s) = \frac{f(s)}{g(s)} \tag{8.8,3}$$

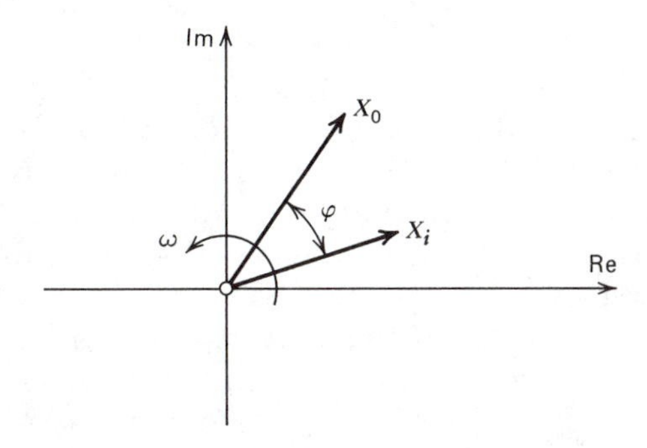

FIG. 8.15 Complex input and output.

so that

$$\bar{X}_o = A_i \frac{f(s)}{(s - i\omega)g(s)} \tag{8.8,4}$$

The roots of the denominator of the right-hand side are

$$i\omega, a_2 \cdots a_n$$

where $a_2 \cdots a_n$ are the roots of $g(s)$. We now apply the expansion theorem, Eq. 8.4,5, to find the complex output

$$\begin{aligned} X_o(t) &= A_i \sum_{r=1}^{n} \left[\frac{(s - a_r)f(s)}{(s - i\omega)g(s)} \right]_{s=a_r} e^{a_r t} \\ &= A_i \left[\frac{f(i\omega)}{g(i\omega)} e^{i\omega t} + c_2 e^{a_2 t} + \cdots + c_n e^{a_r t} \right] \end{aligned} \tag{8.8,5}$$

Let us apply the condition that the system is stable, so that all the roots $a_2 \cdots a_n$ of the characteristic polynomial have negative real parts. Then $e^{a_r t} \to 0$ as $t \to \infty$. Hence the steady-state solution is

$$X_o(t) = A_i \frac{f(i\omega)}{g(i\omega)} e^{i\omega t} \qquad (t \to \infty)$$

which by Eq. 8.8,3 is

$$\begin{aligned} X_o(t) &= A_i G(i\omega) e^{i\omega t} \\ &= G(i\omega) X_i(t) \end{aligned} \tag{8.8,6}$$

or

$$\frac{X_o(t)}{X_i(t)} = G(i\omega) \tag{8.8,7}$$

In general, $G(i\omega)$ is a complex number, varying with the circular frequency ω. Let it be given in polar form by

$$G(i\omega) = KMe^{i\varphi} \tag{8.8,8}$$

Then

$$\frac{X_o(t)}{X_i(t)} = KMe^{i\varphi} \tag{8.8,9}$$

From Eq. 8.8,9 we see that the amplitude ratio of the steady-state output to the input is $|X_o(t)/X_i(t)| = KM$, i.e., that the output amplitude is $A_o = KMA_i$, and that the phase relation is as shown in Fig. 8.15. The output leads the input by the angle φ. The quantity M, which is the modulus of $G(i\omega)$ divided by K, we shall call the *magnification factor*, or *dynamic gain*, and the product KM we call the *total gain*. It is important to note that M and φ are frequency-dependent.

Graphical representations of the frequency response commonly take the form of either vector plots of $Me^{i\varphi}$ (Nyquist diagram) or plots of M and φ as functions of frequency (Bode diagram). Examples of these are shown in Figs. 8.16 and 8.17.

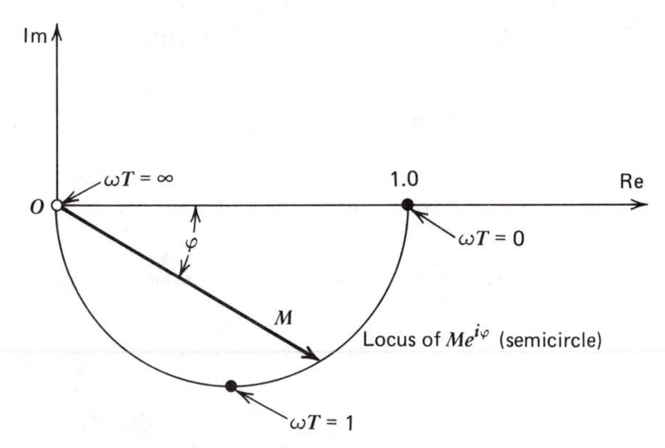

FIG. 8.16 Vector plot of $Me^{i\varphi}$—first-order system.

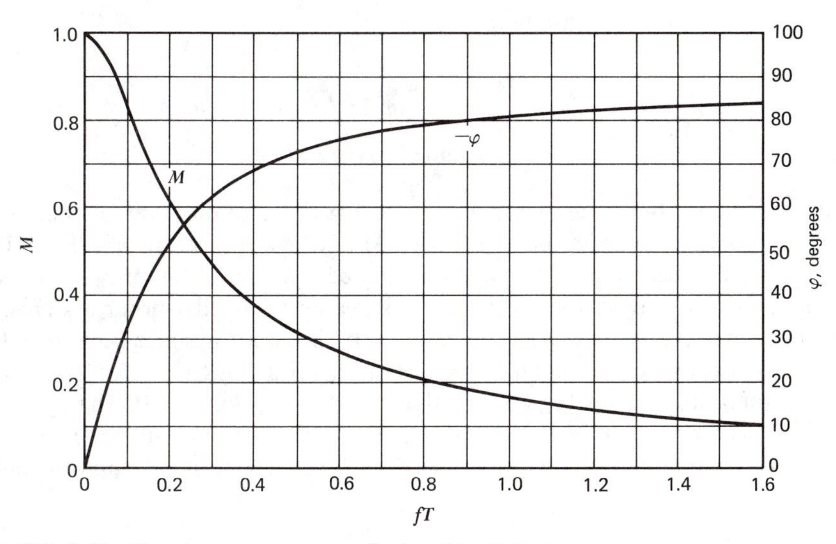

FIG. 8.17 Frequency response—first-order system.

$$M = \text{dynamic gain, } \frac{\text{output amplitude}}{\text{input amplitude}}$$

$$\varphi = \text{phase of output}$$

$$f = \text{input frequency}$$

Frequency Response of First-Order System

The differential equation of a first-order system with time constant T is Eq. 8.5,2. The Laplace transform of the equation, for zero initial conditions is

$$\left(s + \frac{1}{T}\right)\bar{x} = \bar{f}$$

The system input is $f(t)$, and the output is $x(t)$. The transfer function is therefore

$$G(s) = \frac{\bar{x}}{\bar{f}} = \frac{1}{s + 1/T} \tag{8.8,10}$$

and

$$K = \lim_{s \to 0} G(s) = T$$

The frequency response is determined by the vector $G(i\omega)$

$$G(i\omega) = KMe^{i\varphi} = \frac{T}{1 + i\omega T}$$

thus

$$Me^{i\varphi} = \frac{1 - i\omega T}{1 + \omega^2 T^2} \tag{8.8,11}$$

From Eq. 8.8,11, M and φ are found to be

$$\begin{aligned} M &= \frac{1}{(1 + \omega^2 T^2)^{1/2}} \\ -\varphi &= \tan^{-1} \omega T \end{aligned} \tag{8.8,12}$$

A vector plot of $Me^{i\varphi}$ is shown in Fig. 8.16. This kind of diagram is sometimes called the transfer-function locus. Plots of M and φ are given in Fig. 8.17. The abscissa is fT, where $f = \omega/2\pi$, the input frequency. This is the only parameter of the equations, and so the curves are applicable to all first-order systems. It should be noted that at $\omega = 0$, $M = 1$ and $\varphi = 0$. This is always true because of the definitions of K and $G(s)$—it can be seen from Eq. 8.7,2 that $G(s) = K$ when $s = 0$, which automatically gives the above values of M and φ for $\omega = 0$. The dynamic gain M is seen on Fig. 8.17 to decrease continuously as the frequency increases, while at the same time the phase lag increases from zero to its asymptotic value of 90°.

Frequency Response of a Second-Order System

The transfer function of a second-order system was found in Sec. 8.7. From Eqs. 8.7,3 and 4 we obtain the frequency-response vector as

$$Me^{i\varphi} = \frac{\omega_n^2}{(\omega_n^2 - \omega^2) + 2i\zeta\omega_n\omega} \tag{8.8,13}$$

From the modulus and argument of Eq. 8.8,13, we find that

$$\begin{aligned} M &= \frac{1}{\{[1 - (\omega/\omega_n)^2]^2 + 4\zeta^2(\omega/\omega_n)^2\}^{1/2}} \\ -\varphi &= \tan^{-1} \frac{2\zeta\omega/\omega_n}{1 - (\omega/\omega_n)^2} \end{aligned} \tag{8.8,14}$$

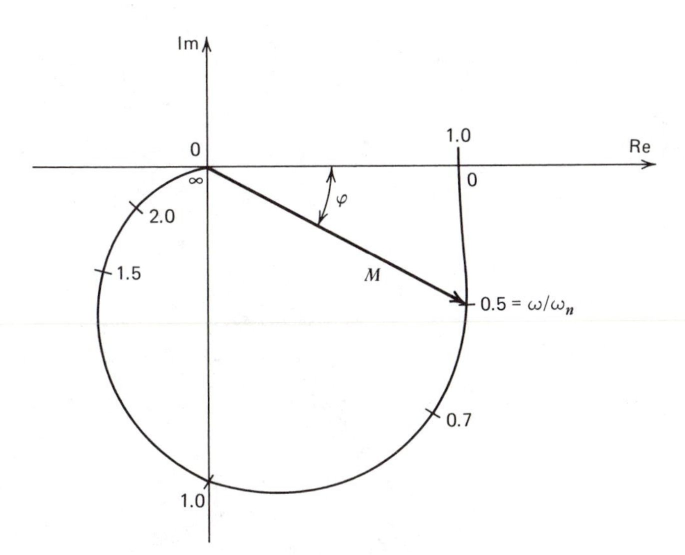

FIG. 8.18 Vector plot of $Me^{i\varphi}$ for second-order system. Damping ratio $\zeta = 0.4$.

A representative vector plot of $Me^{i\varphi}$, for damping ratio $\zeta = 0.4$, is shown in Fig. 8.18, and families of M and φ are shown in Fig. 8.19. Whereas a single pair of curves serves to define the frequency response of all first-order systems (Fig. 8.17), it takes two families of curves, with the damping ratio as parameter, to display the characteristics of all second-order systems. The importance of the damping as a parameter should be noted on Fig. 8.19. It is especially powerful in controlling the magnitude

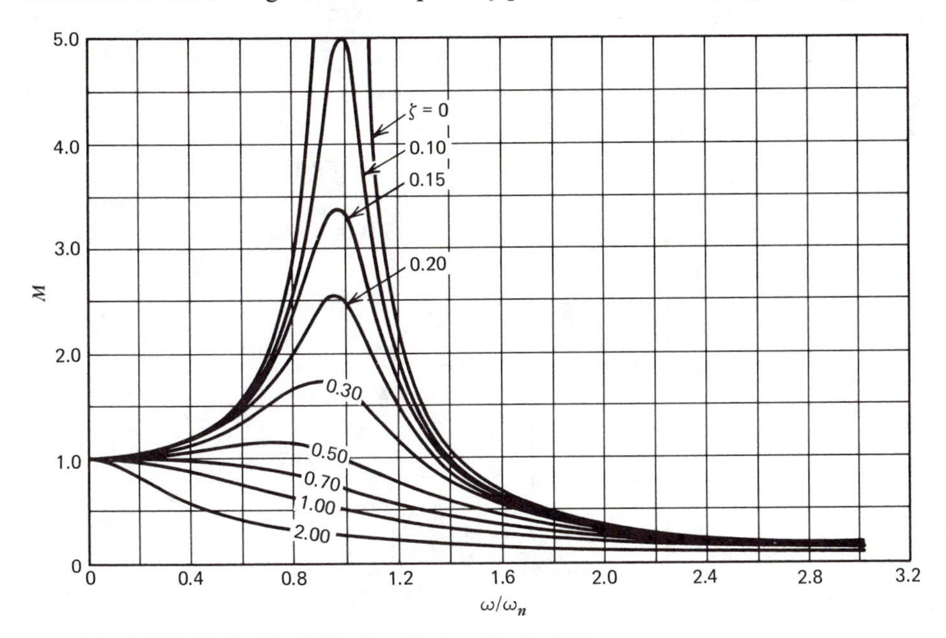

FIG. 8.19a Dynamic gain of second-order system.

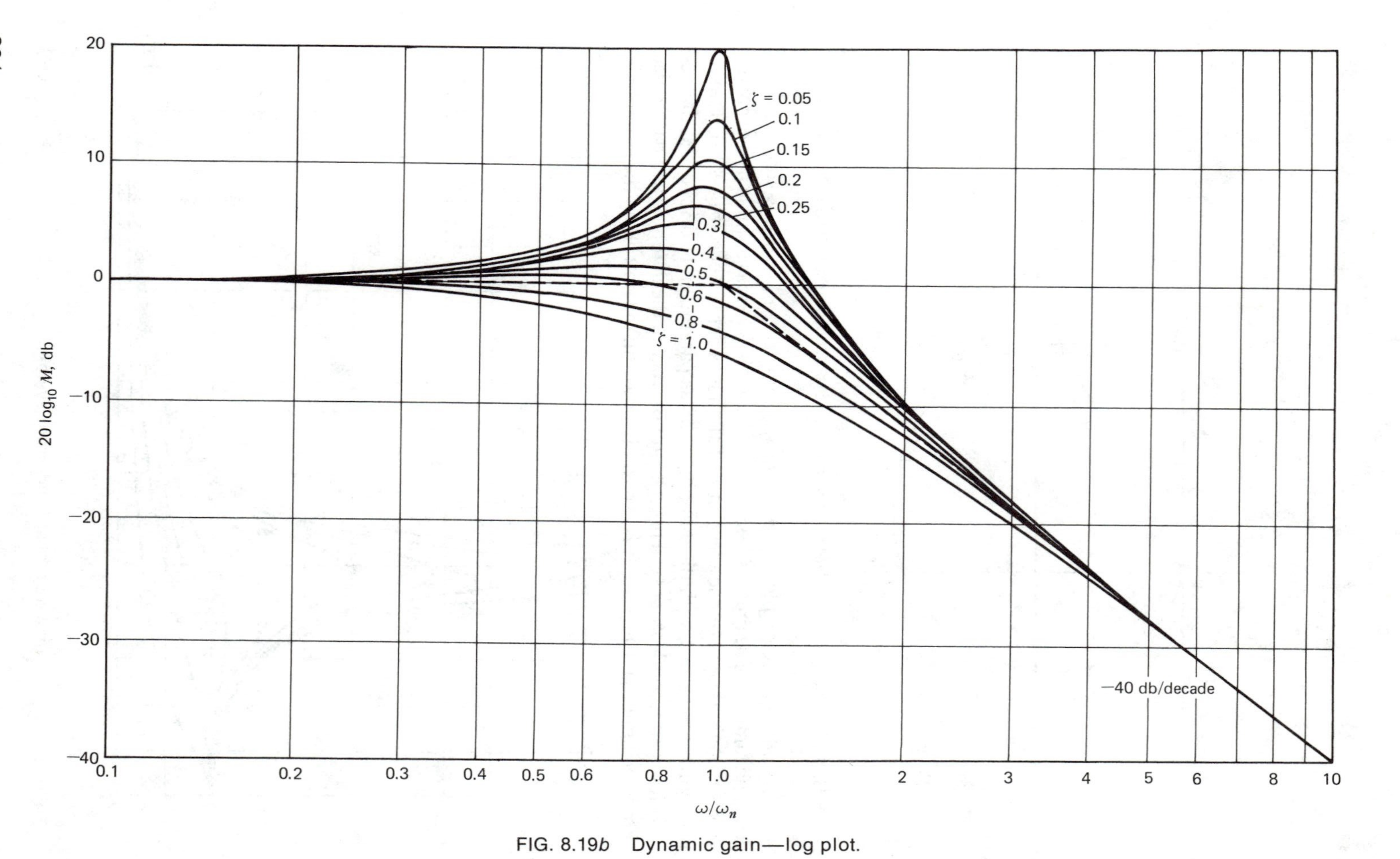

FIG. 8.19*b* Dynamic gain—log plot.

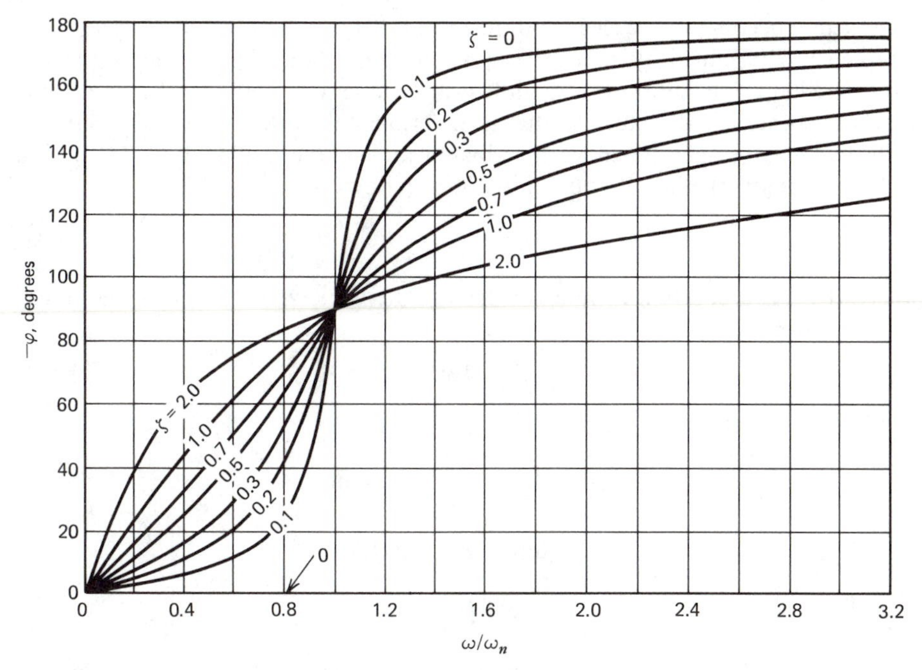

FIG. 8.19*c* Phase lag of second-order system.

of the resonance peak that occurs near unity frequency ratio. At this frequency the phase lag is by contrast independent of ζ, as all the curves pass through $\varphi = -90^\circ$ there. For all values of ζ, $M \to 1$ and $\varphi \to 0$ as $\omega/\omega_n \to 0$. This shows that, whenever a system is driven by an oscillatory input whose frequency is low compared to the undamped natural frequency, the response will be *quasistatic*. That is, at each instant, the output will be the same as though the instantaneous value of the input were applied statically.

The behavior of the output when ζ is near 0.7 is interesting. For this value of ζ, it is seen that φ is very nearly linear with ω/ω_n up to 1.0. Now the phase lag can be interpreted as a time lag, $\tau = -(\varphi/2\pi)T = -\varphi/\omega$ where T is the period. The output wave form will have its peaks retarded by τ sec relative to the input. For the value of ζ under consideration, $-\varphi/(\omega/\omega_n) \doteqdot \pi/2$ or $-\varphi/\omega = \pi/2\omega_n = \frac{1}{4}T_n$, where $T_n = 2\pi/\omega_n$, the undamped natural period. Hence we find that, for $\zeta \doteqdot 0.7$, there is a nearly constant time lag $\tau \doteqdot \frac{1}{4}T_n$, independent of the input frequency, for frequencies below resonance.

8.9 TRANSITION BETWEEN THE TIME AND FREQUENCY DOMAINS

In this section we derive relations that show the connection between the indicial and impulsive admittances (response in the *time domain*) and the frequency response (response in the *frequency domain*). These relations are of great importance, for with them it becomes quite clear that the response dynamics of a linear system are completely determined by *either* the impulsive admittance $h(t)$ *or* the frequency response $G(i\omega)$. Furthermore, straightforward numerical methods of going from one to the other are available, so that in experimental work the more convenient of the two may be determined, with no loss in value of the results.

Let us define the impulse function $\delta(t)$ as follows (see Fig. 8.20):

$$\delta(t) = \lim_{\varepsilon \to 0} f(\varepsilon, t) \tag{8.9,1}$$

where $f(\varepsilon, t)$ is a continuous function having the value zero except in the range zero to ε, and such that

$$\int_0^\varepsilon f(\varepsilon, t)\, dt = 1 \tag{8.9,2}$$

The Laplace transform of $\delta(t)$ is

$$\begin{aligned} \bar{\delta}(s) &= \int_0^\infty e^{-st} \lim_{\varepsilon \to 0} f(\varepsilon, t)\, dt \\ &= \lim_{\varepsilon \to 0} \int_0^\varepsilon e^{-st} f(\varepsilon, t)\, dt \\ &= e^{-s(0)} \int_0^\varepsilon f(\varepsilon, t)\, dt \\ &= 1 \end{aligned} \tag{8.9,3}$$

Hence, if a linear system, with transfer function $G(s)$, be subject to the input $\delta(t)$, then the Laplace transform of the response is

$$\bar{h}(s) = G(s)\bar{\delta}(s) = G(s) \tag{8.9,4}$$

The inverse transform is found from the inversion theorem, Eq. 8.4,7. We assume that the system is stable, so that all poles of $G(s)$ lie in the left half-plane. (This is

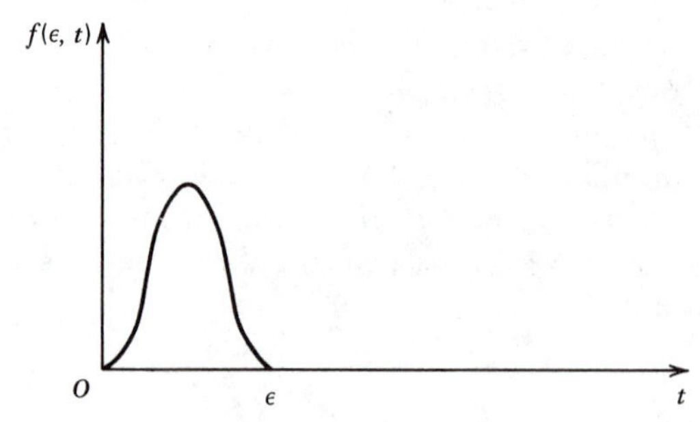

FIG. 8.20 The impulse function.

$$\int_0^\varepsilon f(\varepsilon, t)\, dt = 1.0$$

$$\delta(t) = \lim_{\varepsilon \to 0} f(\varepsilon, t)$$

also the condition that $G(i\omega)$ shall give the frequency response.) We may then take $\gamma = 0$, so that $s = i\omega$, and

$$h(t) = \frac{1}{2\pi i}\int_{-i\infty}^{i\infty} e^{i\omega t}\bar{h}(i\omega)\, d(i\omega)$$

$$= \frac{1}{2\pi}\int_{-\infty}^{\infty} e^{i\omega t}G(i\omega)\, d\omega \tag{8.9,5}$$

The impulsive admittance is thus found to be (for stable systems) the inverse Fourier transform of the frequency response. G. F. Floyd (see ref. 12.1, pp. 334–336) has shown that the integral of Eq. 8.9,5 may be expressed in an alternative form over the frequency range 0 to ∞.

$$h(t) = \frac{2}{\pi}\int_0^{\infty} \cos \omega t\; KM \cos \varphi\, d\omega \tag{8.9,6}$$

where $M(\omega)$ and $\varphi(\omega)$ are the dynamic gain and phase angle. Huss and Donegan (ref. 8.3) have devised a convenient scheme for numerical integration of Eq. 8.9,6.

From Eq. 8.9,4, it also follows that

$$G(s) = \bar{h}(s) = \int_0^{\infty} e^{-st}h(t)\, dt$$

or that

$$G(i\omega) = \int_0^{\infty} e^{-i\omega t}h(t)\, dt \tag{8.9,7}$$

This relation, which is reciprocal to Eq. 8.9,5, shows that the frequency response is the Fourier transform of the impulsive admittance [note that $h(t) = 0$ for $t < 0$]. A convenient numerical method for evaluating this integral is also given by Huss and Donegan.

The reciprocal relations derived above are particularly useful in experimental work (e.g., flight testing), when it is desired to measure the response characteristics, and when the transfer function of the system is unknown. For example, from an experimental determination of the load-factor frequency response to elevator-angle input, Eq. 8.9,6 permits the calculation of $h(t)$. The transient load-factor response to any arbitrary elevator-angle input can then be computed by means of the convolution integral, Eq. 8.6,5 or 8.6,6.

8.10 ADDITIONAL SYMBOLS INTRODUCED IN CHAPTER 8

$A(t)$ indicial admittance

$G(s)$ transfer function

$h(t)$ impulsive admittance

K static gain (see Eq. 8.7,2)

$\mathscr{L}[\ ]$ Laplace transform of [ ]

M dynamic gain (see Eq. 8.8,8)

$\delta(t)$ delta function (see Eq. 8.9,1)

φ phase angle of frequency response (Eq. 8.8,8)

$\overline{(\ )}$ Laplace transform of ( )

8.11 BIBLIOGRAPHY

8.1 H. S. Carslaw and J. C. Jaeger. *Operational Methods in Applied Mathematics*, 2nd ed. Oxford University Press, London, 1947.

8.2 J. C. Jaeger. *An Introduction to the Laplace Transformation*. Methuen & Co. London, 1949.

8.3 C. R. Huss and J. J. Donegon. Method and Tables for Determining the Time Response to a Unit Impulse from Frequency Response Data and for Determining the Fourier Transform of a Function of Time. *NACA TN 3598*, 1956.

8.4 J. J. Donegan and C. R. Huss. Comparison of Several Methods for Obtaining the Time Response of Linear Systems to Either a Unit Impulse or Arbitrary Input from Frequency-Response Data. *NACA TN 3701*, 1956.

8.5 T. von Kármán and M. A. Biot. *Mathematical Methods in Engineering*. McGraw-Hill Book Co., New York, 1940.

8.6 H. S. Carslaw. *Fourier's Series and Integrals*, 3rd ed. Macmillan, London, 1930.

8.7 J. J. Donegan and C. R. Huss. Incomplete Time Response to a Unit Impulse, and Its Application to Lightly Damped Linear Systems. *NACA TN 3897*, 1956.

RESPONSE TO ACTUATION OF THE CONTROLS (OPEN LOOP)

CHAPTER 9

9.1 GENERAL REMARKS

It is convenient to classify vehicle motion according to whether it is *free* or *forced*. Chapters 6 and 7 were devoted to a number of examples of the former, and in this chapter we give some examples of the latter. The particular cases studied are those in which the motion results from nonautonomous actuation of the controls. That is, we exclude those in which the controls are moved *in response to* the vehicle motion in accordance with a prescribed law, as by an autopilot. Such motions are the subject of the following chapter. We should recall that for linear systems with constant coefficients there is really only one *fundamental* response problem. The impulse response, the step response, and the frequency response are all explicitly related to one another and to the transfer function (Chap. 8). In addition, the convolution theorem (8.6,5) enables the response to any arbitrary control variation to be calculated from a knowledge of the response to either an impulse or a step input.

In the examples that follow, we consider the response of the airplane to actuation of its principal controls: the throttle and the three aerodynamic control surfaces—elevator, rudder, and aileron. The examples include both step and frequency response.

As remarked above, the basic items needed for computing response are the transfer functions relating the various control inputs to the various desired responses. These can readily be obtained from the Laplace-transformed equations 4.16,6 and 4.16,7.

Longitudinal Control

The two principal quantities that need to be controlled in symmetric flight are the speed (v_c) and the flight path angle γ ($\gamma = \theta - \alpha$, see Fig. 4.3), that is, the C.G.

velocity vector. To achieve this obviously entails the ability to apply control forces both parallel and perpendicular to the flight path. The former is provided by thrust or drag control, and the latter by control of the lift via elevator deflection or wing flaps. It is evident that the main *initial* response to opening the throttle (increasing the thrust) is a forward acceleration, i.e., change of speed. The main *initial* response to elevator deflection is a rotation in pitch, with consequent change in angle of attack and lift, followed by flight path curvature, i.e., $\dot{\gamma}$. When the transient that follows such control action has ultimately died away (for a stable airplane), the new steady state can be found as in conventional performance analysis. Figure 9.1 shows the basic relations. The steady speed u_0 at which the airplane flies is governed by C_{L_0}, which is in turn fixed by δ_e (Sec. 2.5). Hence a given δ_e implies a given u_0. Now the flight path angle is determined at any given speed by the thrust (Fig. 9.1). Thus the *ultimate* result of moving the throttle at fixed δ_e (when **T** passes through the C.G.) is to change γ without changing speed. But we saw above that the *initial* response to ΔT is just the opposite—*a change in speed without change in* γ. The short- and long-term effects of this control action are clearly quite contrary. Likewise, we saw that the initial effect of moving the elevator is to rotate the vehicle and change γ without significant speed change, whereas the ultimate consequence is substantial change in both speed and γ. The total picture of longitudinal control is obviously not simple, and the transients that connect the initial and final responses require investigation. We shall see that these transients

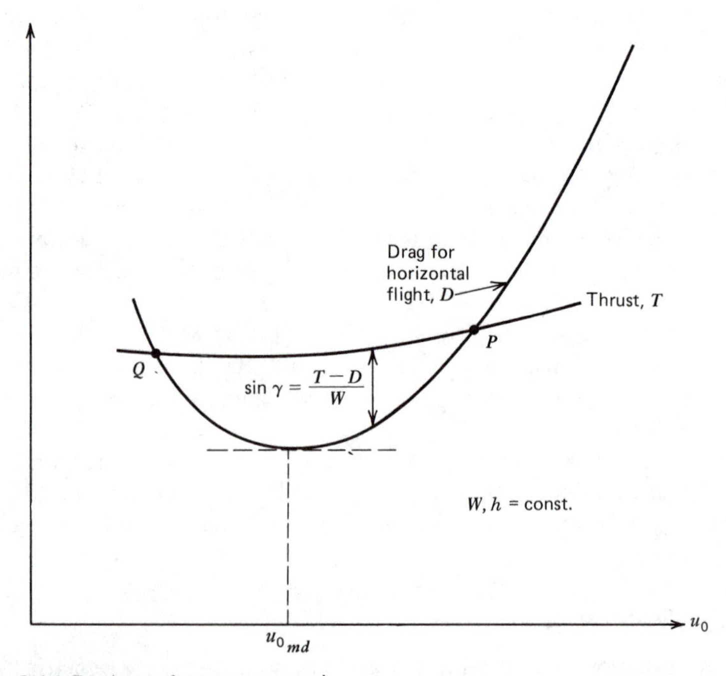

FIG. 9.1 Basic performance graph.

are usually dominated by the long-period lightly damped phugoid oscillation, and that the final steady state may only be reached after a very long time.

Lateral Control

The lateral controls (the aileron and rudder) on a conventional airplane have three principal functions.

1. To provide trim in the presence of asymmetric thrust associated with power plant failure.
2. To provide corrections for unwanted motions associated with atmospheric turbulence or other random events.
3. To provide for turning maneuvers—i.e., rotation of the velocity vector in a horizontal plane.

The first two of these purposes are served by having the controls generate aerodynamic moments about the x and z axes—rolling and yawing moments. For the third a force must be provided that has a component normal to v_c and in the horizontal plane. This is, of course, the component $L \sin \phi$ of the lift when the airplane is banked at angle ϕ. Thus the lateral controls (principally the aileron) produce turns as a secondary result of controlling ϕ.

Ordinarily, the long-term responses to deflection of the aileron and rudder are very complicated, with all the lateral degrees of freedom being excited by each. Solution of the complete equations of motion is the only way to appreciate these fully. Certain useful approximations of lower order are however available.

9.2 RESPONSE TO ELEVATOR INPUT

For the conventional case of cruising flight, Eqs. 4.16,6, with slight alterations, can be used for response to elevator. We let $\theta_0 = 0$, assume that $G_{zu} = G_{mu} = G_{zq} = 0$, note that $s\bar{\theta} = \bar{q}$, equation (*e*), omit equation (*d*), and rewrite the remaining equations in the convenient matrix form:

$$\begin{bmatrix} (2\mu s - G_{xu}) & -G_{x\alpha} & 0 & C_{L_0} \\ 2C_{L_0} & (2\mu s - G_{z\alpha}) & -2\mu & 0 \\ 0 & -G_{m\alpha} & (i_B s - G_{mq}) & 0 \\ 0 & 0 & 1 & -s \end{bmatrix} \begin{bmatrix} \bar{u} \\ \bar{\alpha} \\ \bar{q} \\ \bar{\theta} \end{bmatrix} = \begin{bmatrix} 0 \\ G_{z\eta} \\ G_{m\eta} \\ 0 \end{bmatrix} \bar{\eta} \tag{9.2,1}$$

The aerodynamic transfer functions on the r.h.s. can usually be represented well enough by (see Sec. 4.16)

$$\begin{aligned} G_{z_\eta} &= -C_{L_\delta} \\ G_{m\eta} &= C_{m_\delta} + sC_{m_{\dot{\delta}}} \end{aligned} \tag{9.2,2}$$

and $C_{m_{\dot{\delta}}}$ is furthermore frequently neglected.

If we denote the 4×4 matrix on the l.h.s. of Eq. (9.2,1) by **P** then the equation may be written more compactly as

$$\mathbf{P}\begin{bmatrix}\bar{u}\\ \bar{\alpha}\\ \bar{q}\\ \bar{\theta}\end{bmatrix} = \begin{bmatrix}0\\ G_{z\eta}\\ G_{m\eta}\\ 0\end{bmatrix}\bar{\eta} \tag{9.2,3}$$

and from this

$$\begin{bmatrix}\bar{u}\\ \bar{\alpha}\\ \bar{q}\\ \bar{\theta}\end{bmatrix} = \mathbf{P}^{-1}\begin{bmatrix}0\\ G_{z\eta}\\ G_{m\eta}\\ 0\end{bmatrix}\bar{\eta} \tag{9.2,4}$$

It follows that the four transfer functions for the four response variables are

$$\begin{bmatrix}G_{u\eta}\\ G_{\alpha\eta}\\ G_{q\eta}\\ G_{\theta\eta}\end{bmatrix} = \mathbf{P}^{-1}\begin{bmatrix}0\\ G_{z\eta}\\ G_{m\eta}\\ 0\end{bmatrix} \tag{9.2,5}$$

The transfer functions given of course relate to the nondimensional variables, not dimensional ones, for example,

$$G_{u\eta} = \frac{\mathscr{L}[\hat{u}]}{\mathscr{L}[\eta]} \qquad G_{q\eta} = \frac{\mathscr{L}[\hat{q}]}{\mathscr{L}[\eta]} \tag{9.2,6}$$

The particular transfer functions shown are not the only ones of interest. Other responses, such as flight path angle and load factor, may be wanted. Thus, since $\gamma = \theta - \alpha$, we simply have

$$G_{\gamma\eta} = G_{\theta\eta} - G_{\alpha\eta} \tag{9.2,7}$$

Since the load factor is $n = L/W$ and is unity in the reference flight condition, we have to first order (an exercise for the student):

$$\Delta n = \frac{\Delta L}{W} = \frac{\Delta C_L}{C_{L_0}} + 2\hat{u} \tag{9.2,8}$$

ΔC_L in Eq. 9.2,8 is conveniently eliminated by

$$\Delta\bar{C}_L = G_{Lu}\bar{u} + G_{L\alpha}\bar{\alpha} + G_{Lq}\bar{q} + G_{L\delta}\bar{\eta}$$

After substituting into the Laplace transform of Eq. 9.2,8 and dividing by $\bar{\eta}$, we get

$$G_{n\eta} = \frac{\Delta\bar{n}}{\bar{\eta}} = \frac{1}{C_{L_0}}\left[(G_{Lu} + 2C_{L_0})G_{u\eta} + G_{L\alpha}G_{\alpha\eta} + G_{Lq}G_{q\eta} + G_{L\delta}\right] \tag{9.2,9}$$

The preceding equations can be used directly for machine computation of frequency response functions, which basically requires only routine operations on matrices with complex coefficients; an example of this application is given below. However, for analysis one needs literal expressions for the various transfer functions, and in some applications one also needs their inverses (the impulsive admittances). This is not a practical analytical procedure for the complete system, even with the simplified equation 9.2,1. To obtain exact solutions for the impulse or step responses, the preferred method is to solve the original differential equations on an analog or digital computer. For analytical work in control system design, however, approximate forms of the transfer functions may be very useful. This topic is pursued in more depth in Chap. 10, of ref. 1.10 where such approximations are developed.

Numerical Example—Frequency Response

For this example we take the same hypothetical jet transport used in Sec. 6.5, flying at the same speed and height. For the aerodynamic transfer functions in **P** we use the stability derivative representation, i.e., all but $G_{m\alpha}$ are the same as the corresponding stability derivatives, and $G_{m\alpha} = C_{m_\alpha} + sC_{m_{\dot{\alpha}}}$. In addition, for the control aerodynamics we use

$$G_{z\eta} = -C_{L_\delta} = -0.24/\text{radian}$$
$$G_{m\eta} = C_{m_\delta} = -0.72/\text{radian}$$

The exact frequency response was calculated from (9.2,5) by substituting $s = i\hat{\omega}$ in them. The results are shown on Fig. 9.2 for speed, angle of attack, and flight-path angle.

The exact solutions show that the responses in the "trajectory" variables u and γ are dominated entirely by the large peak at the phugoid frequency. Because of the light damping in this mode, the dynamic gains at resonance are very large. The peak $|G_{u\eta}|$ of about 85 means that a speed amplitude of 10% of u_0 would result from an elevator angle amplitude of only $0.1 \times 57.3/85 = 0.068°$. Similarly at resonance an oscillation of 10° in γ would result from about 1/10° elevator amplitude. For both these variables, the response diminishes rapidly with increasing frequency, becoming negligibly small above the short-period frequency. The phase angle for $\hat{u}$, Fig. 9.2*b*, is zero at low frequency, diminishes rapidly to $-180°$ at the phugoid frequency (very much like the lightly damped, second-order systems of Fig. 8.18) and subsequently at the short-period frequency undergoes a further drop characteristic of a heavily damped, second-order system. The "chain" concept of high-order systems in series (Sec. 8.7) is well exemplified by this graph.

By contrast, the attitude variables α and q show important effects at both the phugoid and short-period frequencies. The complicated behavior of α near the phugoid frequency indicates the sort of thing that can occur with high-order systems. It is associated with a pole and a zero of $G_{\alpha\eta}$ occurring close together in the transfer function. Again, above the short-period frequency, the amplitude of α and q both fall off rapidly.

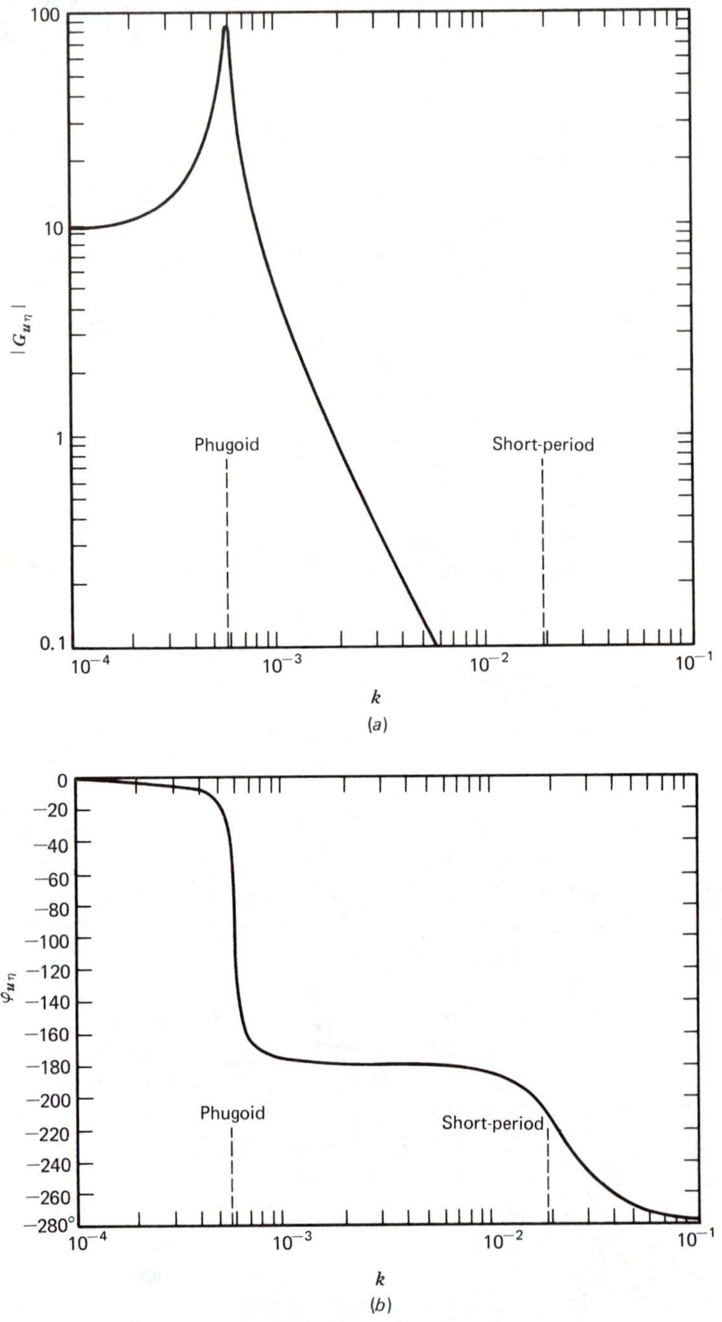

FIG. 9.2 Frequency-response functions, elevator angle input. Jet transport cruising at high altitudes (*a*) Speed amplitude. (*b*) Speed phase. (*c*) Angle of attack amplitude. (*d*) Angle of attack phase. (*e*) Flight-path angle amplitude. (*f*) Flight-path angle phase.

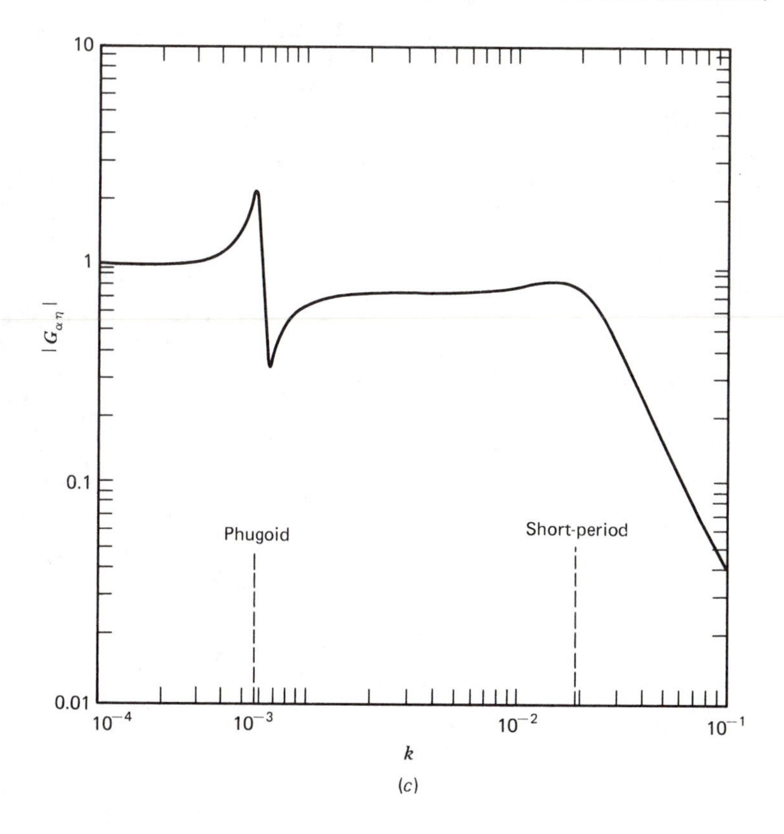

(c)

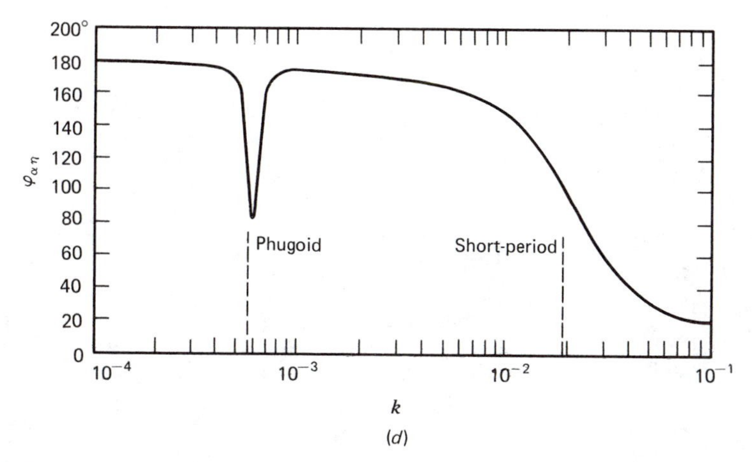

(d)

FIG. 9.2 *(continued)*

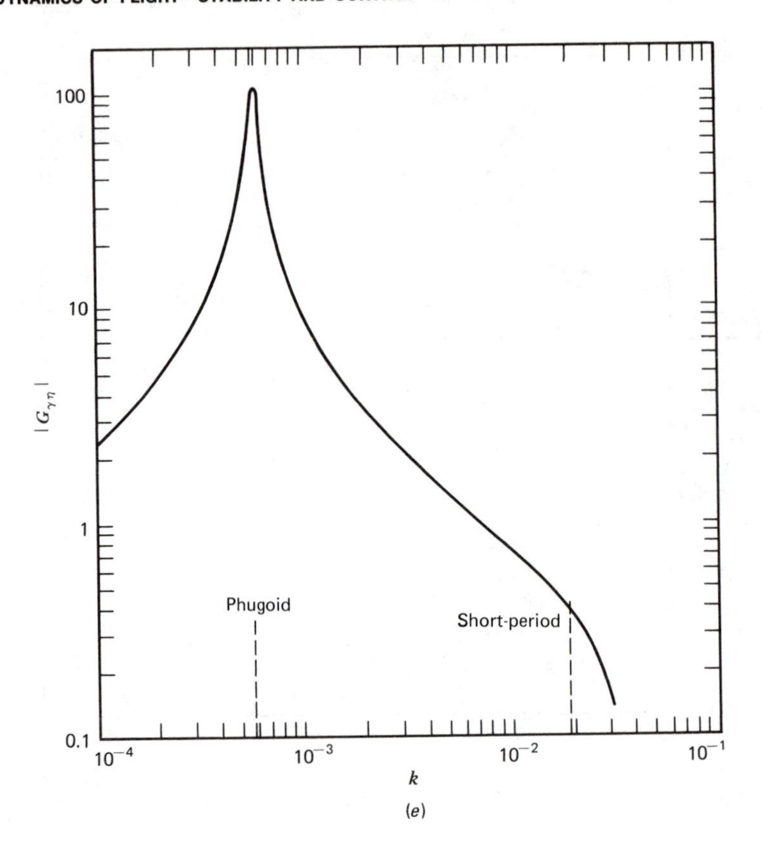

(*e*)

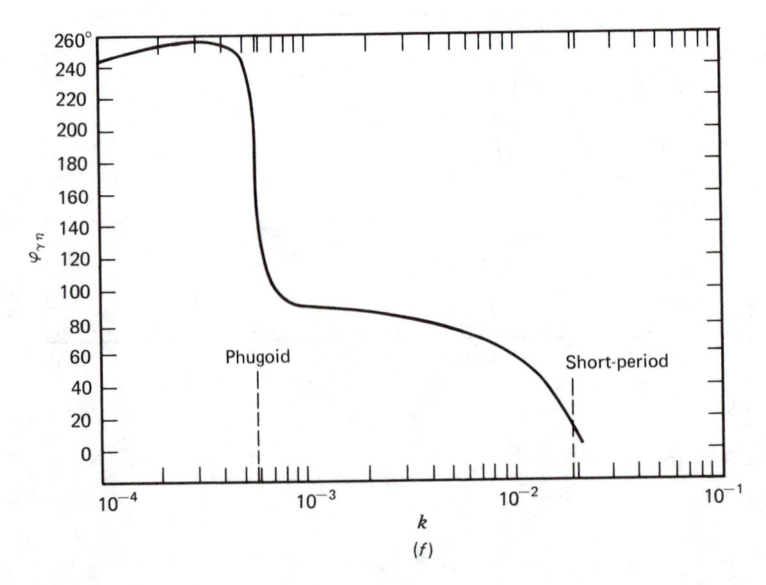

(*f*)

FIG. 9.2 (*continued*)

Numerical Example—Step Response

For the same airplane and flight conditions as in the previous example, the response to a step-function input in the elevator angle was computed, using (4.15,7) mechanized for a 10-volt analog computer. As an aid to readers unfamiliar with analog computation, the details of this one example are set out rather fully. The other examples that follow later were computed in essentially the same way, but the details of scaling and circuits are omitted. With the same assumptions as made previously the differential equations with numerical coefficients are:

$$\begin{aligned}
D\hat{u} &= -6.92 \times 10^{-5}\hat{u} + 2.55 \times 10^{-4}\alpha - 4.60 \times 10^{-4}\theta \\
D\alpha &= -9.20 \times 10^{-4}\hat{u} - 9.00 \times 10^{-3}\alpha + \hat{q} - 4.42 \times 10^{-4}\eta \\
D\hat{q} &= 2.03 \times 10^{-6}\hat{u} - 3.62 \times 10^{-4}\alpha - 1.43 \times 10^{-2}\hat{q} - 3.77 \times 10^{-4}\eta \\
D\theta &= \hat{q}
\end{aligned} \tag{9.2,10}$$

To mechanize them for analog computation we make the following transformation of variables:

$$[u] = s_u\hat{u} \qquad [\alpha] = s_\alpha\alpha; \qquad [q] = s_q\hat{q} \qquad [\theta] = s_\theta\theta \tag{9.2,11}$$

where the quantities in square brackets, $[u]$ etc., denote machine voltages, and s_u etc. are scale factors. Time scaling is by the law

$$\tau = s_t\hat{t}$$

where τ is laboratory clock time, or *macsecs* (for computing machine seconds), and $\hat{t}$ is the nondimensional time variable of the differential equations. To relate the computer results to real flight time t we use

$$\tau = s_t \frac{t}{t^*}, \qquad t^* = 0.0105 \text{ sec} \tag{9.2,12}$$

On recalling that D in Eq. (9.2,10) represents $d/d\hat{t}$, and defining $[\ ]' = (d/d\tau)[\ ]$, the transformation of the equations into differential equations for the voltage yields

$$\begin{aligned}
[u]' &= \frac{s_u}{s_t}\left\{-\frac{6.92 \times 10^{-5}}{s_u}[u] + \frac{2.55 \times 10^{-4}}{s_\alpha}[\alpha] - \frac{4.60 \times 10^{-4}}{s_\theta}[\theta]\right\} \\
[\alpha]' &= \frac{s_\alpha}{s_t}\left\{-\frac{9.20 \times 10^{-4}}{s_u}[u] - \frac{9.00 \times 10^{-3}}{s_\alpha}[\alpha] \right. \\
&\qquad \left. + \frac{1}{s_q}[q] - 4.42 \times 10^{-4}\eta\right\} \\
[q]' &= \frac{s_q}{s_t}\left\{\frac{2.03 \times 10^{-6}}{s_u}[u] - \frac{3.62 \times 10^{-4}}{s_\alpha}[\alpha] \right. \\
&\qquad \left. - \frac{1.43 \times 10^{-2}}{s_q}[q] - 3.77 \times 10^{-4}\eta\right\} \\
[\theta]' &= \frac{s_\theta}{s_t s_q}[q]
\end{aligned} \tag{9.2,13}$$

Note that we have chosen for convenience to give the control angles as η rather than $[\eta]/s_\eta$ since η not $[\eta]$ is to be specified. The scale factors used were as follows:

$$s_u = 10 \text{ v/unit}$$
$$s_\alpha = 10 \text{ v/radian}$$
$$s_q = 1000 \text{ v/unit}$$
$$s_\theta = 10 \text{ v/radian}$$

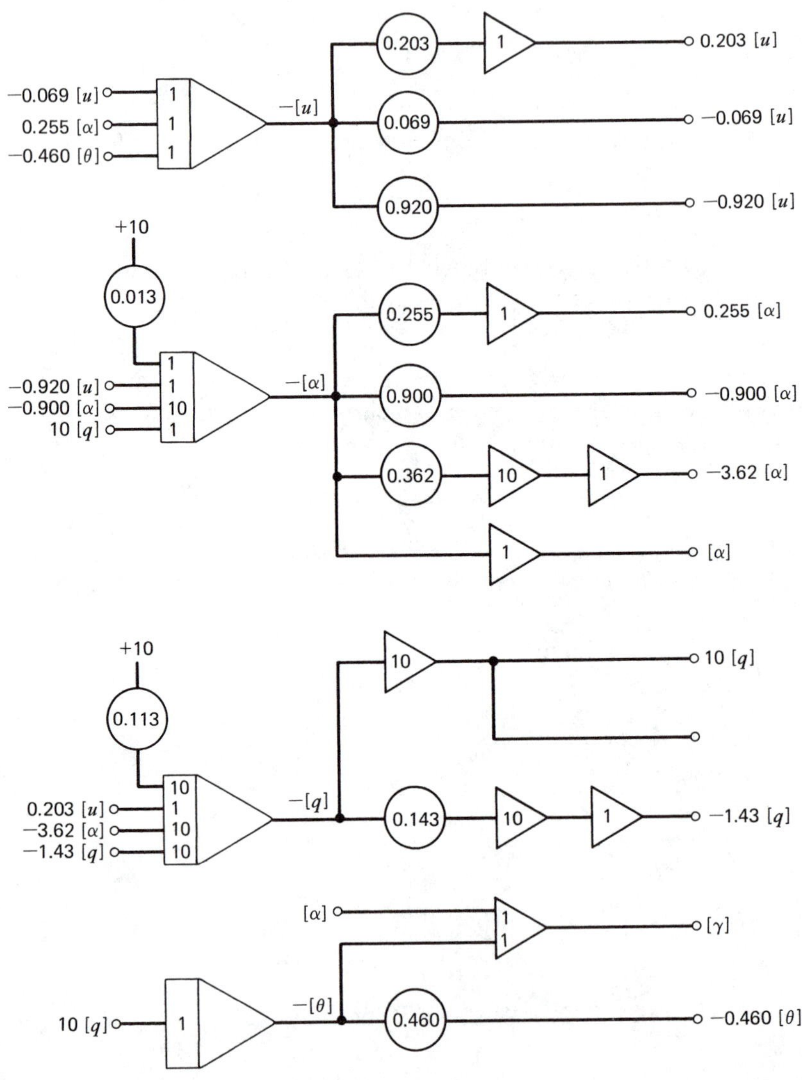

FIG. 9.3 Analog circuit diagram for response to elevator step. $\eta = -0.03$ rad, $s_t = 10^{-3}$.

Since the response shortly after $t = 0$ is governed mainly by the short-period mode, and the long-term response by the phugoid mode, a single time scale is not appropriate for both. Hence two time scales were used:

To show long-term response: $s_t = 10^{-3}$

To show initial response: $s_t = 10^{-2}$

The analog circuit for $s_t = 10^{-3}$ and $\eta = -.03$ rad is shown in Fig. 9.3, using conventional symbols for integrators, summers, etc.

When $s_t = 10^{-3}$, the time relation, from Eq. 9.2,12

$$t = \frac{t^*}{s_t}\tau = 10.5\tau$$

and hence the process proceeds about 10 times faster on the computer than in flight. When $s_t = 10^{-2}$, the process proceeds at nearly real flight time.

The results for $\hat{u}$, α, and γ are shown for both time scales on Figs. 9.4 and 9.5. These curves were recorded by a conventional $x - y$ plotter; the time base was generated on the computer by integrating a constant.

Figure 9.4 shows that α increases rapidly and quickly damps out to its asymptotic value. u and γ, however, make a slow, weakly damped approach to their final values, the initial overshoot being very large for both. If the reason for moving

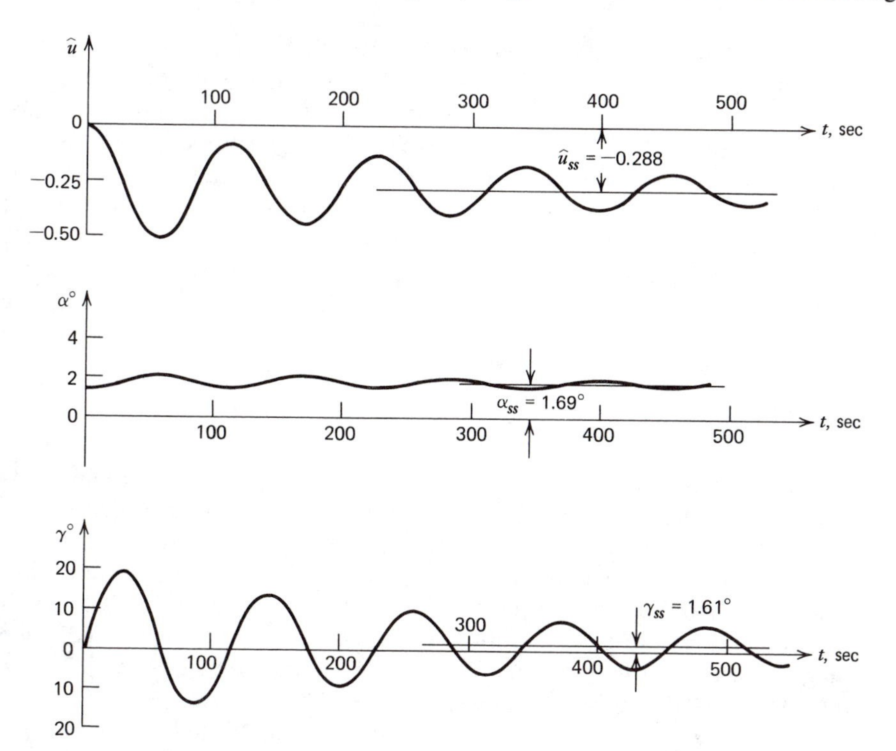

FIG. 9.4 Response to elevator ($\eta = -0.03$ rad). Jet transport cruising at high altitude.

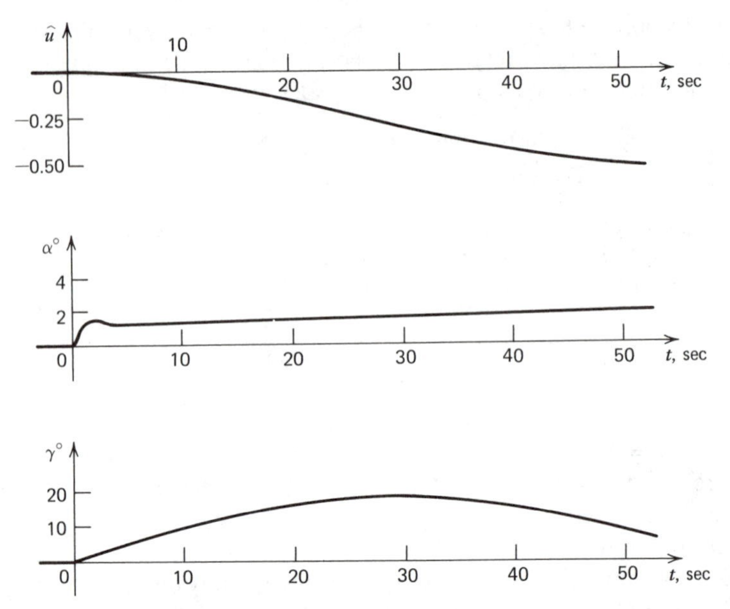

FIG. 9.5 Response to elevator ($\eta = -0.03$ rad). Jet transport cruising at high altitude.

the elevator were to change to a new steady state, the maneuver has not been a very effective one! After 500 sec the oscillations in u and γ have still not disappeared. The behavior near $t = 0$ is shown more clearly on Fig. 9.5; the rapid rise in α is dominated by the short-period mode. It is only after α has changed that the associated increase in lift can act to curve the flight path (via $\Delta L \doteq mu_0\dot{\gamma}$); thus the increase in γ lags that in α. At the same time the increased drag due to α, and the "downhill" component of the weight combine to produce a reduction in speed, which lags still farther behind. The response in γ is not in fact very rapid. It takes about 10 sec to increase γ by about 10° with this elevator deflection. In this time the vehicle has traveled 7330 ft.

9.3 RESPONSE TO THE THROTTLE

The initial response of an airplane to movement of the throttle is actually quite dependent on the details of the engine control system and on the type of propulsion system. For jet engines it takes an appreciable time for the rpm and thrust to increase after opening the throttle, and this can be an important factor in emergency conditions. The response of a propeller, which increases thrust by a change of blade angle, is more rapid. We make the simple assumption here that opening the throttle produces a step change in C_T of amount ΔC_T.

We assume further that the thrust line passes through the C.G. so that there is no moment from the thrust and its only effect is to add ΔT to the aerodynamic force ΔX in Eq. 4.14,9. The net effect of this addition on Eq. 4.16,6 is to add ΔC_T

on the r.h.s. of (*a*). Finally then, the governing equation is the same as Eq. 9.2,1 on the l.h.s., and the r.h.s. is

$$\begin{bmatrix} 1 \\ 0 \\ 0 \\ 0 \end{bmatrix} \Delta C_T \tag{9.3,1}$$

Numerical Example—Step Response

Analog computations were made for the airplane and flight condition of the previous example, with $\Delta C_T = 0.0125$. The results for $\hat{u}$, α, and γ are shown on Fig. 9.6. The motion at this time scale is clearly dominated by the lightly damped phugoid. We see that the speed begins to increase immediately, before the other variables have time to change. It then undergoes a damped oscillation, returning finally to its initial value. The angle of attack varies only slightly, and γ makes an oscillatory approach to its final positive value γ_{ss}. The ultimate steady state is a

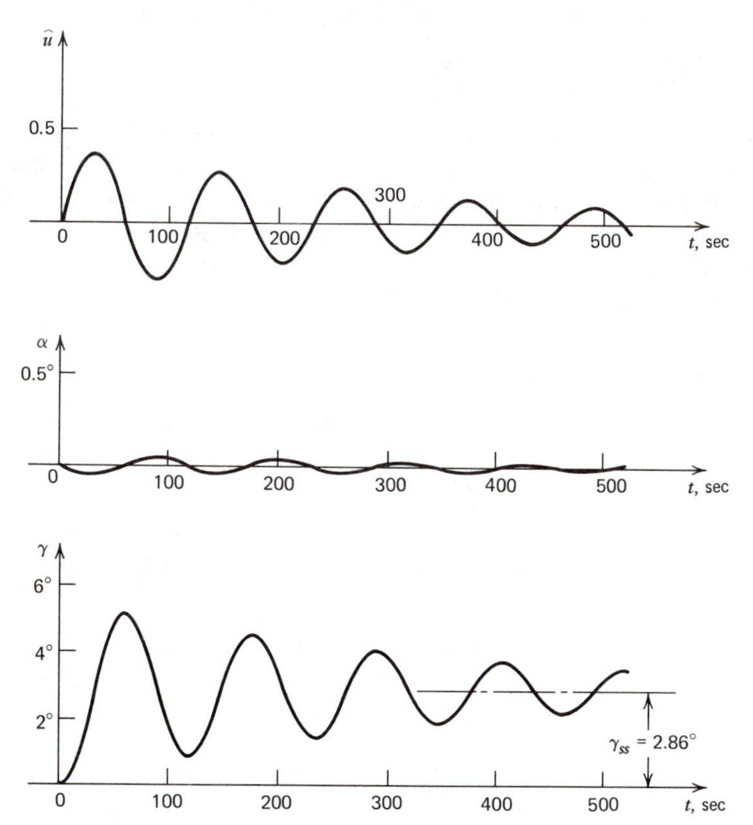

FIG. 9.6 Response to throttle ($\Delta C_T = 0.0125$). Jet transport cruising at high altitude. Thrust line passing through C.G.

climb with $u = \alpha = 0$. When the thrust line does not pass through the C.G., the response is different in several details. Principally, the thrust moment produces a rapid change in α, followed by an oscillatory decay to a new $\alpha_{ss} \neq 0$, and the final speed changes to a new steady state $\hat{u}_{ss} \neq 0$.

9.4 LATERAL STEADY STATES

The basic flight condition is steady symmetric flight, in which all the lateral variables β, p, r, ϕ are identically zero. Unlike the elevator and the throttle, the lateral controls, the aileron and rudder, are not used *individually* to produce changes in the steady state. This is because the steady-state values of β, p, r, ϕ that result from a constant δ_a or δ_r are not generally of interest as a useful flight condition. There are two lateral steady states that are of interest, however, each of which requires the joint application of aileron and rudder. These are the *steady sideslip*, in which the flight path is rectilinear, and the *steady turn*, in which the angular velocity vector is vertical. We look into these below before proceeding to the study of dynamic response to the lateral controls.

The Steady Sideslip

The steady sideslip is a condition of nonsymmetric rectilinear translation. It is sometimes used, particularly with light airplanes, to correct for cross-wind on landing approaches. Glider pilots also use this maneuver to steepen the glide path, since the L/D ratio decreases due to increased drag at large β. In this flight condition $D = d/d\hat{t} \equiv 0$, and $p = r = y' = 0$. Thus, with reference to Eq. 4.15,8, the only nonzero state variables are β, ϕ, and ψ. We take $\theta_0 \equiv 0$, and reduce Eq. 4.15,8 in accordance with the sideslip conditions given. The result is

$$\begin{aligned} C_{y_\beta}\beta + C_{L_0}\phi + C_{y_\zeta}\zeta &= 0 \\ C_{l_\beta}\beta + C_{l_\zeta}\zeta + C_{l_\xi}\xi &= 0 \\ C_{n_\beta}\beta + C_{n_\zeta}\zeta + C_{n_\xi}\xi &= 0 \end{aligned} \tag{9.4,1}$$

The three equations contain the four variables β, ϕ, ζ, ξ. Hence an infinite set of solutions exists, in which any one of the four may be selected arbitrarily. If we choose ϕ to be arbitrary the equations can be solved for the corresponding β, ζ, ξ (provided of course that its matrix is not singular). Thus

$$\begin{bmatrix} \beta \\ \zeta \\ \xi \end{bmatrix} = \mathbf{A}^{-1} \begin{bmatrix} -C_{L_0}\phi \\ 0 \\ 0 \end{bmatrix}$$

where (9.4,2)

$$\mathbf{A} = \begin{bmatrix} C_{y_\beta} & C_{y_\zeta} & 0 \\ C_{l_\beta} & C_{l_\zeta} & C_{l_\xi} \\ C_{n_\beta} & C_{n_\zeta} & C_{n_\xi} \end{bmatrix}$$

As an example, consider the jet transport used previously, at $C_{L_0} = 1.0$, with the β derivatives as in Sec. 7.2. In addition to these we need the control derivatives, for which we use

$$C_{y_\zeta} = 0.067 \qquad C_{l_\zeta} = 0.003 \qquad C_{n_\zeta} = -0.040$$
$$C_{l_\xi} = -0.065 \qquad C_{n_\xi} = 0.005$$

It is evident from Eq. 9.4,2 that β, ζ, and ξ are all proportional to ϕ, hence the ratios of the angles are constant. The numerical result is:

$$\frac{\phi}{\beta} = 0.0558; \qquad \frac{\zeta}{\beta} = 1.675; \qquad \frac{\xi}{\beta} = -1.800$$

so that for a sideslip of 10°, the other angles are $\phi = 0.56°$, $\zeta = 16.75°$, $\xi = -18.00°$. As expected, a slip to the right requires left rudder and right aileron. The control angles are seen to be large; powerful controls are needed to sideslip at large angles. When the matrix **A** is singular, it only indicates that ϕ is zero in the sideslip. In that case the equations can be rearranged to put ϕ on the l.h.s. and β on the r.h.s., in which case the new matrix is very unlikely to be singular.

The Steady Turn

The steady circling flight of an airplane (Fig. 9.7) is a particular case of control response which is of considerable interest, being the commonest of all maneuvers. In this maneuver, the six linear and angular velocity components are constants, and hence the differential equations of motion reduce to algebraic equations. This simplification enables us to retain the nonlinear terms in the angle of bank Φ. This is desirable since we would like to treat cases where it is too large for the linearization to be valid; e.g., $\Phi = 60°$. We shall, however, assume that the angle of climb Θ is small. Θ and Φ are both constant during the maneuver, but Ψ is not. It can be seen from the definitions of the orientation angles (see Sec. 4.5) that $\dot{\Psi}$ is the rate of change of azimuth of the airplane x axis. This we denote by Ω, the *rate of turn*. Ω is positive for a right turn.

Under the conditions stated above, the kinematical equations, Eqs. 4.12,3 give the following values for the angular velocities:

$$\begin{aligned} P &= -\Omega\Theta \\ Q &= \Omega \sin \Phi \\ R &= \Omega \cos \Phi \end{aligned} \tag{9.4,3}$$

Even in a fairly rapid turn, e.g., 360° per min, Ω is quite small (i.e., $2\pi/60 \doteqdot 0.1$), so that the three angular velocities may be treated as small quantities, and their squares and products neglected. Furthermore, we may choose the axes so that the xy plane contains the velocity vector of the C.G., so that $W = 0$. Finally, we shall specify that the turn is truly banked, which is to say that the side force $Y = 0$. Although the velocity component V may not then be exactly zero, it will be small enough to neglect its product with other small quantities. Under these conditions,

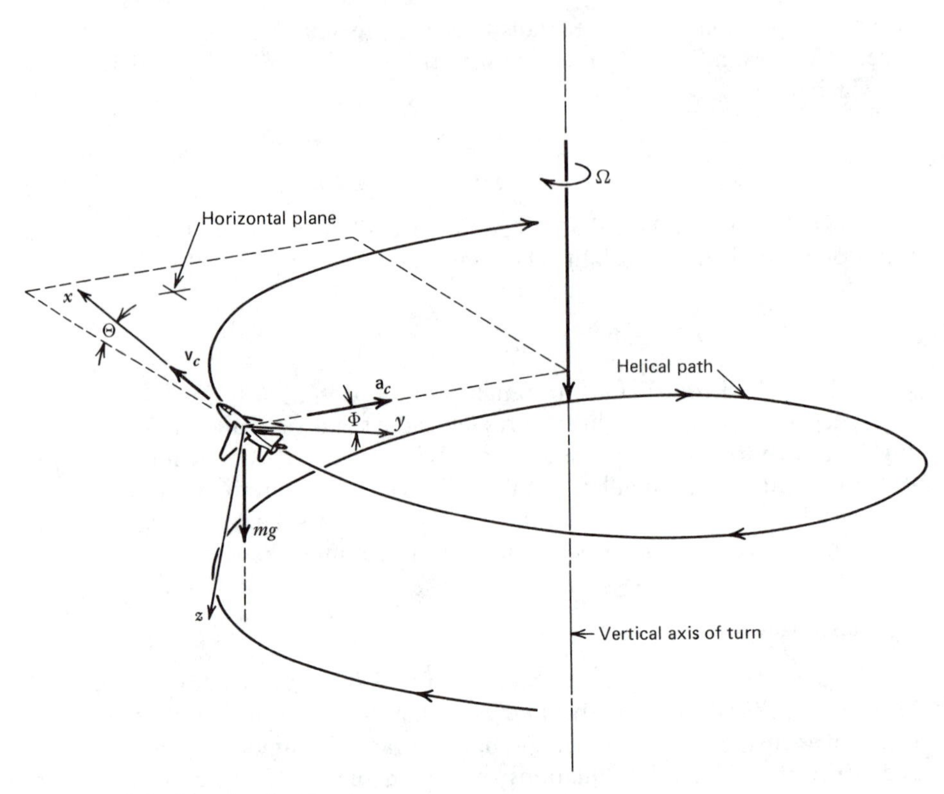

FIG. 9.7 Steady climbing turn.

and in the absence of gyroscopic effects, the dynamical equations 4.12,1 and 4.12,2 reduce to

$$\begin{aligned} X - mg\Theta &= 0 \\ mg \sin \Phi &= m\Omega U \cos \Phi \\ Z + mg \cos \Phi &= -m\Omega U \sin \Phi \\ L = M &= N = 0 \end{aligned} \tag{9.4,4}$$

From Eqs. 9.4,4 we find that the angle of bank is given by

$$\tan \Phi = \frac{U}{g} \Omega$$

and that the Z force is

$$Z = -mg \sec \Phi$$

The normal load factor in the turn is then simply related to the angle of bank; namely

$$n = \frac{-Z}{mg} = \sec \Phi$$

The control angles required to maintain the steady turn are found from the fact that the three aerodynamic moments are zero. Let us introduce a reference flight condition, which is straight-line flight at the given height, speed, and angle of climb, and let the prefix Δ denote changes from this reference condition. Then

$$\begin{aligned} \Delta C_l &= C_{l_\beta}\Delta\beta + C_{l_p}\Delta\hat{p} + C_{l_r}\Delta\hat{r} + C_{l_\xi}\Delta\xi + C_{l_\zeta}\Delta\zeta = 0 & (a)\\ \Delta C_m &= C_{m_\alpha}\Delta\alpha + C_{m_q}\Delta\hat{q} + C_{m\eta}\Delta\eta = 0 & (b)\\ \Delta C_n &= C_{n_\beta}\Delta\beta + C_{n_p}\Delta\hat{p} + C_{n_r}\Delta\hat{r} + C_{n_\xi}\Delta\xi + C_{n_\zeta}\Delta\zeta = 0 & (c) \end{aligned} \qquad (9.4,5)$$

where

$$\begin{aligned} \Delta\beta &\doteqdot 0 & (a)\\ \Delta\hat{p} &= \frac{Pb}{2U} = -\frac{\Omega\Theta b}{2U} & (b)\\ \Delta\hat{r} &= \frac{Rb}{2U} = \frac{\Omega b}{2U}\cos\Phi & (c)\\ \Delta\hat{q} &= \frac{Qc}{2U} = \frac{\Omega c}{2U}\sin\Phi & (d) \end{aligned} \qquad (9.4,6)$$

The value of $\Delta\alpha$ is most simply obtained from the change in C_L. The reference value is, for small angle of climb θ, $C_{L_0} = W/\frac{1}{2}\rho U^2 S$. The value in the turn is $C_L = -Z/\frac{1}{2}\rho U^2 S = nC_{L_0}$. Hence

$$\Delta C_L = (n-1)C_{L_0} = C_{L_\alpha}\Delta\alpha + C_{L_q}\Delta\hat{q} + C_{L_\eta}\Delta\eta \qquad (9.4,7)$$

Elevator Angle in the Turn

From Eqs. 9.4,5*b* and 9.4,7, $\Delta\alpha$ is eliminated to get the change in elevator angle:

$$\Delta\eta = \frac{(n-1)C_{L_0}C_{m_\alpha} + \Delta\hat{q}(C_{L_\alpha}C_{m_q} - C_{L_q}C_{m_\alpha})}{C_{L_\eta}C_{m_\alpha} - C_{L_\alpha}C_{m_\eta}}$$

$\Delta\hat{q}$, which occurs in this expression, is known from the conditions of the turn Eq. 9.4,6*d*:

$$\begin{aligned} \Delta\hat{q} &= \frac{\Omega c}{2U}\sin\Phi = \frac{gc}{2U^2}\frac{\sin^2\Phi}{\cos\Phi}\\ &= \frac{gc}{2U^2}n\left(1 - \frac{1}{n^2}\right) \end{aligned}$$

The factor g/U^2 can be eliminated in terms of C_{L_0}, and the relative density $\mu = m/\frac{1}{2}\rho Sc$ introduced to convert this expression into

$$\Delta\hat{q} = \frac{C_{L_0}}{2\mu}\frac{(n-1)(n+1)}{n}$$

Finally we get, for $\Delta\eta$,

$$\Delta\eta = (n-1)C_{L_0}\frac{C_{m_\alpha} + \dfrac{n+1}{2\mu n}(C_{L_\alpha}C_{m_q} - C_{L_q}C_{m_\alpha})}{C_{L_\eta}C_{m_\alpha} - C_{L_\alpha}C_{m_\eta}} \tag{9.4,8}$$

The signs and magnitudes of the terms in this expression are such that $\Delta\eta$ is negative for $n > 1$; i.e., up elevator is required to hold the turn. The factor $(n-1)C_{L_0}$ indicates that the elevator angle increases with the load factor in the turn, and that it is largest at low speeds.

Example

The information given by Eq. 9.4,8 is plotted on Fig. 9.8 as it applies to the airplane used in the previous examples (see Secs. 6.5 and 9.2). The values of C_{L_α}, C_{L_q}, and C_{L_η} are the negatives of C_{z_α}, C_{z_q} and C_{z_η} respectively.

The elevator angles obtained are seen to be quite small.

Aileron and Rudder Angles in the Turn

The aileron and rudder angles required to maintain the steady turn are obtained by simultaneous solution of Eqs. 9.4,5 *a* and *c*. The result obtained, with $\Delta\beta = 0$, is

$$\begin{aligned}\Delta\xi &= \frac{(C_{l_p}C_{n_\zeta} - C_{l_\zeta}C_{n_p})\Delta\hat{p} + (C_{l_r}C_{n_\zeta} - C_{l_\zeta}C_{n_r})\Delta\hat{r}}{C_{l_\zeta}C_{n_\xi} - C_{l_\xi}C_{n_\zeta}} \\ \Delta\zeta &= \frac{(C_{l_p}C_{n_\xi} - C_{l_\xi}C_{n_p})\Delta\hat{p} + (C_{l_r}C_{n_\xi} - C_{l_\xi}C_{n_r})\Delta\hat{r}}{C_{l_\xi}C_{n_\zeta} - C_{l_\zeta}C_{n_\xi}}\end{aligned} \tag{9.4,9}$$

where $\Delta\hat{p}$ and $\Delta\hat{r}$ are as given by Eq. 9.4,6.

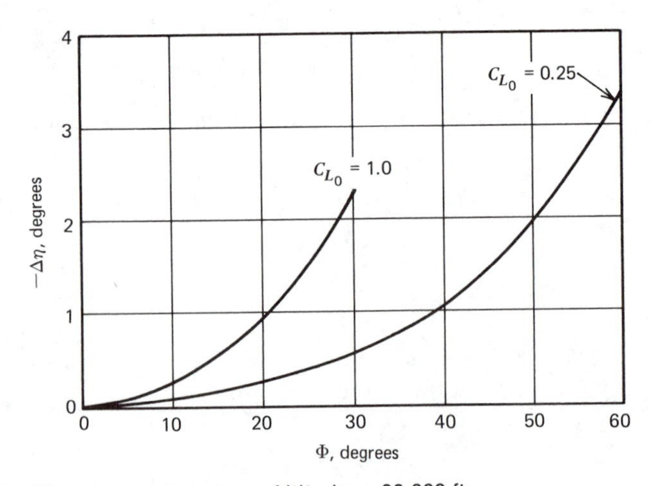

FIG. 9.8 Elevator angle to turn. Altitude = 30,000 ft.

Example

The airplane used in the previous examples was considered to be in turning flight at a speed of 300 mph and altitude of 30,000 ft. In carrying out the calculations, allowance was made for the variation of some of the stability derivatives with C_L (see Sec. 7.2). Two cases were calculated: climbing flight at $\theta = 15°$, and gliding flight at $\theta = -15°$. For each case a range of bank angles was used, and the values of $\Delta\hat{p}$, $\Delta\hat{r}$, $\Delta\xi$, and $\Delta\zeta$ were calculated. Plots of the results are shown in Figs. 9.9 and 9.10. They show that the control angles are generally small, and that

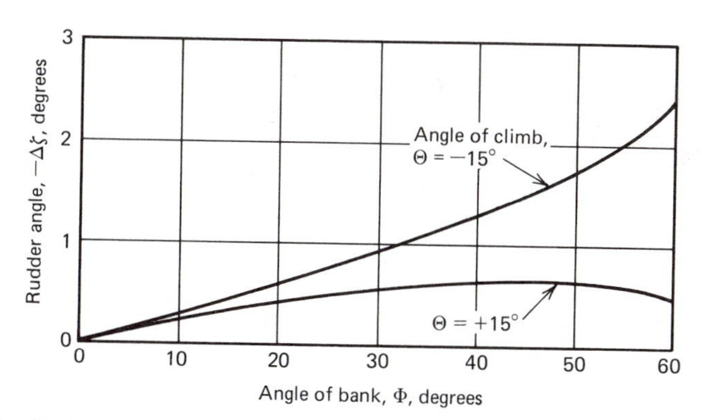

FIG. 9.9 Rudder angle in a steady turn.

$u_0 = 300$ mph
Altitude = 30,000 ft.

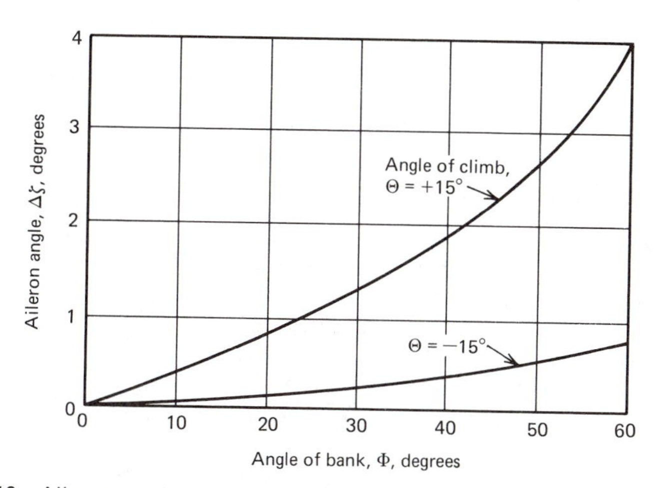

FIG. 9.10 Aileron angle in a steady turn.

$u_0 = 300$ mph
Altitude = 30,000 ft.

they depend very much on the angle of climb θ. The reason for this is that the rolling component of the airplane angular velocity depends on $\theta[\Delta\hat{p} = (\Omega b/2U)\theta]$. Since the derivatives C_{l_p} and C_{n_p} are both large, this rate of roll exerts a considerable influence on the steady-state control-surface angles. Since positive Φ corresponds to a right turn, we see from Figs. 9.9 and 9.10 that to turn right requires right rudder (ζ negative) and "off-bank" ailerons (right aileron down, ξ positive).

9.5 LATERAL FREQUENCY RESPONSE

The computation of the lateral frequency response, with aileron or rudder input, is carried out with a set of transfer functions obtained from Eqs. 4.16,7*a*, *b*, *c*, and *f*. For this purpose we set $\theta_0 = 0$ (for horizontal flight) and rewrite the equation in the compact matrix form

$$\mathbf{Px} = \mathbf{Qu} \tag{9.5,1}$$

where

$$\mathbf{x} = [\bar{\beta}\ \bar{p}\ \bar{r}\ \bar{\phi}]^T$$
$$\mathbf{u} = [\bar{\zeta}\ \bar{\xi}]^T$$

$$\mathbf{P} = \left[\begin{array}{c|c|c|c} (2\mu s - G_{y\beta}) & -G_{yp} & (2\mu - G_{yr}) & -C_{L_0} \\ \hline -G_{l\beta} & (i_A s - G_{lp}) & -(i_E s + G_{lr}) & 0 \\ \hline -G_{n\beta} & -(i_E s + G_{np}) & (i_C s - G_{nr}) & 0 \\ \hline 0 & 1 & 0 & -s \end{array}\right]$$

and

$$\mathbf{Q} = \begin{bmatrix} G_{y\zeta} & 0 \\ G_{l\zeta} & G_{l\xi} \\ G_{n\zeta} & G_{n\xi} \\ 0 & 0 \end{bmatrix}$$

The required transfer functions are then given by the matrix

$$\mathbf{G} = \mathbf{P}^{-1}\mathbf{Q} \tag{9.5,2}$$

where

$$\mathbf{G} = \begin{bmatrix} G_{\beta\zeta} & G_{\beta\xi} \\ G_{p\zeta} & G_{p\xi} \\ G_{r\zeta} & G_{r\xi} \\ G_{\phi\zeta} & G_{\phi\xi} \end{bmatrix}$$

In addition to those indicated by the transfer functions contained in Eq. 9.5,2, there are two other responses of interest, i.e., yaw angle ψ and lateral displacement y'. These are found from Eq. 4.16,7*g*, and the Laplace transform of the nondimensional form of Eq. 4.14,6*b* as

$$\bar{\psi} = \frac{1}{s}\bar{r}$$

and (9.5,3)

$$\bar{y} = \frac{1}{s}(\bar{\psi} + \bar{\beta})$$

It follows that the corresponding transfer functions for rudder input are

$$G_{\psi\zeta} = \frac{\bar{\psi}}{\bar{\zeta}} = \frac{1}{s}G_{r\zeta} \qquad (a)$$

(9.5,4)

$$G_{y\zeta} = \frac{\bar{y}}{\bar{\zeta}} = \frac{1}{s}(G_{\psi\zeta} + G_{\beta\zeta}) \qquad (b)$$

with two similar relations for the aileron transfer functions.

Numerical Example

The frequency-response functions for the jet transport in horizontal flight at 30,000 ft altitude and $C_{L_0} = 0.25$ were calculated from the above equations by setting $s = i\hat{\omega}$. All the aerodynamic transfer functions were replaced by the corresponding derivatives, i.e., $\bar{G}_{y\beta} = C_{y_\beta}$, $G_{l\xi} = C_{l_\xi}$ etc. Thus we have neglected terms such as $C_{l_{\dot{\xi}}}$. The numerical values are the same ones used in the previous examples. The results for some of the state variables are shown in Figs. 9.11 and 9.12.[1] Figure 9.11 shows the responses in β, ϕ, and r to rudder input. The principal feature is the peak at the frequency of the Dutch roll, which because of the relatively light damping of this mode, is substantial. For example, a 1° rudder amplitude at the Dutch roll frequency produces about $4\frac{1}{2}°$ β amplitude and $6\frac{1}{2}°$ roll amplitude. At zero frequency β, p, and r are finite, but ϕ and ψ are infinite. That is, the computed steady state associated with rudder input is a constant rotational motion $\omega_{ss} = ip_{ss} + kr_{ss}$. Since the equations were linearized with respect to ϕ and are therefore not valid for large ϕ, this steady state is spurious. The slopes of the high-frequency asymptotes can be predicted from the structure of the general transfer function matrix. For the given rudder input it yields slopes of -2 for β, ϕ, and -1 for r. These slopes are reached approximately by $k = 0.1$ for r and β, but not for ϕ. This is because the coefficient of a cubic term in the numerator of $G_{\beta\zeta}$ contains the small aerodynamic derivative C_{y_ζ}.

Figure 9.12 shows similar results for aileron angle input. The absence of the control term C_{y_ξ} makes the high-frequency asymptote of $|G_{\beta\xi}|$ a line of slope -2 instead of -1.

All the amplitude curves on both figures show a rapid reduction of response once the frequency exceeds that of the lateral oscillation mode.

The sharp dip in $|G_{r\zeta}|$ at $k \doteq 0.0025$ is characteristic of a zero in the transfer function lying close to the imaginary axis at this frequency.

[1] Note that the abscissa on the graphs is $\omega\bar{c}/2u_0$, not $\omega b/2u_0$.

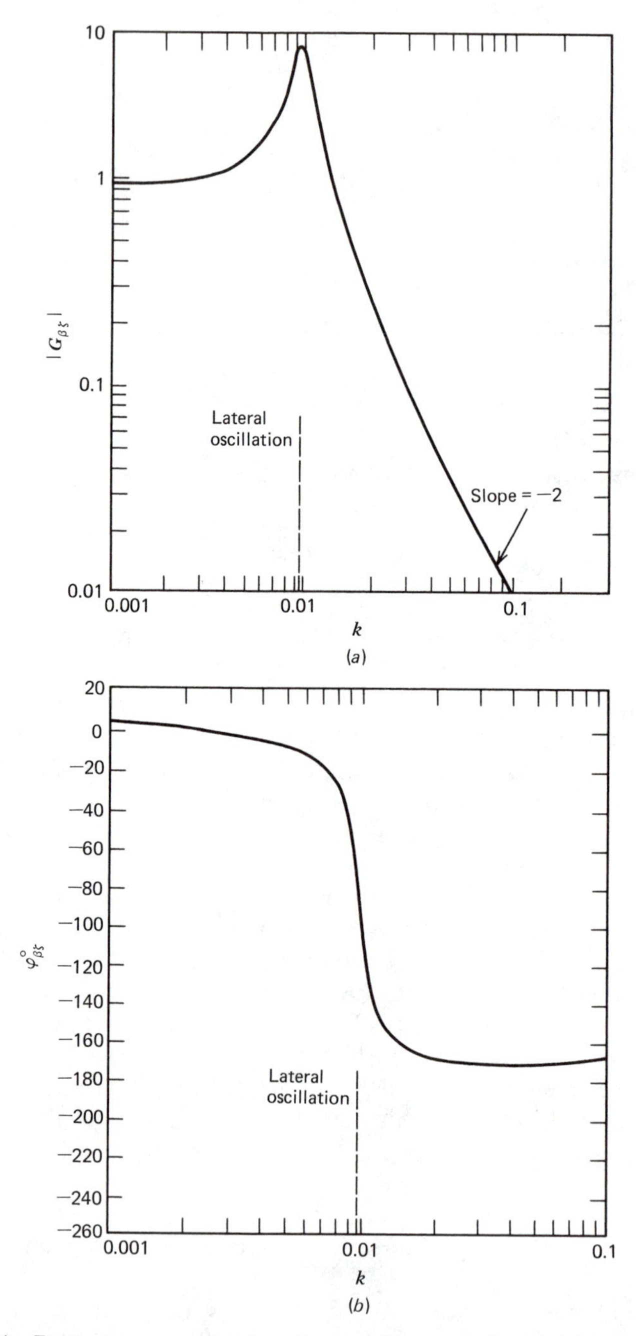

FIG. 9.11 Frequency-response functions, rudder angle input. Jet transport cruising at high altitude. (*a*) Sideslip amplitude. (*b*) Sideslip phase. (*c*) Roll amplitude. (*d*) Roll phase. (*e*) Yaw-rate amplitude. (*f*) Yaw-rate phase.

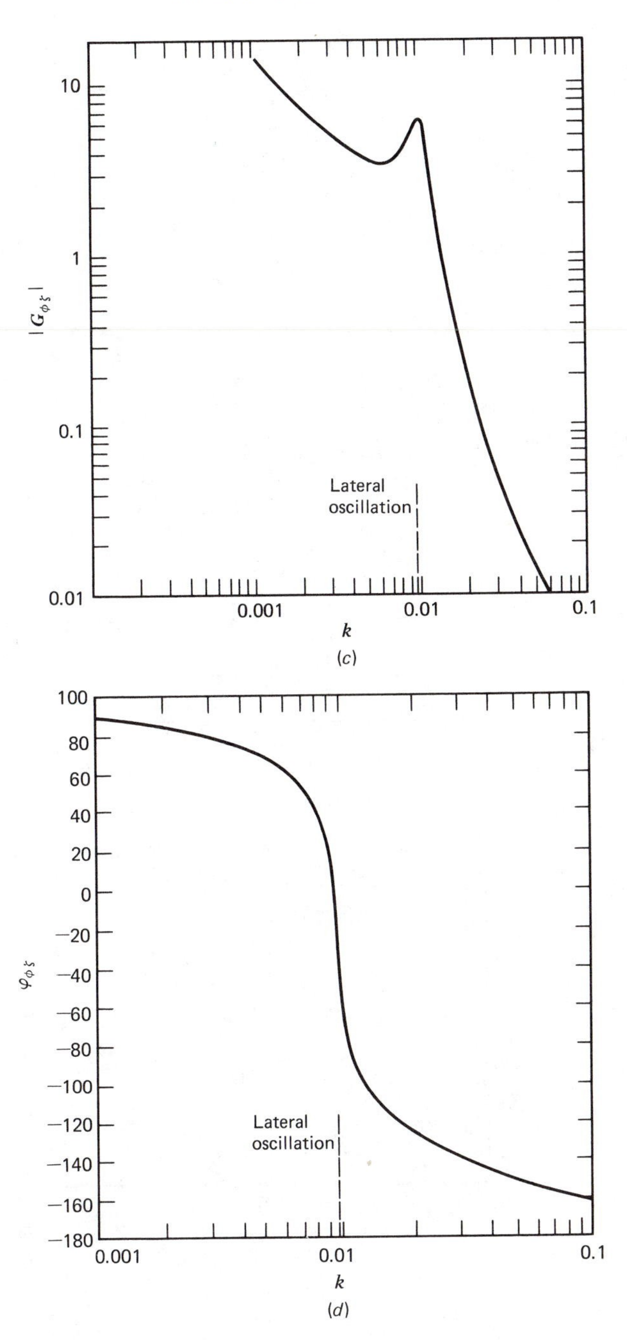

FIG. 9.11 (*continued*)

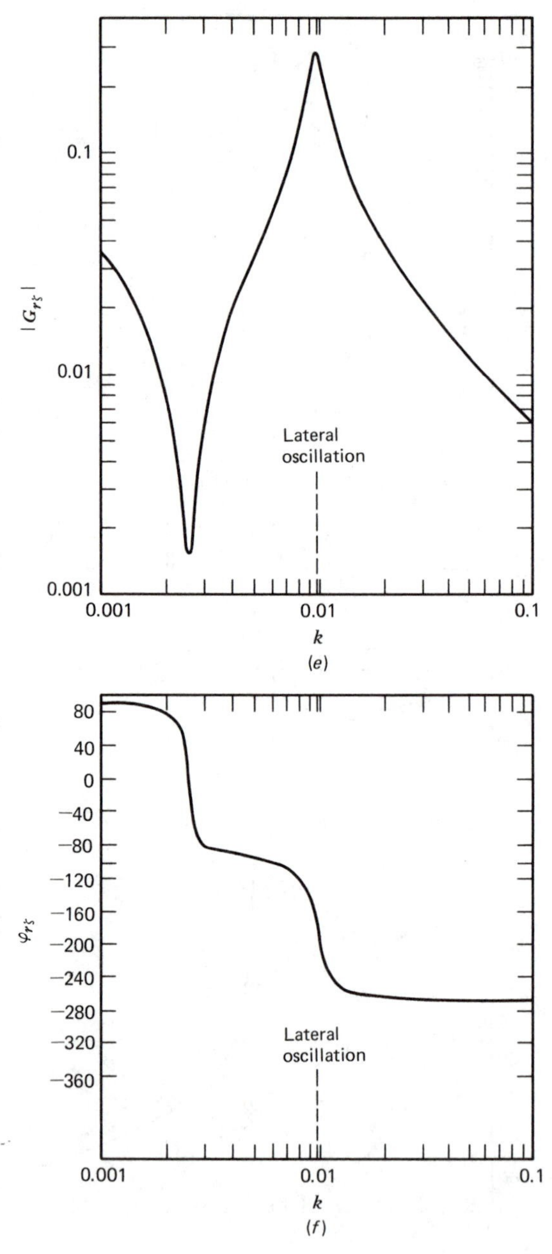

FIG. 9.11 (*continued*)

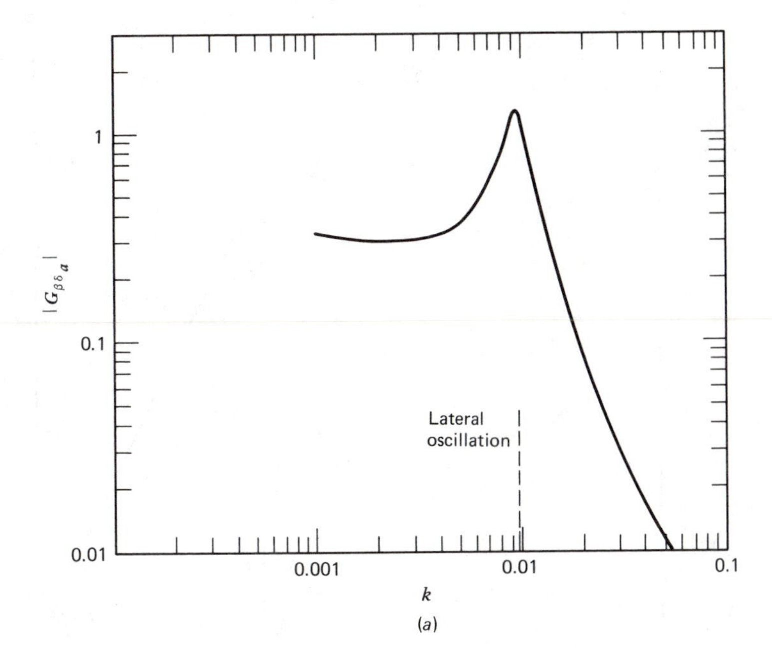

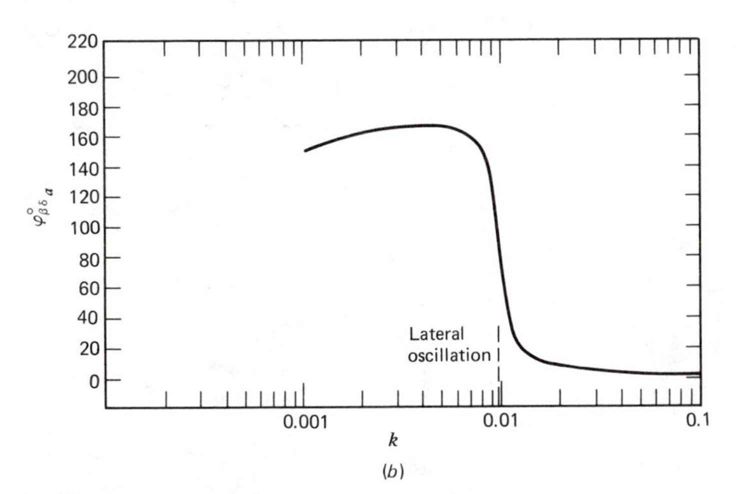

FIG. 9.12 Frequency-response functions, aileron angle input. Jet transport cruising at high altitude. (*a*) Sideslip amplitude. (*b*) Sideslip phase. (*c*) Roll amplitude. (*d*) Roll phase. (*e*) Yaw-rate amplitude. (*f*) Yaw-rate phase.

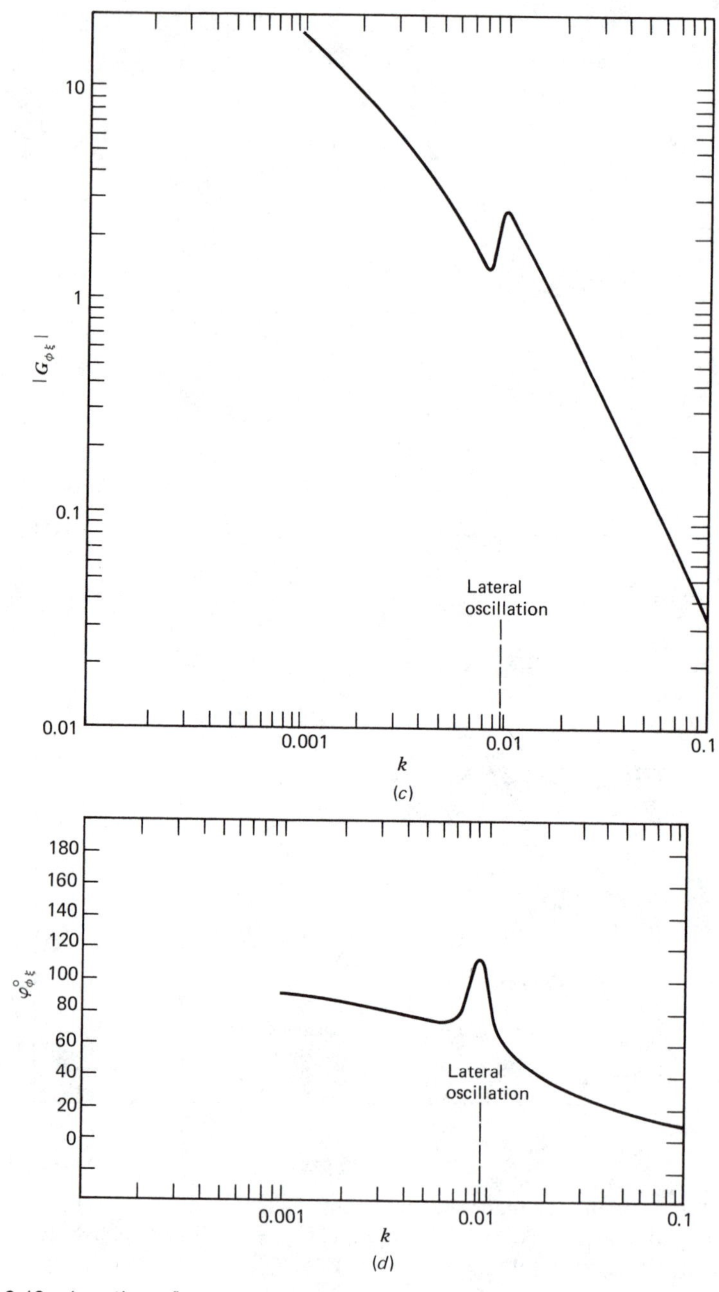

FIG. 9.12 (*continued*)

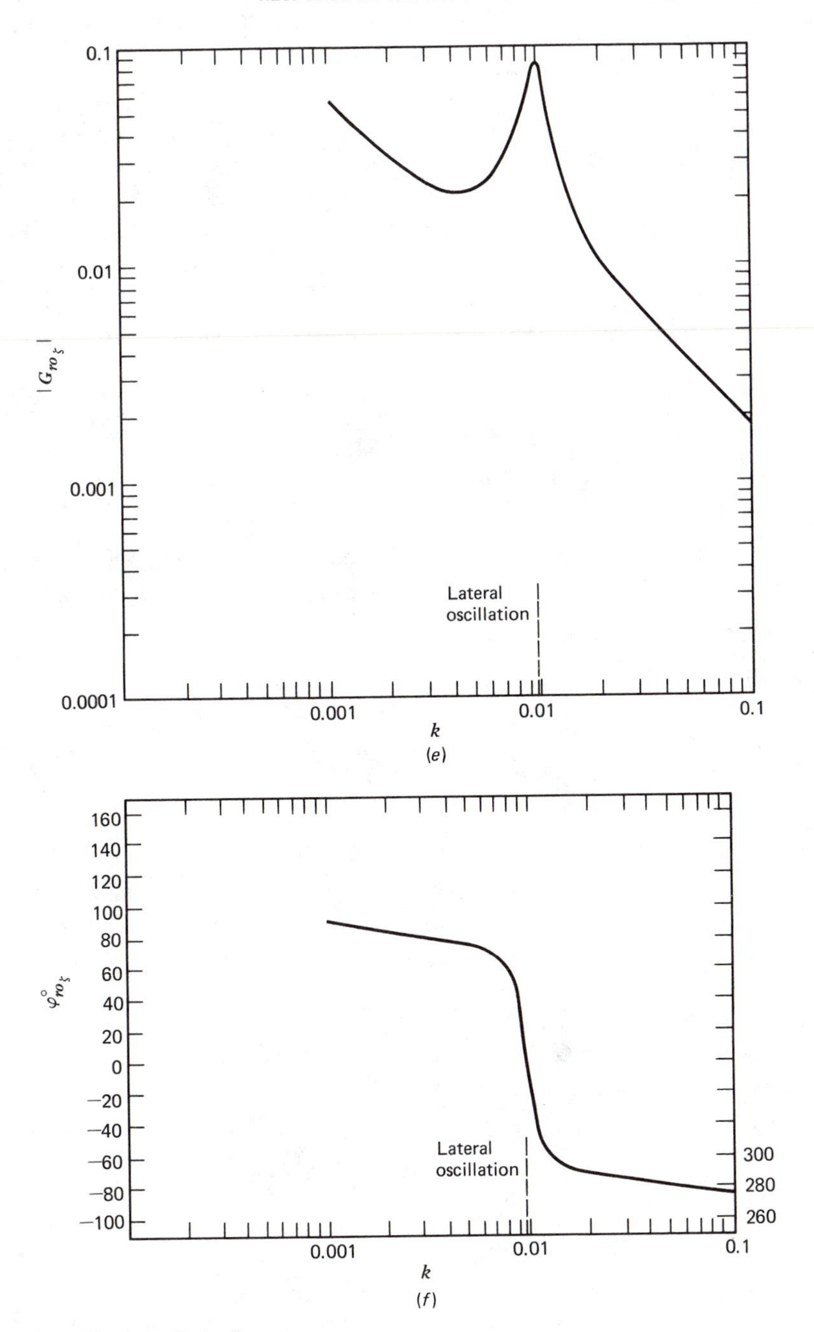

FIG. 9.12 (*continued*)

9.6 TRANSIENT RESPONSE TO AILERON AND RUDDER

We have seen that useful lateral steady states are produced only by certain definite combinations of the control deflections. It is evident then that our interest in the response to a single lateral control should be focused primarily on the initial behavior. The equations of motion provide some insight on this question directly. Following a step input of one of the two controls the state variables at $t = 0^+$ are all still zero, and from Eq. 4.15,8 we can deduce that their initial rates of change, neglecting $C_{l_{\dot{\xi}}}$ and $C_{n_{\dot{\xi}}}$, are related to the control angles by

$$\begin{aligned} 2\mu D\beta &= C_{y_\zeta}\zeta && (a) \\ i_A D\hat{p} - i_E D\hat{r} &= C_{l_\xi}\xi + C_{l_\zeta}\zeta && (b) \quad (9.6,1) \\ -i_E D\hat{p} + i_c D\hat{r} &= C_{n_\xi}\xi + C_{n_\zeta}\zeta && (c) \end{aligned}$$

The initial sideslip rate $\dot{\beta}$ is thus seen to be governed solely by the rudder and is easily seen from Eq. 9.6,1a to be positive (slip to the right) when ζ is positive (left rudder). Of somewhat more interest is the rotation generated. This can be found by solving Eqs. 9.6,1b, c for $\hat{p}$ and $\hat{r}$, from which the initial angular acceleration is the vector

$$D\hat{\omega} = \mathbf{i}D\hat{p} + \mathbf{k}D\hat{r} \tag{9.6,2}$$

The direction of this vector is the initial axis of rotation, and this is of interest. It lies in the xz plane, the plane of symmetry of the airplane, as illustrated in Fig. 9.13a. The angle δ it makes with the x axis is, of course,

$$\delta = \tan^{-1}\frac{\hat{r}}{\hat{p}} \tag{9.6,3}$$

Let us consider the case of "pure" controls, i.e., those with no aerodynamic cross-coupling, so that $C_{l_\zeta} = C_{n_\xi} = 0$. The ailerons then produce pure rolling moment and the rudder produces pure yawing moment. In that case Eq. 9.6,1 yields

$$\tan\delta = \frac{\hat{r}}{\hat{p}} = \frac{i_A C_{n_\zeta}\zeta + i_E C_{l_\xi}\xi}{i_C C_{l_\xi}\xi + i_E C_{n_\zeta}\zeta} \qquad (a) \quad (9.6,4)$$

and for zero aileron angle (response to rudder)

$$\delta_R = \tan^{-1}\frac{i_A}{i_E} \qquad (b)$$

For zero rudder angle (response to aileron)

$$\delta_A = \tan^{-1}\frac{i_E}{i_C} \qquad (c)$$

The angles δ_A, δ_R are seen to depend very much on the product of inertia i_E. When it is zero, the result is as intuitively expected, the rotation that develops is about either the x axis (aileron deflected) or the z axis (rudder deflected). For a vehicle such as the jet transport of previous examples, with $A' = 0.4C'$, the values of A, C, E given by Eq. 4.5,7 yield the results shown in Fig. 9.14. The relations are also

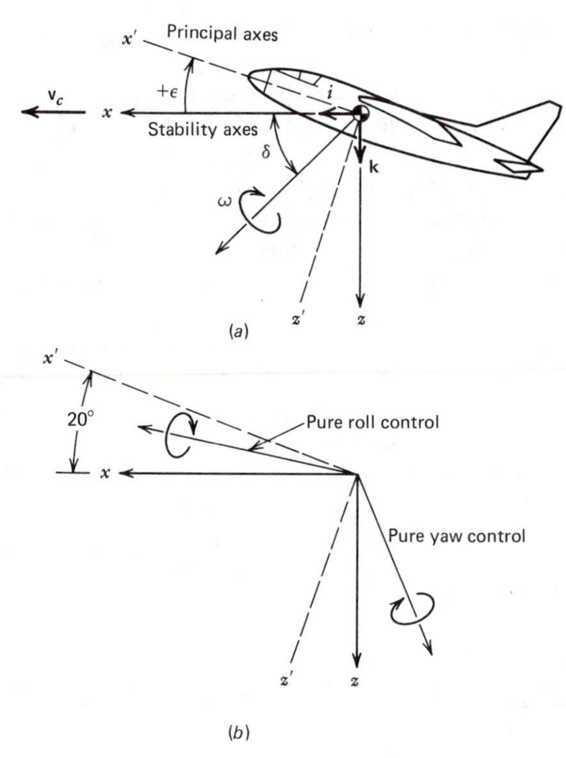

FIG. 9.13 Initial response to lateral control. (*a*) General. (*b*) Example jet transport.

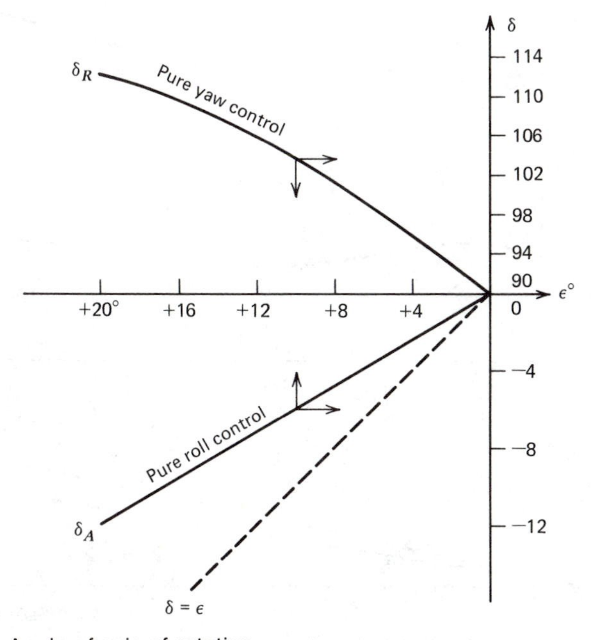

FIG. 9.14 Angle of axis of rotation.

shown to scale in Fig. 9.13b for $\varepsilon = 20°$ (high angle of attack). It can be seen that there is a tendency for the vehicle to rotate about the principal x axis, rather than about the axis of the aerodynamic moment. This is simply because A/C is appreciably less than unity. Now the jet transport of our example is by no means "slender," in that it is of large span and has wing-mounted engines. For an SST or a slender missile, the trend shown is much accentuated, until in the limit as aspect ratio $\rightarrow 0$, both tan δ_R and tan δ_A tend to -tan ε, and the vehicle rotates initially about the x' axis no matter what control is used!

The above analysis tells us how the motion starts but not how it continues. For that we need solutions of Eqs. 4.15,8. Solutions for the example jet transport at $C_{L_0} = 0.25$ at 30,000 ft altitude were obtained by analog simulation of these equations, and the results for β, $\hat{p}$, ϕ, and ψ are shown in Figs. 9.15 and 9.16. Figure 9.15 shows the response to negative aileron angle (corresponding to entry into a right turn). The main feature is the rapid acquisition of roll rate, and its integration

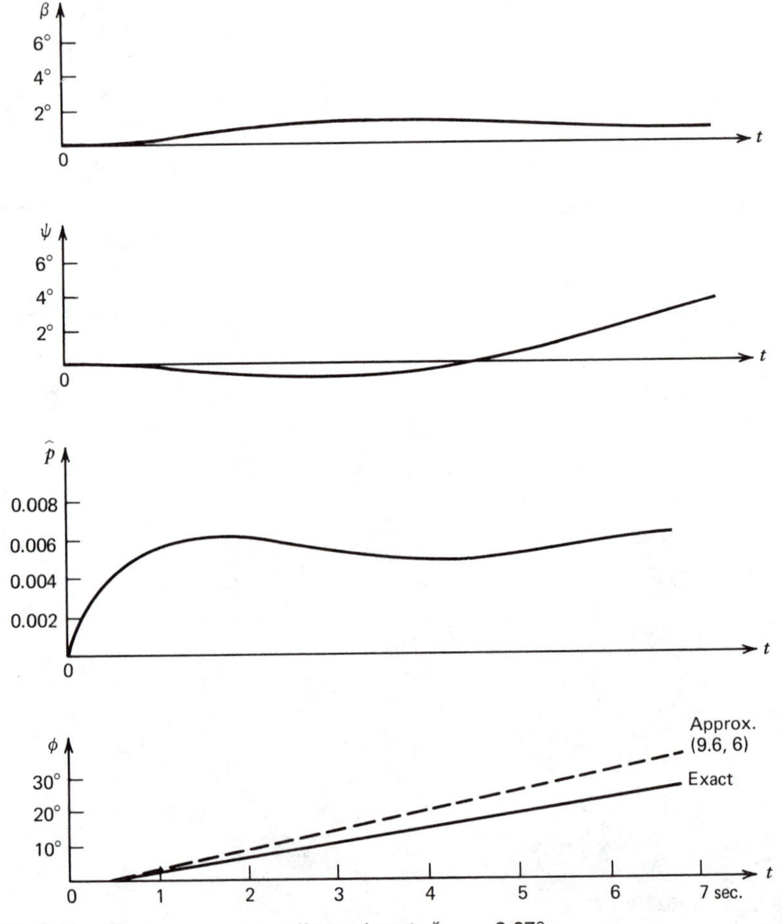

FIG. 9.15 Step response to aileron input. $\xi = -2.87°$.

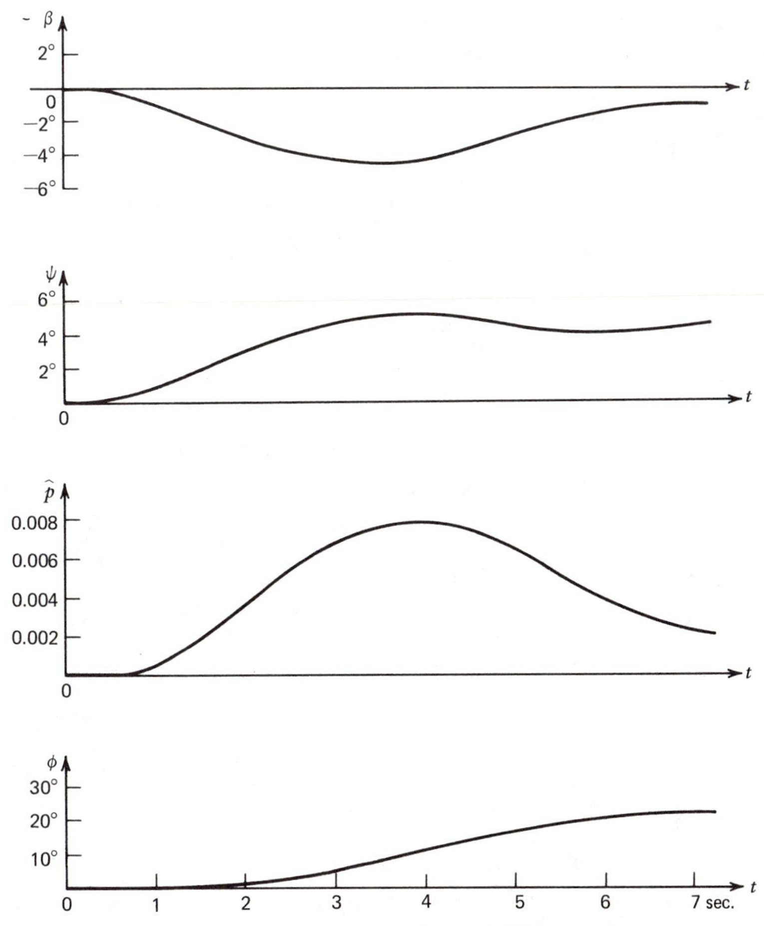

FIG. 9.16 Step response to rudder input. $\zeta = -2.87°$.

to produce bank angle ϕ. The maximum roll rate is achieved in about $1\frac{1}{2}$ sec, and a bank angle of about 25° at the end of 6 sec. Because of the aileron adverse yaw derivative $C_{n_\xi} > 0$, the initial yawing moment is negative, causing the nose to swing to the left, with consequent negative ψ and positive β. The positive β, via the dihedral effect $C_{l_\beta} < 0$ produces a negative increment in C_l, opposing the rolling motion. More than 4 sec elapse before the nose swings into the desired right turn.

Figure 9.16 shows the response to a negative (right) rudder angle of the same magnitude as the aileron angle on Fig. 9.15. This causes the nose to swing rapidly to the right, β being initially roughly equal and opposite to ψ indicating virtually no change in the direction of the velocity vector. The result of $\beta < 0$ (because of $C_{l_\beta}\beta$) is a positive rolling moment and buildup of positive ϕ.

Right rudder, like right aileron, is seen to produce a transition into a turn to the right, but neither does so optimally. A correct transition into a truly banked turn requires the coordinated use of both controls, and if there is to be no loss of altitude (see Sec. 9.4) of the elevator as well.

An approximation to the ϕ response to ξ can be obtained from a single-degree-of-freedom roll analysis.

From Eq. 4.15,8*b*, by assuming that only $\hat{p}$ and ξ differ from zero, and neglecting $C_{l_{\dot{\xi}}}$, we get

$$(i_A D - C_{l_p})\hat{p} = C_{l_\xi}\xi$$

or

$$i_A D^2\phi - C_{l_p}D\phi = C_{l_\xi}\xi \tag{9.6,5}$$

For zero initial conditions, the solution to Eq. 9.6,5 is

$$\frac{\phi}{\xi} = -\frac{C_{l_\xi}}{C_{l_p}}\left\{\hat{t} + \frac{i_A}{C_{l_p}}\left[1 - \exp\frac{C_{l_p}\hat{t}}{i_A}\right]\right\} \tag{9.6,6}$$

This result is compared with the exact solution on Fig. 9.15 and is seen to give a good approximation to ϕ over the most important first few seconds. This simple analysis supplies a useful criterion for roll control. It yields as the steady-state roll rate

$$\hat{p}_{ss} = -\frac{C_{l_\xi}}{C_{l_p}}\xi \tag{9.6,7}$$

A specification on $\hat{p}_{ss}$ then leads to an aileron design to provide the necessary $C_{l_\xi}\xi$.

9.7 ADDITIONAL SYMBOLS INTRODUCED IN CHAPTER 9

- k reduced frequency, $\omega\bar{c}/2u_0$
- $G_{\mu\nu}(s)$ transfer function, $\mu(t)$ output, $\nu(t)$ input
- $A_{\mu\nu}(t)$ indicial admittance, $\mu(t)$ output, $\nu(t) = 1(t)$ input
- $\hat{\omega}$ ωt^*
- Ω rate of turn in a steady turn

See also Secs. 2.11, 3.11, 4.18, 5.17, 6.13, and 8.12.

CLOSED-LOOP CONTROL

CHAPTER 10

10.1 GENERAL PRINCIPLES

Although open-loop responses of the kind studied in some depth in Chap. 9 are very revealing in bringing out inherent vehicle dynamics, they do not in themselves usually represent real operating conditions. Every phase of the flight of an airplane can be regarded as the accomplishment of a set task—i.e., flight on a specified trajectory. That trajectory may simply be a straight horizontal line traversed at constant speed, or it may be a turn, a transition from one symmetric flight path to another, a landing flare, following an ILS or navigation radio beacon, homing on a moving target, etc. All of these situations are characterized by a common feature, namely, the presence of a *desired state*, steady or transient, and of departures from it that are designated as errors. These errors are of course a consequence of the unsteady nature of the real environment and of the imperfect nature of the physical system comprising the vehicle, its instruments, its controls, and its guidance system (whether human or automatic). The correction of errors implies a knowledge of them, i.e., of error-measuring (or state-measuring) devices, and the consequent actuation of the controls in such a manner as to reduce them. This is the case whether control is by human or by automatic pilot. In the former case—the human pilot—the state information sensed is a complicated blend of visual and motion cues, and instrument readings. The logic by which this information is converted into control action is only imperfectly understood, but our knowledge of the physiological "mechanism" that intervenes between logical output and control actuation is somewhat better. In the latter case—the automatic control—the sensed information, the control logic, and the dynamics of the control components are usually well known, so that system performance is in principle quite predictable. The process of using state information to govern the control inputs is known as *closing the loop*, and the resulting system as a *closed-loop control* or *feedback control*. The terms *regulator* and *servomechanism* describe particular applications of the feedback principle. Figure 8.14 shows a general block diagram describing the feedback situation. In the present context we regard $\mathbf{y}$ as the state vector, $H(s)$ as an operator (linear in the figure, but of course not necessarily so) and $\boldsymbol{\varepsilon}$ as the control vector. Clearly, since real flight situations virtually always entail closed-loop control, a study of the consequences of closing the loop is in order.

Another factor that cannot be separated from those referred to above is the force amplification or power amplification common in the control systems of large aircraft. The control forces needed on large high-speed aircraft may exceed the capabilities of human pilots. Thus another dynamic system—powered controls—intervenes between the pilot and the aerodynamic surfaces. Such subsystems are themselves commonly servomechanisms—closed-loop systems that drive the surfaces in response to pilot commands. Thus we are frequently concerned with "loops within loops," a very common situation. For example, the "outermost" loop might be a guidance loop that controls the error in vehicle position relative to an ILS beam. An inner loop might be a *stability augmentation system* (treated later in Sec. 10.3) whose purpose is to improve the inherent lateral dynamics of the vehicle and, finally, within this one there may be still another loop associated with the control-surface servo.

Although flight dynamicists (who usually come from an aerospace engineering background) and control engineers (who frequently have a background in electrical engineering) usually communicate adequately on problems of mutual concern, there is often understandably some difference in their points of view. This is illustrated somewhat facetiously in Fig. 10.1. At one extreme, the control engineer

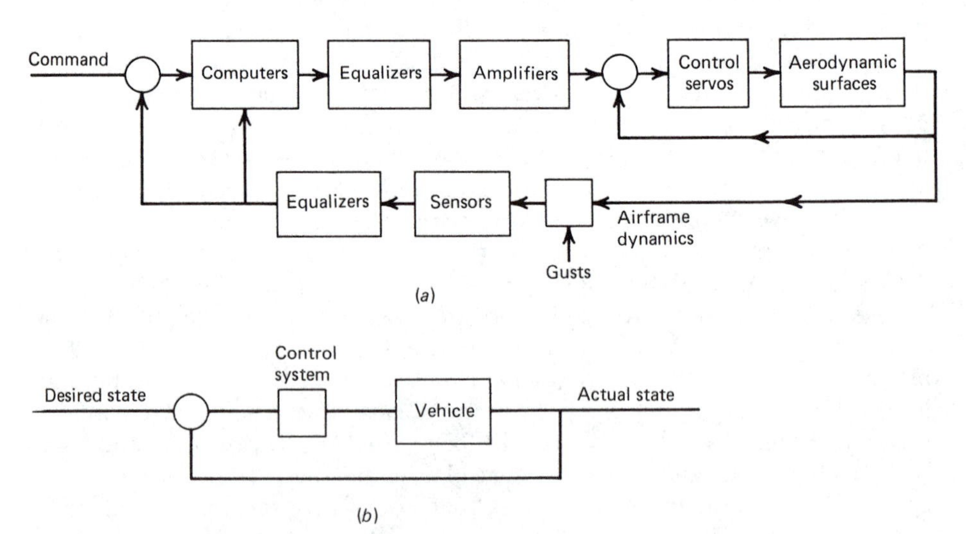

FIG. 10.1 Closed-loop control—two extreme views. (*a*) The control engineer's viewpoint. (*b*) The flight dynamicist's viewpoint.

may overemphasize the many elements that comprise the control system, and tend to minimize the role of the dynamics of the vehicle itself—perhaps replacing all its rich and varied detail with oversimplified approximate transfer functions. At the other extreme, the flight dynamicist may substitute some simple algebraic relations for the entire control system. Neither extreme is right for the final solution of real problems, but both may have their merits for certain purposes. We naturally tend here to the flight dynamicist's view of the system in the illustrations that follows. For example, it is sometimes very helpful to consider the loop closure as simply modifying some of the existing aerodynamic derivatives, or adding new ones. Specifically let y be any nondimensional state variable, and let a control surface be displaced in response to this variable according to the law

$$\Delta\delta = k\,\Delta y; \qquad k = \text{const}$$

(Here k is a simplified representation of all the sensor and control system dynamics!) Then a typical aerodynamic force or moment coefficient C_a will be incremented by

$$\begin{aligned} \Delta C_a &= C_{a_\delta}\Delta\delta \\ &= C_{a_\delta}k\,\Delta y \end{aligned} \tag{10.1,1}$$

This is the same as adding a synthetic increment

$$\Delta C_{a_y} = kC_{a_\delta} \tag{10.1,2}$$

to the aerodynamic derivative C_{a_y}. Thus if y be yaw rate and δ be rudder angle, then the synthetic increment in the yaw-damping derivative is

$$\Delta C_{n_r} = kC_{n_{\delta r}} \tag{10.1,3}$$

which might be the kind of change required to correct a lateral dynamics problem. This example is in fact the basis of the often-applied "yaw damper," a stability-augmentation feature. Again, if y be the roll angle and δ the aileron, we get the *entirely new* derivative

$$C_{l_\phi} = kC_{l_{\delta_a}} \tag{10.1,4}$$

the presence of which can profoundly change the lateral characteristics.

Sensors

We have already alluded to the general nature of feedback control, and the need to provide *sensors* that ascertain the *state* of the vehicle. When human pilots are in control, their eyes and kinesthetic senses, aided by the standard flight information displayed by their instruments, provide this information. (In addition, of course, their brains supply the logical and computational operations needed, and their neuro-muscular systems all or part of the actuation.) In the absence of human control, when the vehicle is under the command of an autopilot, the sensors must, of course, be physical devices. As already mentioned, some of the state information needed is measured by the standard flight instruments—air speed, altitude, rate of climb, heading, etc. This information may or may not be of a quality and in a form suitable for incorporation into an automatic control system. In any event it is not generally enough. When both guidance and attitude-stabilization needs are considered, the state information needed may include:

Position and velocity vectors relative to a suitable reference frame.

Vehicle attitude (θ, ϕ).

Rotation rates (p, q, r).

Aerodynamic angles (α, β).

Acceleration components of a reference point in the vehicle.

The above is not an exhaustive list. A wide variety of devices are in use to measure these variables, from Pitot-static tubes to sophisticated inertial-guidance platforms. Gyroscopes, accelerometers, magnetic and gyro compasses, angle-of-attack and sideslip vanes, and other devices all find applications as sensors. The most common form of sensor output is an electrical signal, but fluidic devices have also been used. Although in the following examples we tend to assume that the desired variable can be measured independently, linearly, and without time lag, this is of course an idealization that is only approached but never reached in practice. Every sensing device, together with its associated transducer and amplifier, is itself a dynamic system with characteristic frequency response, noise, nonlinearity, and cross-coupling. These attributes cannot finally be ignored in the design of real systems, although one can usefully do so in preliminary work. As an example of cross-coupling effects, consider the sideslip sensor assumed to be available in the stability augmentation system of Sec. 10.3. Assume, as might well be the case, that it consists of a sideslip vane mounted on a boom projecting forward from the nose. Such a device would in general respond not only to $\beta = \sin^{-1} (v/v_c)$ but also to atmospheric turbulence (side gusts), to roll and yaw rates, and to lateral acceleration a_y at the vane hinge. Thus the output signal would in fact be a complicated mathematical function of several state variables, representing several feedback loops, rather than being simply proportional to β as assumed in the example. The objective in sensor design is, of course, to minimize all the unwanted extraneous effects, and to provide sufficiently high frequency response and low noise in the sensing system.

This brief discussion serves only to draw attention to the important design and analytical problems related to sensors, and to point out that their real characteristics, as opposed to their idealizations, need finally to be taken into account in design.

10.2 EXAMPLE—SUPPRESSION OF THE PHUGOID

The characteristic lightly damped, low-frequency oscillation in speed, pitch attitude, and altitude that was identified in Chapter 6, was seen in Chapter 9 to lead to large peaks in the frequency-response curves (Fig. 9.2) and long transients

(Figs. 9.4 and 9.6). Similarly, in the control-fixed case, there are large undamped responses in this mode to disturbances such as atmospheric turbulence. These variations in speed, height, and attitude are in fact not in evidence in actual flight; the pilot (human or automatic) effectively suppresses them, maintaining flight at more or less constant speed and height. The logic by which this process of suppression takes place is not unique. In principle it can be achieved by using feedback signals derived from any one or a combination of pitch attitude θ, altitude h, speed v_c, and their derivatives. In practice, the availability and accuracy of the state information determines what feedback is used. We shall see that a simple negative feedback of pitch attitude suffices effectively to eliminate the phugoid. Pitch attitude is instantly and accurately available from either the real or artificial horizon. We shall also see that operating on speed error can produce pitch maneuvers free of phugoid oscillations.

Consider the system shown in Fig. 10.2, in which θ_C is the pitch command, $G_p(s)$ is the overall transfer function of the control system, and g is a disturbance (gust) input. The pitch attitude is given by

$$\bar{\theta} = G_{\theta g}\bar{g} + G_{\theta\eta}\bar{\eta} \tag{10.2,1}$$

and we readily find the overall transfer functions

$$\begin{aligned} \frac{\bar{\theta}}{\bar{\theta}_c} &= \frac{G_p G_{\theta\eta}}{1 + G_p G_{\theta\eta}} \\ \frac{\bar{\theta}}{\bar{g}} &= \frac{G_{\theta g}}{1 + G_p G_{\theta\eta}} \end{aligned} \tag{10.2,2}$$

The stability with respect to θ_c or g inputs is given by the roots of the characteristic equations of these two overall transfer functions. As long as η and g are both inhomogeneous inputs to the linear aircraft system, it can be seen that the denominators of $G_{\theta\eta}$ and $G_{\theta g}$ are the same, each being the characteristic polynomial of the system. Thus we may write

$$G_{\theta\eta} = \frac{N_1}{D} \qquad G_{\theta g} = \frac{N_2}{D} \tag{10.2,3}$$

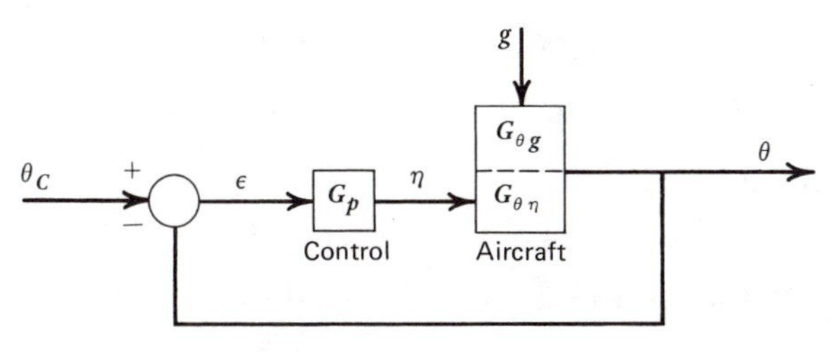

FIG. 10.2 Phugoid suppression system.

where N_1, N_2, D are polynomials in s, and the overall transfer functions are

$$\frac{\bar{\theta}}{\bar{\theta}_c} = \frac{G_p N_1}{D + G_p N_1}$$
$$\frac{\bar{\theta}}{\bar{g}} = \frac{N_2}{D + G_p N_1} \tag{10.2,4}$$

The poles of these transfer functions, which are the roots of the characteristic equations, will be the same if $G_p N_1$ and N_2 have no poles (or the same poles), and in that case the stability with respect to gust inputs will be the same as that for pitch command inputs. A reasonably general form for $G_p(s)$ for this application is

$$G_p(s) = \frac{k_1}{s} + k_2 + k_3 s$$

For obvious reasons, the three terms on the r.h.s. are called, respectively, *integral control*, *proportional control* and *rate control*, because of the way they operate on the error ε. The particular form of the controlled system, here $G_{\theta\eta}(s)$, determines which of k_1, k_2, k_3 need to be nonzero, and what their magnitudes should be for good performance. Integral control has the characteristic of a *memory*, and steady-state errors cannot persist when it is present. Rate control has the characteristic of *anticipating* the future values of the error and thus generates *lead* in the control actuation. It turns out that all we need here is proportional control, so we choose $G_p(s) = K$, a constant, and the characteristic equation is

$$D(s) + KN_1(s) = 0 \tag{10.2,5}$$

To proceed further, we need explicit expressions for N_1 and D. To obtain these, we turn to Eq. 4.16,6 and reduce them to correspond to the phugoid approximation of Sec. 6.7. In this approximation, the pitching moment equation is assumed to be satisfied by maintaining the quasistatic relation between α and η, i.e.,

$$G_{m\alpha}\bar{\alpha} + G_{m\eta}\bar{\eta} = 0 \tag{10.2,6}$$

This makes α a simple linear function of η. The remaining equations needed are Eqs. 4.16,6*a*, *b*. We make some further assumptions, namely,

$$\begin{aligned} \theta_0 = 0, \qquad G_{zq} = G_{z\eta} = G_{zu} = 0, \qquad G_{xu} = -2C_{D_0}, \\ G_{z\alpha} = -C_{L_\alpha}, \qquad G_{x\alpha} = C_{L_0} - C_{D_\alpha}, \\ G_{m\alpha} = C_{m_\alpha}, \qquad G_{m\eta} = C_{m_\delta} \end{aligned} \tag{10.2,7}$$

With these, we can rewrite the two equations as follows:

$$\begin{bmatrix} (2\mu s + 2C_{D_0}) & C_{L_0} \\ 2C_{L_0} & -2\mu s \end{bmatrix} \begin{bmatrix} \bar{u} \\ \bar{\theta} \end{bmatrix} = \begin{bmatrix} (C_{L_0} - C_{D_\alpha}) \\ -(2\mu s + C_{L_\alpha}) \end{bmatrix} \frac{-C_{m_\delta}}{C_{m_\alpha}} \bar{\eta} \tag{10.2,8}$$

The required transfer function is readily found from Eq. 10.2,8 to be

$$G_{\theta\eta} = \frac{N_1(s)}{D(s)} = \frac{n_2 s^2 + n_1 s + n_0}{c_2 s^2 + c_1 s + c_0} \tag{10.2,9}$$

In the expression for n_1 we further neglect C_{D_0} compared to C_{L_α}, with the result:

$$
\begin{aligned}
n_2 &= -4\mu^2 \frac{C_{m_\delta}}{C_{m_\alpha}} \\
n_1 &= -2\mu C_{L_\alpha} \frac{C_{m_\delta}}{C_{m_\alpha}} \qquad (10.2,10) \\
n_0 &= -2[C_{L_\alpha} C_{D_e} + C_{W_e}(C_{W_e} - C_{D_\alpha})] \frac{C_{m_\delta}}{C_{m_\alpha}}
\end{aligned}
$$

$$
\begin{aligned}
c_2 &= 4\mu^2 \\
c_1 &= 4\mu C_{D_e} \qquad (10.2,11) \\
c_0 &= 2C_{L_0}{}^2
\end{aligned}
$$

The characteristic equation is, from Eqs. 10.2,5 and 10.2,9,

$$(c_2 + Kn_2)s^2 + (c_1 + Kn_1)s + (c_0 + Kn_0) = 0 \qquad (10.2,12)$$

and the feedback is seen to affect every term in the equation. We also observe that the numerator of the open-loop transfer function $G_{\theta\eta}$ plays a decisive role in determining the characteristics of the closed-loop system.

The frequency and damping of the system are now obtained from Eq. 10.2,12 as

$$
\begin{aligned}
\omega_n' &= \left(\frac{c_0 + Kn_0}{c_2 + Kn_2}\right)^{1/2} = \omega_n \left(\frac{1 + Kn_0/c_0}{1 + Kn_2/c_2}\right)^{1/2} \qquad (10.2,13) \\
2\zeta' &= \frac{c_1 + Kn_1}{\sqrt{(c_2 + Kn_2)(c_0 + Kn_0)}} = 2\zeta \frac{(1 + Kn_1/c_1)}{\sqrt{(1 + Kn_2/c_2)(1 - Kn_0/c_0)}}
\end{aligned}
$$

where $\omega_n = (c_0/c_2)^{1/2}$ and $2\zeta = c_1/\sqrt{c_2 c_0}$ are the fixed-control phugoid parameters. Using the data for the jet transport cruising at 30,000 ft altitude given in Sec. 6.1, and $C_{m_\delta} = C_{m_\alpha}$ we get the numerical values

$$\frac{n_0}{c_0} = -2.02, \qquad \frac{n_1}{c_1} = -130, \qquad \frac{n_2}{c_2} = -1.0$$

from which

$$
\begin{aligned}
\frac{\omega_n'}{\omega_n} &= \left(\frac{1 - 2.02K}{1 - K}\right)^{1/2} \qquad (10.2,14) \\
\frac{\zeta'}{\zeta} &= \frac{1 - 130K}{\sqrt{(1 - K)(1 - 2.02K)}}
\end{aligned}
$$

Even with small gain K the damping of the phugoid is very much increased. The original value was $\zeta = 0.0535$, so to produce a dead-beat transient for which $\zeta = 1$, we require $\zeta'/\zeta = 18.7$, which is produced by a gain $-K = 0.17$. Note that the gain is negative, since a positive error ε (Fig. 10.2) indicates the nose is too low, and up-elevator ($\delta_e < 0$) is required to correct. With the gain needed for $\zeta = 1.0$, we get $\omega_n'/\omega_n = 1.07$, so the frequency has been increased by only 7%, and the phugoid approximation for $G_{\theta\eta}$ is clearly adequate.

This calculation shows how a human or automatic pilot could eliminate the phugoid oscillations quite simply, using readily available state information. The exact control law by which a human pilot actually achieves this result may in fact be somewhat different from that assumed here, but it is probable that θ is the prime variable on which he operates.

Change of Flight-Path Angle

The phugoid makes its presence known not only in the form of transient perturbations from a steady state, but also in maneuvers, as illustrated in Sec. 9.3. We saw there for example that in changing from level to climbing flight by opening the throttle (Fig. 9.6) there results a protracted, weakly damped approach to the new state that would take some 10 min to complete. Transitions from one value of γ to another are obviously not made in this manner, and the pilot suppresses the oscillation in this case as well. Provided that the correct θ is known for the climb condition, the same technique as discussed above would work, i.e., proportional control operating on pitch-attitude error. We illustrate an alternative concept that does not require any knowledge of the final correct pitch attitude, but that uses speed error alone. Figure 10.3*a* shows the system. In this case it is found that proportional control is not adequate—it serves mainly to shorten the period of the oscillation, but has little effect on the damping. To improve damping needs rate control, so the control law used is

$$G_p(s) = k_1 + k_2 s; \qquad k_1, k_2 > 0 \tag{10.2,15}$$

where the signs of the gains have been chosen to give the required corrections.

Just as in the case of θ feedback above, the characteristic equation can be obtained from the approximate transfer function, in this case $G_{u\eta}$. It is found from Eq. 10.2,8 with the same approximations as used above, i.e.,

$$G_{u\eta} = \frac{N_3(s)}{D(s)} \tag{10.2,16}$$

where $D(s)$ is given by Eqs. 10.2,9 and 10.2,11 and

$$\begin{aligned} N_3(s) &= m_1 s + m_0 \\ m_1 &= 2\mu C_{D_\alpha} C_{m_\delta}/C_{m_\alpha} \\ m_0 &= C_{L_0} C_{L_\alpha} C_{m_\delta}/C_{m_\alpha} \end{aligned} \tag{10.2,17}$$

The characteristic equation (cf. Eq. 10.2,5) is

$$D(s) + (k_1 + k_2 s)N_3(s) = 0$$

which becomes

$$(c_2 + m_1 k_2)s^2 + (c_1 + m_1 k_1 + m_0 k_2)s + (c_0 + k_1 m_0) = 0 \tag{10.2,18}$$

The new characteristic equation is again second order, being the sum of the original one and additional terms. When the signs of the quantities in Eq. 10.2,17 are taken into account, the modifications to the three original coefficients can be

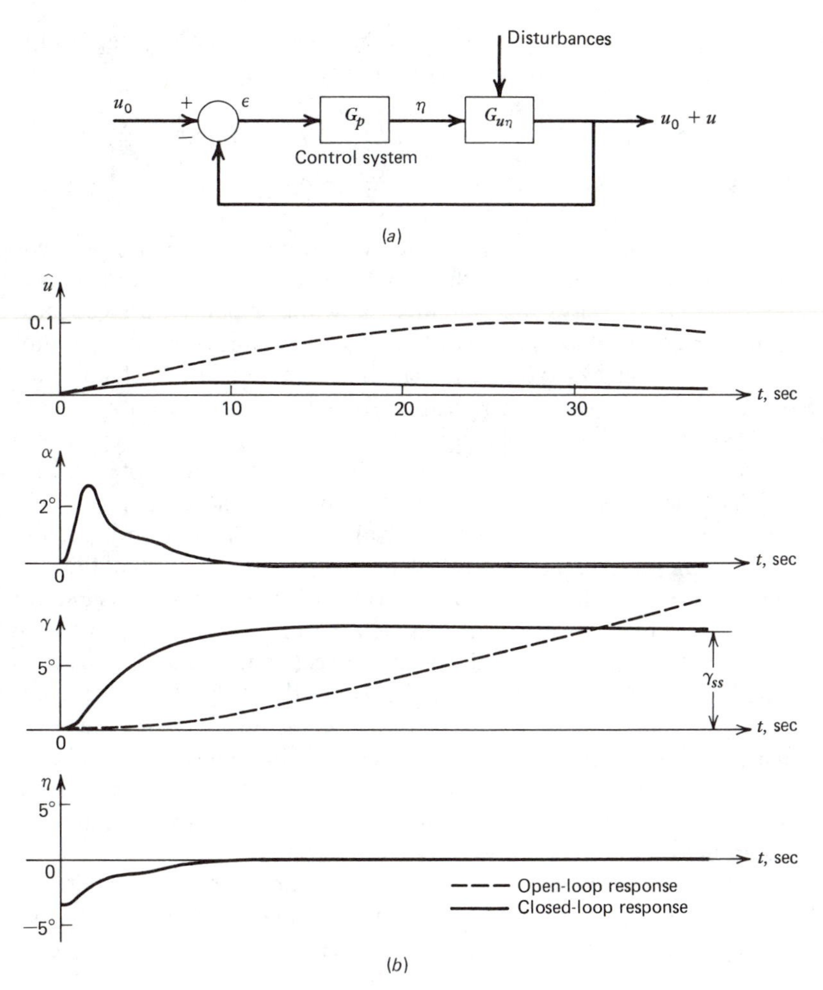

FIG. 10.3 (*a*) System with speed feedback. (*b*) Suppression of phugoid by closed-loop control—response to thrust change.

summarized thus

c_2: increased by amount proportional to k_2

c_1: increased by amounts proportional to k_1 and k_2

c_0: increased by an amount proportional to k_1

Since there are two free constants, k_1 and k_2, we can analytically satisfy two conditions by means of Eq. 10.2,18—one on the period, and one on the damping of the closed-loop system. This procedure is fairly obvious, and is not elaborated on here. The values of the constants finally chosen have to be constrained of course by practical considerations related to sensor and control hardware limitations. Finally, the approximate analysis has to be verified with the complete system of equations. As an example, Fig. 10.3 shows the response to a step input of thrust obtained using analogue computation of the full system of equations. The constants used were

$$k_1 = 0.30 \text{ rad/unit}; \qquad k_2 = 1000 \text{ rad/unit}$$

The first corresponds to an elevator deflection of 0.172° per 1% change in speed, and the second to 25.3° per g of forward acceleration. The airplane and flight condition of the figure are the same as those for Fig. 9.6. The dashed lines show the beginning of the response that would exist without feedback. This would take about 10 min to decay. The solid lines show the response with feedback, and we see that for all practical purposes the transition is completed smoothly and rapidly—within about 15 sec. There is a small overshoot in γ, and small errors in $\hat{u}$ and η that die out rather slowly. This feature could be eliminated at the cost of some additional complexity by introducing some integral control. The elevator angle variation required to accomplish the transition is seen to consist of an initial step (up-elevator) followed by a gradual reduction of the deflection. The conditions near $t = 0$ are, of course, somewhat artificial because of the step input used. A gradual thrust increase would have resulted in a gradual deflection of the elevator. It should be noted that the error in $\hat{u}$, the primary quantity sensed, is indeed kept quite small. The role of α is worth commenting on. At the scale of the figure, there is practically no α change in the open-loop case within the time span shown. The "pulse" in α in the closed-loop case clearly has the effect of producing a corresponding pulse in lift that rotates the velocity vector through the required angle.

Finally, it should be observed that in theory a human pilot has all the state information that we have assumed was available. Speed and its derivative could be obtained from an airspeed indicator, and additional information about $\dot{u}$ can be felt as an inertia force (a "seat-of-the-pants" input). An autopilot could readily have u supplied in electronic form by a conventional transducer, but $\dot{u}$ would be somewhat more troublesome. The two principal alternatives would be differentiation of u, or an acceleration signal from an inertial platform.

10.3 EXAMPLE—STABILITY AUGMENTATION SYSTEM FOR A STOL AIRPLANE

In this example, we consider a situation in which an inherent instability is cured by means of a feedback control. The case is that of a STOL airplane which has an unstable spiral mode at low speed. Its lateral stability characteristics are shown in Figs. 10.4 and 10.5. The spiral mode is seen to have a dangerously short time to double at the highest C_{L_0}, corresponding to a flight condition of powered lift. We postulate that a feedback control for stability augmentation might be useful.

How should we proceed to synthethize such a system? We can choose any of (β, p, r) as variables to sense, and feed back functions of them (cf. Eq. 10.2,1) to produce command signals for the aileron and/or rudder. But which variables shall we choose and what functions of them shall we use? Here the "flight dynamicist's approach" of looking at the feedback control system as a way of modifying the

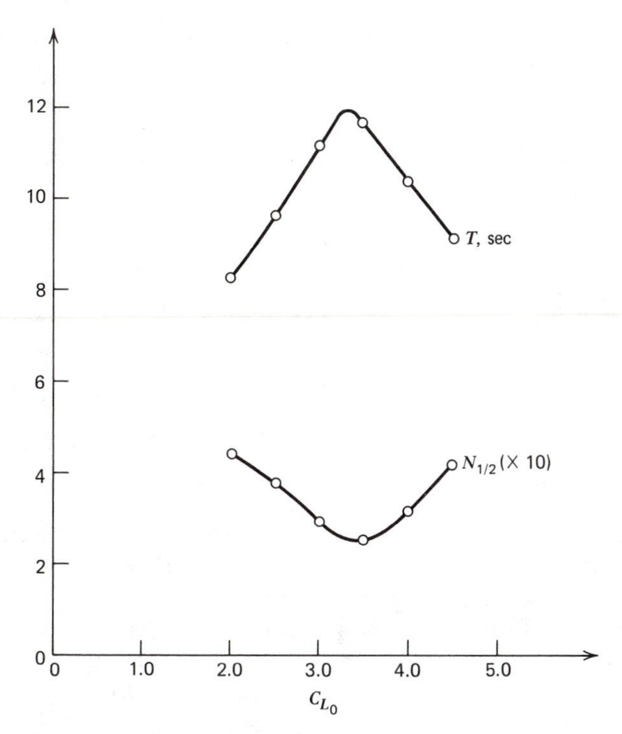

FIG. 10.4 Characteristics of lateral oscillation—STOL airplane.

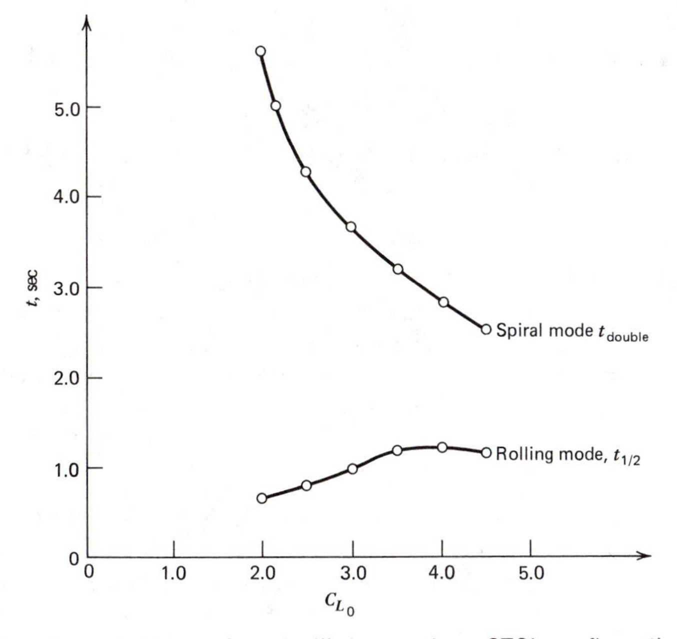

FIG. 10.5 Time constants of nonoscillatory modes—STOL configuration.

aerodynamic derivatives (Sec. 10.1) is helpful. The full set of synthetic changes that can be made in the six lateral moment derivatives is described by the relations

$$\Delta\begin{bmatrix} L_\beta & N_\beta \\ L_p & N_p \\ L_r & N_r \end{bmatrix} = [k_{ij}]^T \begin{bmatrix} L_\xi & N_\xi \\ L_\zeta & N_\zeta \end{bmatrix} \tag{10.3,1}$$

where $[k_{ij}]$ is the 2×3 matrix of feedback gains, i.e.,

$$\begin{bmatrix} \xi \\ \zeta \end{bmatrix} = [k_{ij}] \begin{bmatrix} \beta \\ p \\ r \end{bmatrix} \tag{10.3,2}$$

Thus for example,

$$\xi = k_{11}\beta + k_{12}p + k_{13}r$$

and

$$\Delta N_p = k_{12}N_\xi + k_{22}N_\zeta \tag{10.3,3}$$

Equations 10.3,2 are written in dimensional rather than nondimensional form, since the sensing devices used to generate the feedback signals would ordinarily operate on the dimensional physical variables.

These relations must now be applied with good engineering judgment. Stumbling about blindly in the six-dimensional parameter space of the k_{ij} is not a satisfactory way to find the solution. First, the number of nonzero k_{ij} must be kept to a minimum, since each one entails extra hardware or circuitry, adding to weight, cost, complexity, and failure probability. Second, the engineer must take advantage of his or her understanding of the system and of the fault to be corrected. Here the fault is that the spiral mode is unstable, the other two modes being stable. We know that the criterion for spiral stability in horizontal flight is (Sec. 7.4)

$$(C_{l_\beta}C_{n_r} - C_{l_r}C_{n_\beta}) > 0 \tag{10.3,4}$$

and that it must be the violation of this criterion that is the cause of the instability. For this configuration, the relevant derivatives are, at $C_{L_0} = 4.0$,

$$C_{l_\beta} = 0.010; \qquad C_{n_r} = -0.25; \qquad C_{l_r} = 0.67; \qquad C_{n_\beta} = 0.120$$

which when substituted into Eq. 10.3,4 show the criterion to be violated. It is clear that there is no hope of correcting this situation without changing the sign of one of the four derivatives. In fact the one to which our attention is naturally directed is C_{l_β}, which is here positive, but is ordinarily negative for "well-behaved" airplanes. A "synthetic" C_{l_β} of the required sign can be introduced by aileron feedback of the form

$$\xi = k_{11}\beta, \qquad k_{11} > 0$$

In fact, an attempt at a solution based on this sideslip feedback for $C_{L_0} = 4.0$ was unsuccessful. When k_{11} was made large enough to stabilize the spiral mode, the lateral oscillation was driven unstable. Now it can be concluded from (7.3,5) that C_{n_r} is a powerful factor available to control the damping of the lateral oscillation, and hence an increase in $|C_{n_r}|$ is indicated. This is also beneficial in meeting (Eq.

10.3,4) when combined with a change of sign of C_{l_β}. We therefore choose a second nonzero gain, k_{23}, so that the control deflections are given by

$$\begin{aligned} \xi &= k_{11}\beta \qquad k_{11} > 0 \\ \zeta &= k_{23}r \qquad k_{23} > 0 \end{aligned} \tag{10.3,5}$$

The control derivatives assumed for this example, representative of those that pertain to a deflected slipstream configuration, are

$$\begin{aligned} C_{l_\xi} &= -0.13/\text{rad} \qquad C_{n_\zeta} = -0.30/\text{rad} \\ C_{n_\xi} &= +0.04/\text{rad} \qquad C_{l_\zeta} = +0.04/\text{rad} \end{aligned}$$

With these derivatives, and a control law given by Eq. 10.3,5, values of k_{11} and k_{23} can readily be found that eliminate the instability in the spiral mode while maintaining a stable lateral oscillation. In fact it is only a little more difficult in this case to incorporate a more realistic feedback law than the simple gains of Eq. 10.3,5. Consequently the example has not been computed with Eq. 10.3,5 but rather by assuming that each control actuator is a first-order dynamic system of fast response time. The corresponding control equations used were

$$\begin{bmatrix} \dot{\xi} \\ \dot{\zeta} \end{bmatrix} = -\begin{bmatrix} 10\xi \\ 12\zeta \end{bmatrix} + \begin{bmatrix} K_{11} & 0 \\ 0 & K_{23} \end{bmatrix} \begin{bmatrix} \beta \\ r \end{bmatrix} \tag{10.3,6}$$

which implies that the time constants of the aileron and rudder position servos are, respectively, $\frac{1}{10}$ and $\frac{1}{12}$ sec, that there are zero time lags in the β and r sensors, and that the steady-state gains are

Aileron: $k_{11} = K_{11}/10$ deg/deg
Rudder: $k_{23} = K_{23}/12$ deg/(deg/sec)

Equations 10.3,6 are now incorporated into the basic lateral equations of motion 4.15,8 to yield the final mathematical system. After converting Eq. 10.3,6 to nondimensional form and neglecting $C_{l_{\dot{\xi}}}$ and $C_{n_{\dot{\zeta}}}$ we get

$$\begin{bmatrix} (2\mu D - C_{y_\beta}) & -C_{y_p} & (2\mu - C_{y_r}) & -C_{L_0} & 0 & -C_{y_\zeta} \\ -C_{l_\beta} & (i_A D - C_{l_p}) & -(i_E D + C_{l_r}) & 0 & -C_{l_\xi} & -C_{l_\zeta} \\ -C_{n_\beta} & -(i_E D + C_{n_p}) & (i_C D - C_{n_r}) & 0 & -C_{n_\xi} & -C_{n_\zeta} \\ 0 & -1 & 0 & 0 & D & 0 \\ -K_{11}t^* & 0 & 0 & 0 & (D + 10t^*) & 0 \\ 0 & 0 & -K_{23}t^* & 0 & 0 & (D + 12t^*) \end{bmatrix} \begin{bmatrix} \beta \\ \hat{p} \\ \hat{r} \\ \phi \\ \xi \\ \zeta \end{bmatrix} = 0 \tag{10.3,7}$$

In addition to the numerical values given above, the following pertain to the example, at $C_{L_0} = 4.0$.

$C_{y_\beta} = -2.45$; $C_{y_p} = -0.031$; $C_{l_p} = -0.26$ $C_{np} = -0.140$; $C_{y_r} = 0.56$;
$W = 40{,}000$ lb; $S = 1000$ ft^2; $\bar{c} = 12.4$ ft; $A = 6.5$; $u_0 = 91.7$ fps
$\mu = 12.95$; $i_A = 1.55$; $i_c = 3.89$; $i_E = 0$; $t^* = 0.425$ sec

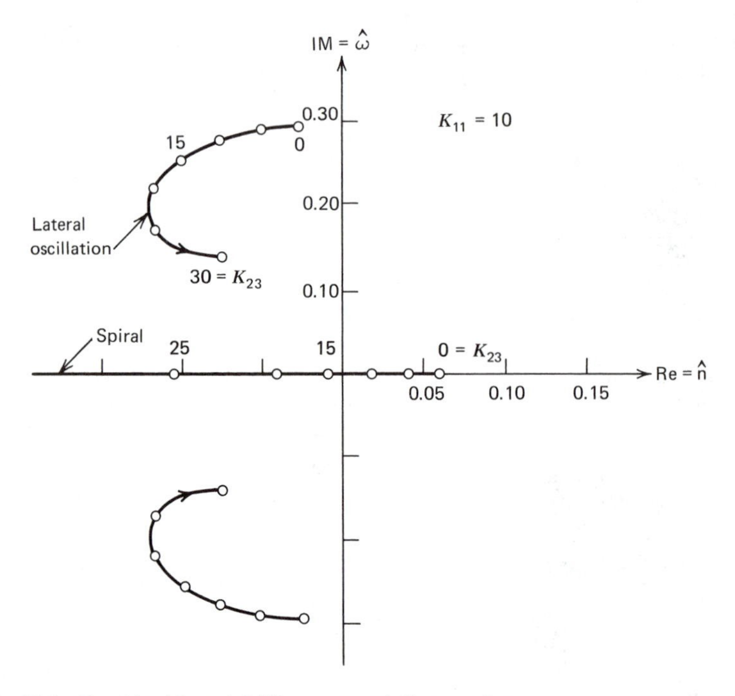

FIG. 10.6 Root loci for stability augmentation system.

With these data, the eigenvalues of the system were found, for ranges of K_{11} and K_{23}, and a typical root locus is shown on Fig. 10.6. There is a substantial range of practical gains for which stability is achieved. For example for $K_{11} = 10, K_{23} = 20$, the spiral and Dutch-roll characteristics are

Spiral: $t_{1/2} = 7.4$ sec

Oscillation: $T = 12.4$ sec, $N_{1/2} = 0.21$ cycles

The corresponding control gains are, respectively, 1 deg/deg for the aileron, and 1.67 deg/(deg/sec) for the rudder. These are both quite modest, and would not likely present any exceptional problems of control design.

10.4 EXAMPLE—ALTITUDE AND GLIDE PATH CONTROL

One of the most important problems in the control of flight path is that of following a prescribed line in space, as defined for example by a radio beacon. This is crucial in the landing situation under poor visibility when the airplane flies down the ILS glide slope. We discuss this case by considering first a simple approximate model that reveals the main features, and then examining a more realistic, and hence more complicated case.

Flight at Exactly Constant Height—Speed Stability

The first mathematical model we consider can be regarded as that corresponding to horizontal flight when a "perfect" autopilot controls the angle of attack in such a way as to keep the height error exactly zero. The result will show that the speed variation is stable at high speeds, but unstable at speeds below a critical value near the minimum drag speed. Neumark (10.2) recounts that this criterion was first discovered *in 1910* by Painlevé, and that it was at first accepted by aeronautical engineers and scientists, but later, on the basis of the theory of the phugoid, which showed no such effect, was rejected as false. In fact, to the extent that pilots can control height error by elevator control alone, i.e., to the extent that they approximate the ideal autopilot we have postulated, the instability at low speed will be experienced in manual flight. Since speed variation is the most noticeable feature of this phenomenon, it is commonly referred to as *speed stability*.

The analysis that follows is essentially that of Neumark, but adapted to the notation and methods of this book. The basic assumption that the flight path is *exactly* horizontal implies $\gamma = 0$, or $\theta = \alpha$. An exactly horizontal flight path also implies $L = W$. The pitching moment equation is assumed to be identically satisfied by means of an appropriate but unspecified control device that supplies the pitching moment as required. The equations for the system are then (4.15,7*a*, *b*, *e*) with $\alpha \equiv \theta$. We also choose $\theta_0 = 0$, and neglect the $\dot{\alpha}$ and q derivatives because the motion under consideration is one that changes slowly. The result is

$$\begin{bmatrix} (2\mu D - C_{x_u}) & -C_{x_\alpha} & C_{L_0} \\ \\ (2C_{L_0} - C_{z_u}) & (2\mu D - C_{z_\alpha}) & -2\mu D \end{bmatrix} \begin{bmatrix} \hat{u} \\ \alpha \\ \alpha \end{bmatrix} = 0 \tag{10.4,1}$$

For the stability derivatives in Eq. 10.4,1 we neglect Mach number effects, and assume $C_{D_0} \ll C_{L_\alpha}$ to get (see Table 5.1):

$$\begin{aligned} C_{x_u} &= -2C_{D_0}; \qquad & C_{x_\alpha} &= C_{L_0} - C_{D_\alpha} \\ C_{z_u} &= 0; & C_{z_\alpha} &\doteq -C_{L_\alpha} \end{aligned}$$

Equations 10.4,1 then reduce to

$$\begin{aligned} 2\mu D\hat{u} + 2C_{D_0}\hat{u} + C_{D_\alpha}\alpha &= 0 \\ 2C_{L_0}\hat{u} + C_{L_\alpha}\alpha &= 0 \end{aligned} \tag{10.4,2}$$

Eliminating α yields a first-order equation for the speed,

$$2\mu D\hat{u} + 2\left(C_{D_0} - \frac{C_{D_\alpha}}{C_{L_\alpha}} C_{L_0}\right)\hat{u} = 0 \tag{10.4,3}$$

for which the solution is the exponential function

$$\hat{u} = \hat{u}_0 e^{-\hat{t}/\hat{T}} \tag{10.4,4}$$

where the "time constant" $\hat{T}$ is given by

$$\hat{T}^{-1} = \frac{1}{\mu}\left(C_{D_0} - \frac{C_{D_\alpha}}{C_{L_\alpha}} C_{L_0}\right) \tag{10.4,5}$$

Equation 10.4,5 leads to the significant conclusion—when $\hat{T}$ is positive, the speed variation is stable, in that an initial error $\hat{u}_0$ decays with time; when $\hat{T}$ is negative, the error grows with time. Now we can rewrite Eq. 10.4,5 as

$$\begin{aligned}\hat{T}^{-1} &= \frac{C_{L_0}}{\mu}\left(\frac{C_{D_0}}{C_{L_0}} - \frac{C_{D_\alpha}}{C_{L_\alpha}}\right)\\ &= \frac{C_{L_0}}{\mu}\left(\frac{C_{D_0}}{C_{L_0}} - \frac{dC_D}{dC_L}\right)\end{aligned} \tag{10.4,6}$$

where dC_L/dC_D is the slope of the tangent to the drag polar (l_2) and C_{L_0}/C_{D_0} is the slope of the secant (l_1), see Fig. 10.7. Thus we see that the factor in parentheses on the r.h.s. of Eq. 10.4,6 passes through zero at point A, when the two lines coincide. When $C_{L_0} > (C_{L_0})_A$, $\hat{T}$ is negative, and vice versa. Point A of course corresponds to max L/D, and hence the important conclusion: *the speed perturbation is stable at speeds higher than that corresponding to* $(L/D)_{max}$, *and unstable for lower speeds.* The unstable region, corresponding to flight at $C_L > C_{L_{opt}}$ is referred to as the "backside of the polar."

FIG. 10.7 Drag polar, **M** = const.

Numerical Example

The jet transport of Sec. 6.5 is used for the example, in horizontal flight at sea level. The data needed for the calculation is as follows:

$$C_{D_0} = 0.016 + \frac{C_{L_0}{}^2}{7\pi}; \qquad C_{L_\alpha} = 4.88; \qquad \frac{dC_D}{dC_L} = \frac{2}{7\pi} C_L$$

$$W/S = 60 \text{ psf}; \qquad \mu = 101.8; \qquad \rho = 0.002378 \text{ slugs/ft.}^3$$
$$u_0 = [2(W/S)/\rho C_{L_0}]^{1/2}; \qquad t^* = 7.70/u_0$$

With these data, the values of C_{L_0} and u_0 at $(L/D)_{max}$ are, respectively, $C'_L = 0.595$ and $u'_0 = 290$ fps. The result of the calculation with Eq. 10.4,6 is shown in Fig. 10.8. There is positive "speed stability" above 290 fps, but the characteristic time to half is large, in excess of 75 sec. In the low-speed range (on "the backside of the polar"), the motion is unstable, with time to double falling as low as 30.5 sec at $C_{L_0} = 1.6$. A low-speed landing approach with this speed characteristic is undesirable from a handling-qualities standpoint. On the other hand, the example corresponds to cruising flight, not landing, since wheels and flaps are retracted.

The speed stability is in fact quite sensitive to the drag characteristics of the airplane. Thus, suppose that undercarriage and flaps have been lowered on the jet transport, with large increases in parasite and induced drag reflected in the polar equation

$$C_D = 0.20 + \frac{1.2C_L{}^2}{7\pi} \tag{10.4,7}$$

The results for this case, also shown on Fig. 10.8, are very different. The divergence time to double is now greater than 30 sec for all speeds above about 99 mph.

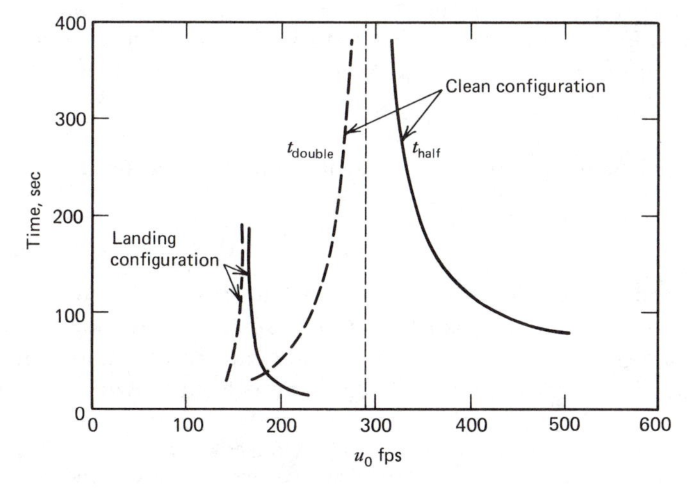

FIG. 10.8 Speed stability of jet transport at sea level.

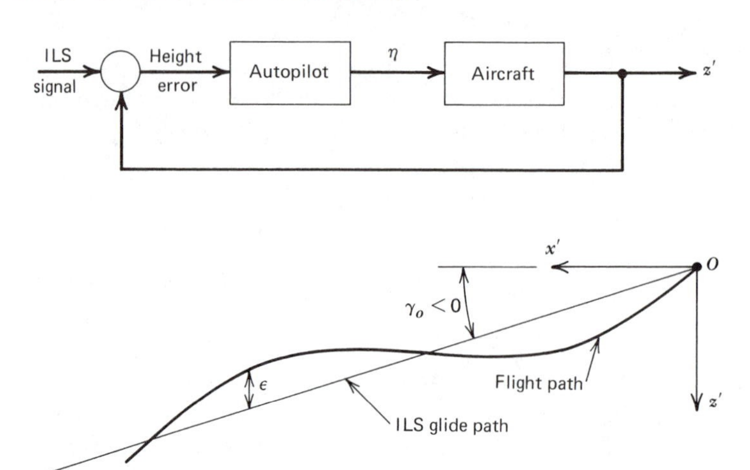

FIG. 10.9 Automatic control of glide path.

Flight on ILS Glide Slope

In the above analysis, we assumed that the airplane was under the control of an ideal autopilot that kept the height error exactly zero. A more realistic model incorporates a feedback control that senses height error and actuates the elevator[1] in response (see Fig. 10.9). The time lag associated with response of height to elevator input may be expected to lead to stability characteristics significantly different from those of the simple model.

Let us assume then that the airplane is making an automatically controlled approach on ILS. That is, a radio beam defines the glide path, and the pitch autopilot is coupled to the radio signal in such a way that height error is sensed and actuates the elevator. The autopilot and control system are relatively fast-acting compared to the pitch response of the vehicle, so we may reasonably assume a simple gain for the transfer function of these elements. Thus the mathematical model is obtained from Eq. 4.15,7 with the additional control law

$$\eta = K_1\varepsilon + K_2\dot{\varepsilon} \qquad (10.4,8)$$

where ε is the height error and we have included both proportional and rate terms.

The height error is defined as

$$\varepsilon = z'_c - z' \qquad (10.4,9)$$

where z'_c is the commanded altitude. We shall not require the commanded flight to be on a straight line, so z'_c is left arbitrary. Since normal glide slopes for civil transports are of the order of 3°, we can readily accommodate the flight path in this case within the range of small perturbations from horizontal flight, and so

[1] A still more sophisticated system uses control of thrust as well as of elevator. This is capable of producing better system performance provided that thrust responds quickly enough to the control command.

take $\theta_0 \equiv 0$. Equation 4.14,6 then yields

$$\frac{dz'}{dt} = -u_0\theta + w \tag{10.4,10}$$

With Eqs. 10.4,9 and 10.4,10 the control equation 10.4,8 becomes

$$\eta = K_1 z_c' + K_2 \dot{z}_c' - K_1 z' - K_2(w - u_0\theta) \tag{10.4,11}$$

Defining $\hat{z} = 2z'/\bar{c}$, we get the nondimensional form of Eq. 10.4,11 as

$$\eta = K_1 \frac{\bar{c}}{2}\hat{z}_c + K_2 u_0 D\hat{z}_c - K_1 \frac{\bar{c}}{2}\hat{z} - K_2 u_0(\alpha - \theta) \tag{10.4,12}$$

We now form the system equations from Eqs. 4.15,7 and 10.4,12. In doing so we note that Eq. 4.15,7*d* is not needed, and we make the following assumptions:

$$C_{z_\eta} = C_{m_{\dot{\eta}}} = C_{z_u} = C_{z_{\dot{\alpha}}} = C_{z_q} = C_{m_u} = 0; \qquad C_{x_u} = -2C_{D_0}.$$

The result is given in Eq. 10.4,13.

$$\begin{bmatrix} (2\mu D + 2C_{D_0}) & -(C_{L_0} - C_{D_\alpha}) & 0 & C_{L_0} & 0 \\ 2C_{L_0} & (2\mu D + C_{L_\alpha} + C_{D_0}) & -2\mu & 0 & 0 \\ 0 & \begin{matrix} -(C_{m_{\dot{\alpha}}} D + C_{m_\alpha}) \\ + C_{m_\eta} K_2 u_0 \end{matrix} & (i_B D - C_{m_q}) & -C_{m_\eta} K_2 u_0 & C_{m_\eta} K_1 \frac{\bar{c}}{2} \\ 0 & 0 & -1 & D & 0 \\ 0 & -1 & 0 & 1 & D \end{bmatrix} \begin{bmatrix} \hat{u} \\ \alpha \\ \hat{q} \\ \theta \\ \hat{z} \end{bmatrix} = \begin{bmatrix} 0 \\ 0 \\ C_{m_\eta}\left(K_1 \frac{\bar{c}}{2} + K_2 u_0 D\right)\hat{z}_c \\ 0 \\ 0 \end{bmatrix} \tag{10.4,13}$$

Numerical Example

Computations of the stability and performance were carried out with Eq. 10.4,13 for the same jet transport airplane used in preceding examples, flying at sea level. The drag polar is Eq. 10.4,7 corresponding to the landing configuration. The data that differ from those of Sec. 6.5 are as follows:

$$C_{D_\alpha} = 0.959, \qquad t^* = 0.0460 \text{ sec}, \qquad \mu = 101.8,$$
$$u_0 = 167.4 \text{ fps}, \qquad C_{L_0} = 1.8, \qquad C_{D_0} = 0.377$$

The eigenvalues corresponding to a range of K_1 and K_2 are shown on Fig. 10.10 in the form of root loci. Point *A* corresponds to the uncontrolled phugoid, and

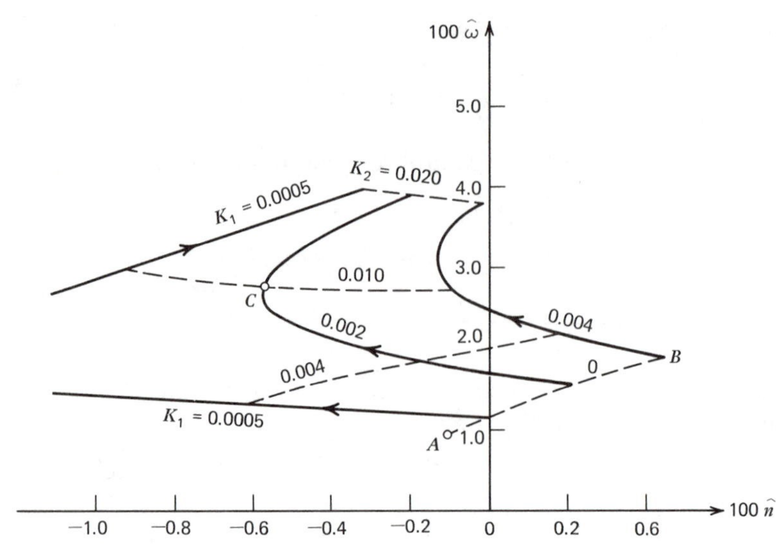

FIG. 10.10 Root locus of glide-path controller.

increasing proportional gain K_1 with zero rate gain produces the branch *AB* of the locus. The system rapidly goes unstable without error-rate control, but is easily stabilized with a modest value of K_2. For example, at point *C* on Fig. 10.10, with $K_1 = 0.002$ (about 12° elevator per 100 ft of height error) and $K_2 = 0.010$ (about 12° elevator per 20 ft/sec height error-rate), the eigenvalue characteristics are:

Phugoid: period = 10.4 sec

$N_{1/2} = 0.54$

Three real roots: $t_{\text{half}} = 94.0, 1.68, 0.86$ sec

The short-period mode has disappeared, being replaced by a pair of real roots, and the third real root is associated with the extra degree of freedom.

The performance of the system, i.e., its ability to track the glide slope, can be in part inferred from the frequency response associated with z_c' input and z' output. This is computed by taking the Laplace transform of Eq. 10.4,13 (which simply changes D to s wherever it occurs), replacing s by $i\hat{\omega}$, and solving the resulting complex algebraic equations for the ratio $\hat{z}/\hat{z}_c$ as a function of $\hat{\omega}$. The result is shown on Fig. 10.11. The system is seen to be able to follow waves in the ILS beam fairly closely down to wavelengths of the order of $\frac{1}{2}$ mile ($\hat{\omega} = 2 \times 10^{-2}$) at which point a phase lag of 40° has developed. This calculation is not, of course, sufficient to decide on the acceptability of the chosen gains. For that purpose one should calculate actual flight paths in the presence of wind shear and turbulence, and relate the dispersions to what is acceptable for a given mission.

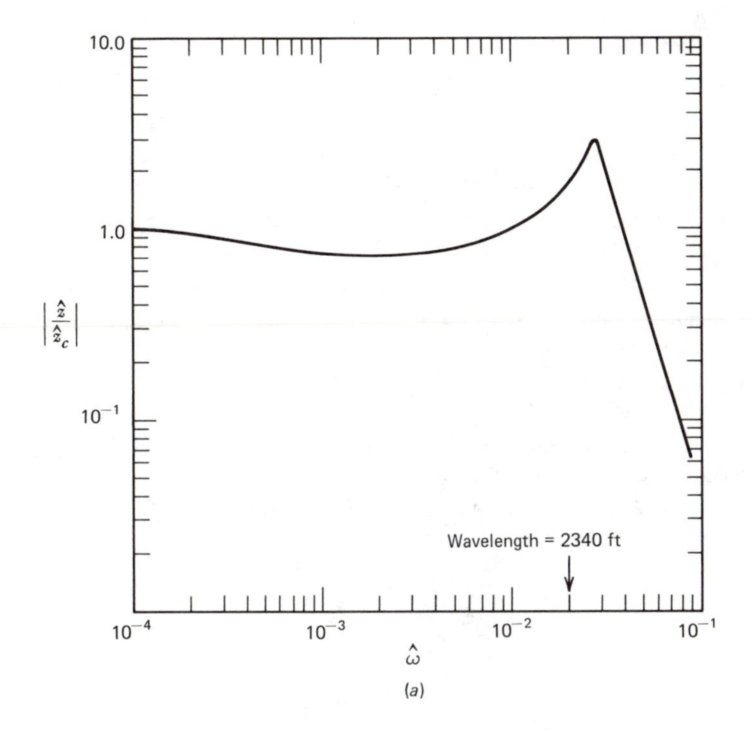

(a)

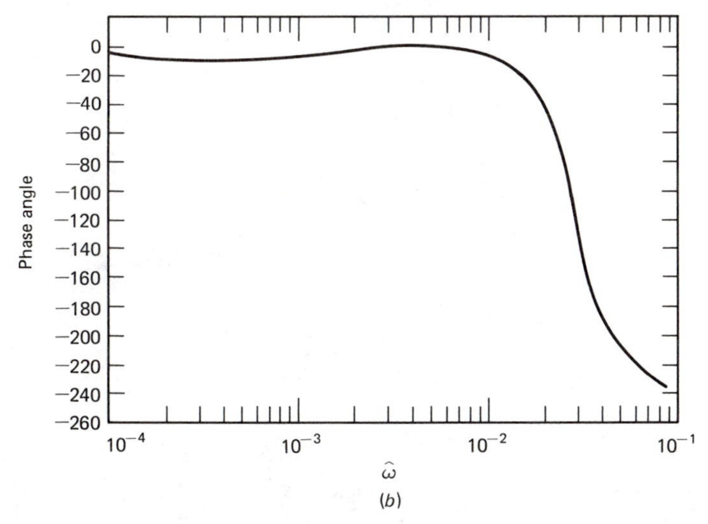

(b)

FIG. 10.11 Response of automatic glide-path controller. (a) Amplitude. (b) Phase angle.

10.5 BIBLIOGRAPHY

10.1 J. W. Tanney. Fluidies. *Progress in Aeronautical Sciences*, vol. 10, Pergamon Press, Oxford, 1970.

10.2 S. Neumark. Longitudinal Stability Below Minimum Drag Speed, and Theory of Stability under Constraint. *ARC R & M 2983*, 1957.

10.3 P. Painlevé. Étude sur le Régime Normal d'un Aeroplane. *La Technique Aeronautique* vol. 1, pp. 3–11, Paris, 1910.

10.4 W. R. Evans. *Control System Dynamics*. McGraw-Hill Pub. Co., New York, 1954.

10.5 D. Graham and D. McRuer. *Analysis of Nonlinear Control Systems*. John Wiley & Sons, Inc., New York, 1961.

10.6 J. H. Blakelock. *Automatic Control of Aircraft and Missiles*. John Wiley & Sons, Inc., New York, 1965.

10.7 S. B. Anderson, H. C. Quigley, and R. C. Innis. Stability and Control Considerations for STOL Aircraft. *CASI Jour.*, vol. 12, no. 5, May 1966.

10.8 A. C. Robinson. Survey of Dynamic Analysis Methods for Flight Control Design. *J. Aircraft*, vol. no. 2, 1969.

RÉSUMÉ OF VECTOR ANALYSIS

APPENDIX A

A.1 DEFINITION AND NOTATION

A *vector quantity* is one that has magnitude and direction (e.g., velocity, force, acceleration).

A *vector* is a directed line segment which, by its length and direction, represents a vector quantity. It is denoted by a boldface symbol, or by an arrow (see Fig. A.1)

$$\mathbf{A}, \qquad \overrightarrow{AB}$$

The magnitude is denoted by plain type, or the absence of the arrow:

$$A, \qquad AB$$

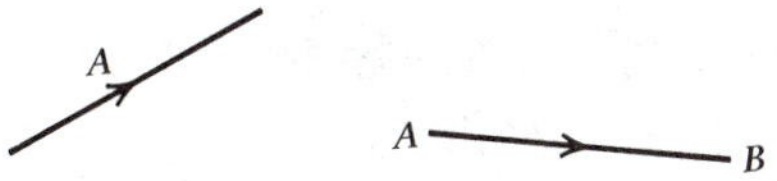

FIG. A.1 Notation for vectors.

A.2 EQUALITY, ADDITION, SUBTRACTION

Two vectors are equal when they have the same length and the same direction (Fig. A.2).

Addition of vectors is performed by the triangle or parallelogram rule (Fig. A.3). Note that

$$\mathbf{A} + \mathbf{B} = \mathbf{B} + \mathbf{A}$$

and that

$$(\mathbf{A} + \mathbf{B}) + \mathbf{C} = \mathbf{A} + (\mathbf{B} + \mathbf{C}) = \mathbf{A} + \mathbf{B} + \mathbf{C}$$

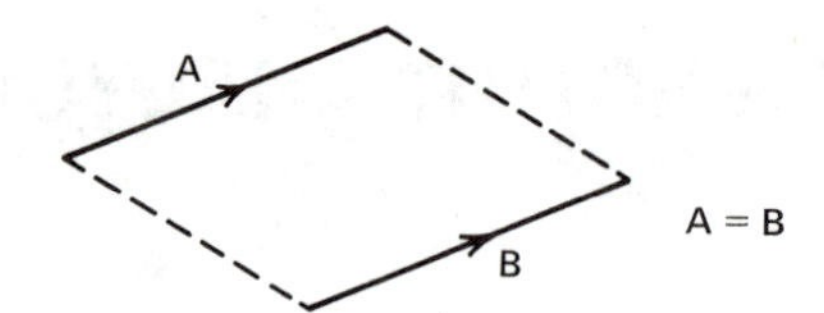

FIG. A.2 Equality of vectors.

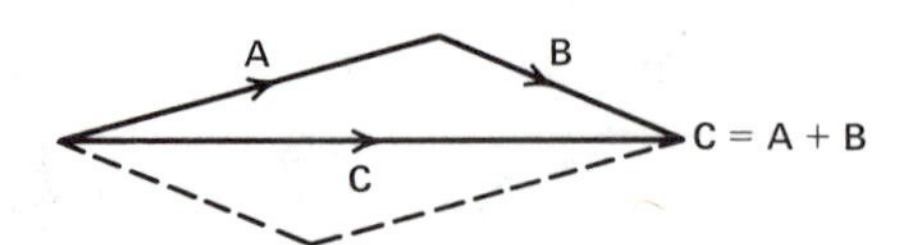

FIG. A.3 Addition of vectors.

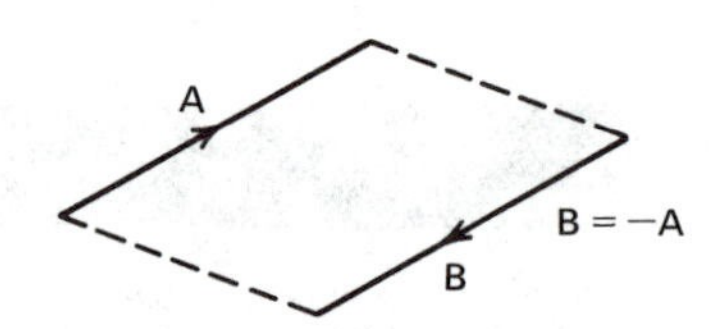

FIG. A.4 Negative of a vector.

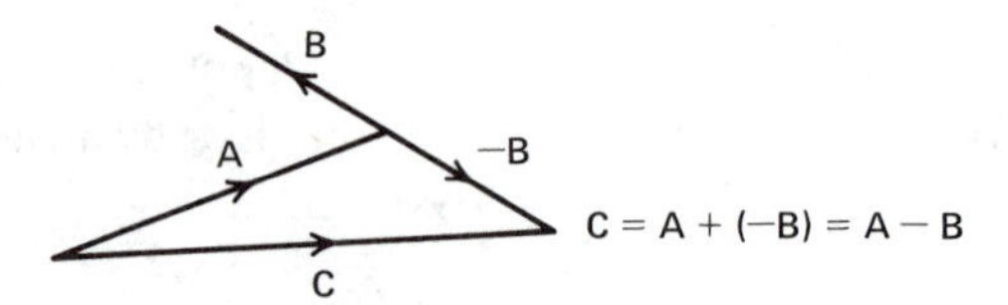

FIG. A.5 Subtraction of vectors.

The negative of a vector is one that has the same magnitude but opposite direction (Fig. A.4).

The difference of two vectors is formed by adding the negative (Fig. A.5).

A.3 COMPONENTS AND UNIT VECTORS

The component of a vector $\overrightarrow{AB}$ on a line l is $\overrightarrow{A'B'}$ where $A'A$ and $B'B$ are normals to l (Fig. A.6).

A unit vector is a vector of unit length in a specified direction. Thus let $\mathbf{e}$ be a unit vector on l. Then the component of $\overrightarrow{AB}$ may also be written as

$$\overrightarrow{A'B'} = \mathbf{e}A'B'$$

A unit orthogonal triad is a set of three unit vectors, mutually perpendicular. They are identified with the three coordinate axes (Fig. A.7). The system is *right-handed* when as shown. That is, a rotation of the x axis into the y axis would drive a screw having a right-hand thread in the positive z direction.

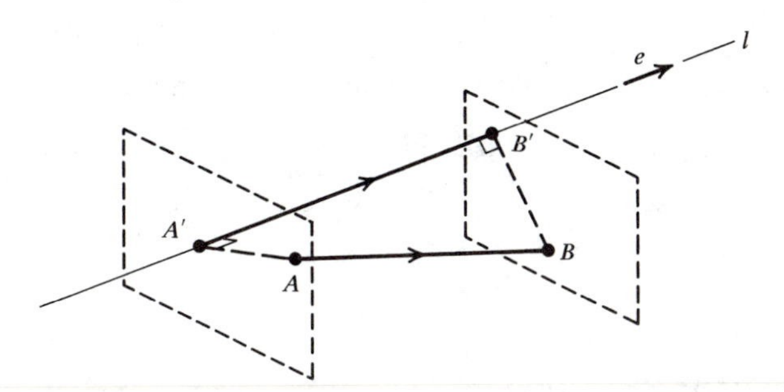

FIG. A.6 Component of a vector.

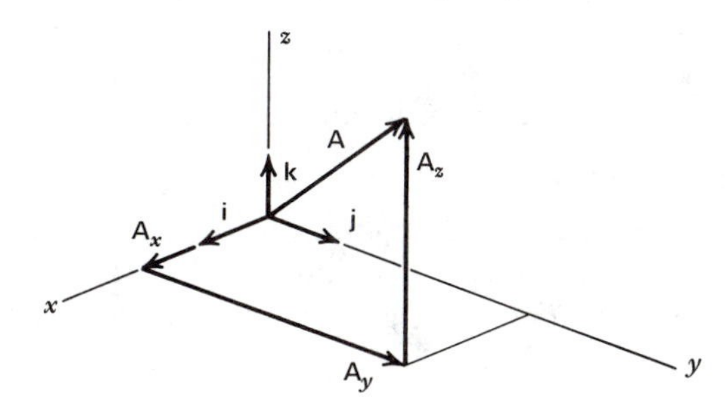

FIG. A.7 Unit vector triad.

A general vector **A**, as in Fig. A.7, has the three components $\mathbf{A}_x$, $\mathbf{A}_y$, $\mathbf{A}_z$. It follows that it may be written as the sum

$$\mathbf{A} = \mathbf{i}A_x + \mathbf{j}A_y + \mathbf{k}A_z \tag{A.3,1}$$

A.4 SCALAR PRODUCT

The scalar product of two vectors **A** and **B** is written

$$C = \mathbf{A} \cdot \mathbf{B} \tag{A.4,1}$$

and is read "*A* dot *B*." *C* is a scalar, and has the magnitude

$$C = AB \cos \theta \tag{A.4,2}$$

where θ is the angle between the two vectors. The following special cases are of interest:

$$\theta = 0, \qquad \mathbf{A} \cdot \mathbf{B} = AB, \qquad \mathbf{A} \cdot \mathbf{A} = A^2$$
$$\theta = \pi/2, \qquad \mathbf{A} \cdot \mathbf{B} = 0$$
$$\theta = \pi, \qquad \mathbf{A} \cdot \mathbf{B} = -AB$$

It follows that

$$\mathbf{i} \cdot \mathbf{i} = \mathbf{j} \cdot \mathbf{j} = \mathbf{k} \cdot \mathbf{k} = 1$$
$$\mathbf{i} \cdot \mathbf{j} = \mathbf{j} \cdot \mathbf{k} = \mathbf{k} \cdot \mathbf{i} = 0$$

The scalar product is distributive, i.e., $\mathbf{A} \cdot (\mathbf{B} + \mathbf{C}) = \mathbf{A} \cdot \mathbf{B} + \mathbf{A} \cdot \mathbf{C}$. From these rules it follows that, when **A** and **B** are given by expressions like Eq. A.3,1, then

$$\begin{aligned} \mathbf{A} \cdot \mathbf{B} &= (\mathbf{i}A_x + \mathbf{j}A_y + \mathbf{k}A_z) \cdot (\mathbf{i}B_x + \mathbf{j}B_y + \mathbf{k}B_z) \\ &= \mathbf{i}A_x \cdot \mathbf{i}B_x + \mathbf{i}A_x \cdot \mathbf{j}B_y + \cdots \\ &= A_xB_x + A_yB_y + A_zB_z \end{aligned} \tag{A.4,3}$$

A.5 VECTOR PRODUCT

The vector product of **A** and **B** is written

$$\mathbf{C} = \mathbf{A} \times \mathbf{B}$$

and is read "*A* cross *B*." **C** is a vector having the following properties (Fig. A.8):

(1) Its magnitude is $C = AB \sin \theta$.

(2) Its direction is perpendicular to the plane of **A** and **B**, and is given by the right-hand rule. That is, the rotation of **A** into **B** by the shortest path would drive a screw having a right-hand thread along the vector **C**.

The following special cases are of interest.

$$\begin{aligned} &\theta = 0 \text{ or } \pi, \qquad \mathbf{A} \times \mathbf{B} = 0 \\ &\theta = \pi/2, \qquad C = AB \\ &\mathbf{A} \times \mathbf{B} = -\mathbf{B} \times \mathbf{A} \\ &\mathbf{A} \times \mathbf{A} = 0 \\ &\mathbf{i} \times \mathbf{i} = \mathbf{j} \times \mathbf{j} = \mathbf{k} \times \mathbf{k} = 0 \\ &\mathbf{i} \times \mathbf{j} = \mathbf{k}, \qquad \mathbf{j} \times \mathbf{k} = \mathbf{i}, \qquad \mathbf{k} \times \mathbf{i} = \mathbf{j} \end{aligned}$$

The vector product also is distributive: i.e.,

$$\mathbf{A} \times (\mathbf{B} + \mathbf{C}) = \mathbf{A} \times \mathbf{B} + \mathbf{A} \times \mathbf{C} \tag{A.5,1}$$

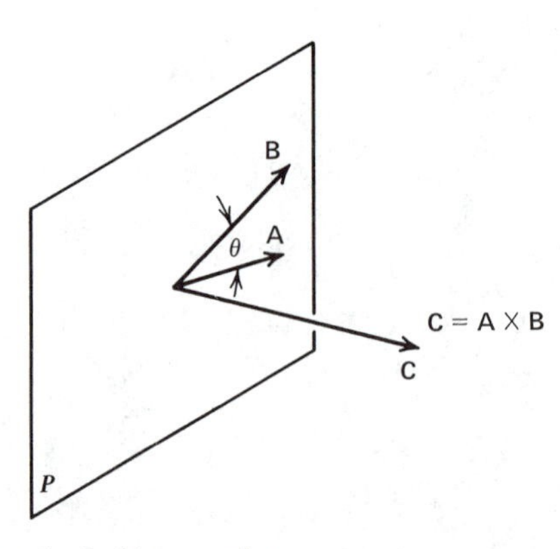

FIG. A.8 Vector product of two vectors.

From the above rules, it follows that, when **A** and **B** are given in component form, then

$$\begin{aligned}\mathbf{A} \times \mathbf{B} &= (\mathbf{i}A_x + \mathbf{j}A_y + \mathbf{k}A_z) \times (\mathbf{i}B_x + \mathbf{j}B_y + \mathbf{k}B_z) \\ &= \mathbf{i}A_x \times \mathbf{i}B_x + \mathbf{i}A_x \times \mathbf{j}B_y + \cdots \\ &= \begin{vmatrix} \mathbf{i} & \mathbf{j} & \mathbf{k} \\ A_x & A_y & A_z \\ B_x & B_y & B_z \end{vmatrix}\end{aligned} \tag{A.5,2}$$

The rule for a vector triple product is

$$\mathbf{A} \times (\mathbf{B} \times \mathbf{C}) = \mathbf{B}(\mathbf{A} \cdot \mathbf{C}) - \mathbf{C}(\mathbf{A} \cdot \mathbf{B}) \tag{A.5,3}$$

Applications in Mechanics

1. The moment of a force **F** about a point O is given by the vector product (Fig. A.9)

$$\mathbf{G} = \mathbf{r} \times \mathbf{F}$$

   where r is any vector from O to the line of action of **F**. The vector **G** has the correct magnitude and direction to represent the moment.

2. The velocity of a point P in a spinning body is given by

$$\mathbf{v} = \boldsymbol{\omega} \times \mathbf{r} \tag{A.5,4}$$

   where ω is the angular velocity vector, and **r** is a vector from any point on the axis to P (Fig. A.10). If, in addition, the point O has the velocity $\mathbf{v}_o$, then the resultant velocity of P is

$$\mathbf{v} = \mathbf{v}_o + \boldsymbol{\omega} \times \mathbf{r} \tag{A.5,5}$$

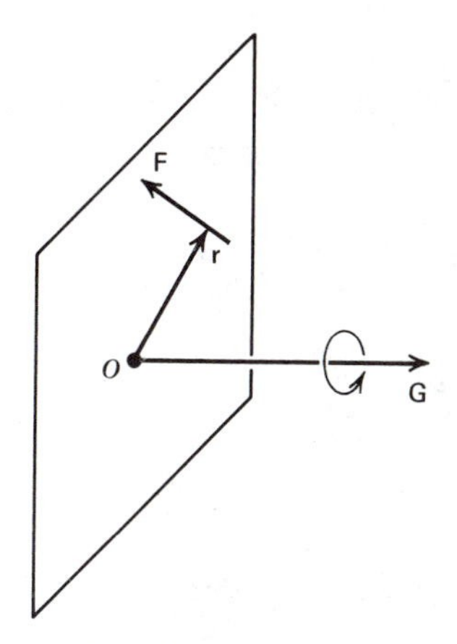

FIG. A.9 Moment of a force.

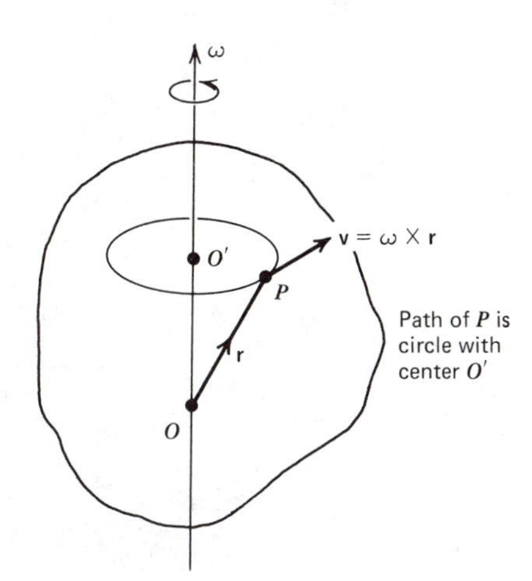

FIG. A.10 Velocity of a point in a spinning body.

A.6 DERIVATIVE OF A VECTOR

Let the vector $\mathbf{A}(t)$ be a function of time. When drawn at successive times from the same origin, the end points trace out a curve called the hodograph (Fig. A.11). (This is a generalization of the usual meaning of the word. Ordinarily it is applied only to the curve associated with the velocity vector.) The derivative is then defined to be

$$\frac{d\mathbf{A}}{dt} = \lim_{\Delta t \to 0} \frac{\mathbf{A}(t + \Delta t) - \mathbf{A}(t)}{\Delta t} = \lim_{\Delta t \to 0} \frac{\Delta \mathbf{A}}{\Delta t} \tag{A.6,1}$$

The limiting direction of $\Delta\mathbf{A}$ is tangent to the hodograph at P, and the magnitude of the limit is

$$\lim_{\Delta t \to 0} \frac{PQ}{\Delta t}$$

which is the speed of P along the hodograph. Thus it follows that the *derivative of a vector is the velocity of its end point along the hodograph.*

Application of the definition A.6,1 leads to the following rules:

(1) $$\frac{d}{dt}(\mathbf{A} + \mathbf{B}) = \frac{d\mathbf{A}}{dt} + \frac{d\mathbf{B}}{dt}$$

(2) $$\frac{d}{dt}(m\mathbf{A}) = m\frac{d\mathbf{A}}{dt} + \mathbf{A}\frac{dm}{dt} \quad (m = \text{scalar function of } t)$$

(3) $$\frac{d}{dt}(\mathbf{A} \cdot \mathbf{B}) = \mathbf{A} \cdot \frac{d\mathbf{B}}{dt} + \frac{d\mathbf{A}}{dt} \cdot \mathbf{B}$$

(4) $$\frac{d}{dt}(\mathbf{A} \times \mathbf{B}) = \frac{d\mathbf{A}}{dt} \times \mathbf{B} + \mathbf{A} \times \frac{d\mathbf{B}}{dt}$$

(5) $$\frac{d\mathbf{A}}{dt} = \mathbf{i}\frac{dA_x}{dt} + \mathbf{j}\frac{dA_y}{dt} + \mathbf{k}\frac{dA_z}{dt} \quad (\mathbf{i}, \mathbf{j}, \mathbf{k} \text{ have fixed directions})$$

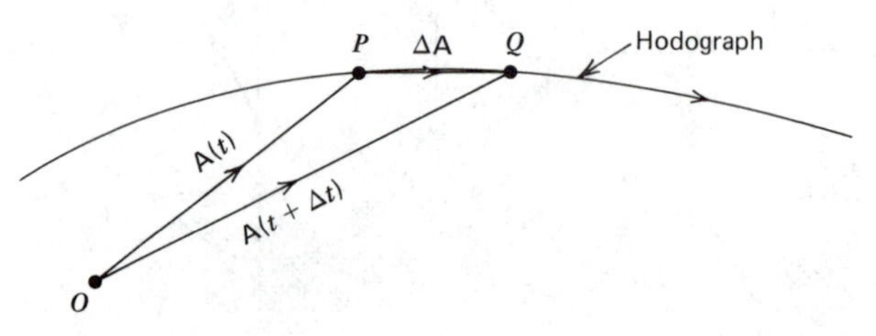

FIG. A.11 Derivative of a vector.

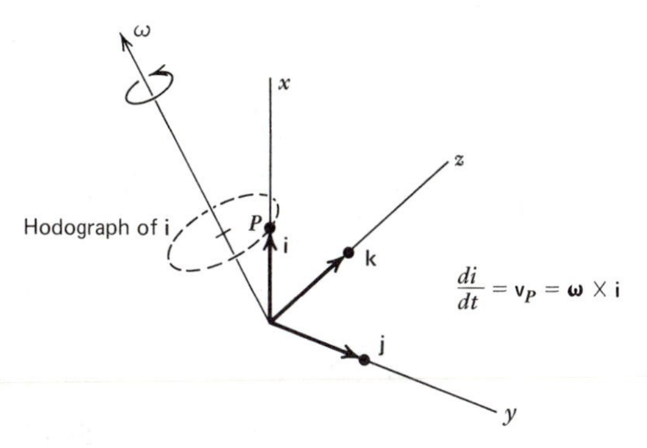

FIG. A.12 Rotating frame of reference.

Derivative in a Rotating Frame of Reference

Rule 5 above does not apply when the unit vector triad is rotating. In that case

$$\frac{d}{dt}(\mathbf{i}A_x + \mathbf{j}A_y + \mathbf{k}A_z) = \mathbf{i}\frac{dA_x}{dt} + \mathbf{j}\frac{dA_y}{dt} + \mathbf{k}\frac{dA_z}{dt} + A_x\frac{d\mathbf{i}}{dt} + A_y\frac{d\mathbf{j}}{dt} + A_z\frac{d\mathbf{k}}{dt}$$

Let the angular velocity of the reference frame be $\boldsymbol{\omega}$ (Fig. A.12). The derivatives of the unit vectors are not zero because of their rotation. To calculate them we note that

1. The derivative (of **i** for example) is the velocity of the end point (of **i**) along the hodograph (of **i**).
2. The hodograph is a circle with center on the $\boldsymbol{\omega}$ axis (Fig. A.12). It follows that

$$\frac{d\mathbf{i}}{dt} = \boldsymbol{\omega} \times \mathbf{i}, \quad \frac{d\mathbf{j}}{dt} = \boldsymbol{\omega} \times \mathbf{j}, \quad \frac{d\mathbf{k}}{dt} = \boldsymbol{\omega} \times \mathbf{k}$$

and hence that

$$\frac{d\mathbf{A}}{dt} = \frac{\delta\mathbf{A}}{\delta t} + \boldsymbol{\omega} \times \mathbf{A}$$

where

$$\frac{\delta\mathbf{A}}{\delta t} = \mathbf{i}\frac{dA_x}{dt} + \mathbf{j}\frac{dA_y}{dt} + \mathbf{k}\frac{dA_z}{dt}$$

A.7 ACCELERATION OF A PARTICLE IN A ROTATING FRAME OF REFERENCE

Let $S'(o'x'y'z')$ be a fixed frame of reference, and $S(oxyz)$ be a second frame of reference moving relative to S'. Its motion is described by $\mathbf{v}_0$, the velocity vector of its origin, and by its angular velocity $\boldsymbol{\omega}$.

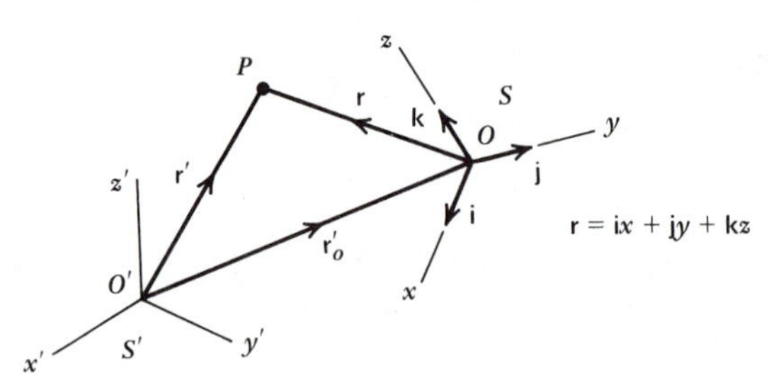

FIG. A.13 Rotating frame of reference.

Let **r** and **r**′ be the position vectors of P relative to the two frames as shown in Fig. A.13. Then

$$\mathbf{r}' = \mathbf{r}'_0 + \mathbf{r}$$

and the velocity of P relative to the fixed frame S' is

$$\mathbf{v} = \frac{d\mathbf{r}'}{dt} = \frac{d\mathbf{r}'_0}{dt} + \frac{d\mathbf{r}}{dt}$$

$$= \mathbf{v}_0 + \frac{d\mathbf{r}}{dt}$$

where $\mathbf{v}_0$ is the velocity of O relative to S'. By the theorem of Sec. A.6 this becomes

$$\mathbf{v} = \mathbf{v}_0 + \frac{\delta \mathbf{r}}{\delta t} + \boldsymbol{\omega} \times \mathbf{r}$$

The acceleration of P is

$$\mathbf{a} = \frac{d\mathbf{v}}{dt} = \frac{\delta \mathbf{v}}{\delta t} + \boldsymbol{\omega} \times \mathbf{v}$$

$$= \frac{\delta \mathbf{v}_0}{\delta t} + \frac{\delta^2 \mathbf{r}}{\delta t^2} + \frac{\delta}{\delta t}(\boldsymbol{\omega} \times \mathbf{r}) + \boldsymbol{\omega} \times \mathbf{v}_0 + \boldsymbol{\omega} \times \frac{\delta \mathbf{r}}{\delta t} + \boldsymbol{\omega} \times (\boldsymbol{\omega} \times \mathbf{r})$$

The first and fourth terms combine to give the acceleration of O relative to S', i.e., $\mathbf{a}_0 = d\mathbf{v}_0/dt = \delta \mathbf{v}_0/\delta t + \boldsymbol{\omega} \times \mathbf{v}_0$, thus

$$\mathbf{a} = \mathbf{a}_0 + \frac{\delta^2 \mathbf{r}}{\delta t^2} + \frac{\delta \boldsymbol{\omega}}{\delta t} \times \mathbf{r} + 2\boldsymbol{\omega} \times \frac{\delta \mathbf{r}}{\delta t} + \boldsymbol{\omega} \times (\boldsymbol{\omega} \times \mathbf{r})$$

Expanding into scalars, we find the three components of **a** to be

$$\begin{aligned}
a_x &= a_{0_x} + \ddot{x} + 2\omega_y \dot{z} - 2\omega_z \dot{y} \\
&\quad - x(\omega_y{}^2 + \omega_z{}^2) + y(\omega_x \omega_y - \dot{\omega}_z) + z(\omega_x \omega_z + \dot{\omega}_y) \\
a_y &= a_{0_y} + \ddot{y} + 2\omega_z \dot{x} - 2\omega_x \dot{z} \\
&\quad + x(\omega_x \omega_y + \dot{\omega}_z) - y(\omega_x{}^2 + \omega_z{}^2) + z(\omega_y \omega_z - \dot{\omega}_x) \\
a_z &= a_{0_z} + \ddot{z} + 2\omega_x \dot{y} - 2\omega_y \dot{x} \\
&\quad + x(\omega_x \omega_z - \dot{\omega}_y) + y(\omega_y \omega_z + \dot{\omega}_x) - z(\omega_x{}^2 + \omega_y{}^2)
\end{aligned}$$

AERODYNAMIC DATA

G. K. DIMOCK AND B. ETKIN

APPENDIX B

B.1 LIFT-CURVE SLOPE, C_{L_α}

Notation

$(a_1)_{0T}$	theoretical two-dimensional, inviscid, incompressible lift-curve slope
$(a_1)_0$	actual two-dimensional incompressible lift-curve slope
$(a_1)_M = C_{L_\alpha}$	lift-curve slope for a finite wing at Mach number M
β	compressibility parameter, $\sqrt{1 - M^2}$
κ	ratio of the experimental section lift-curve slope to the theoretical value $2\pi/\beta$, both taken at the same Mach number $= (a_1)_0/2\pi \quad \text{at } M = 0$; $= \dfrac{(a_1)_{0M}}{2\pi} \cdot \beta \quad \text{at } 0 < M < 1$
A	aspect ratio, b^2/S.
Λ	sweep-back angle of wing quarter-chord line
Λ_β	compressible sweep parameter, $\tan^{-1}(\tan \Lambda/\beta)$
λ	taper ratio, (tip chord/root chord)
$(a_F)_B$	lift-curve slope of fin and rudder in the presence of the body, including the side force induced on the body
$(a_F)_{BT}$	lift-curve slope of fin and rudder, allowing for body and tail plane effects

Notes

Figure B.1,1: $(a_1)_0$. The values of $(a_1)_{0T}$ were determined using potential flow theory. The difference between $(a_1)_{0T}$ and $(a_1)_0$ depends on the development of the boundary layer toward the trailing edge of the airfoil. The curves apply only when there is no separation of flow over the airfoil, and should only be used for airfoils at low incidence. The accuracy of the data is within $\pm 5\%$.

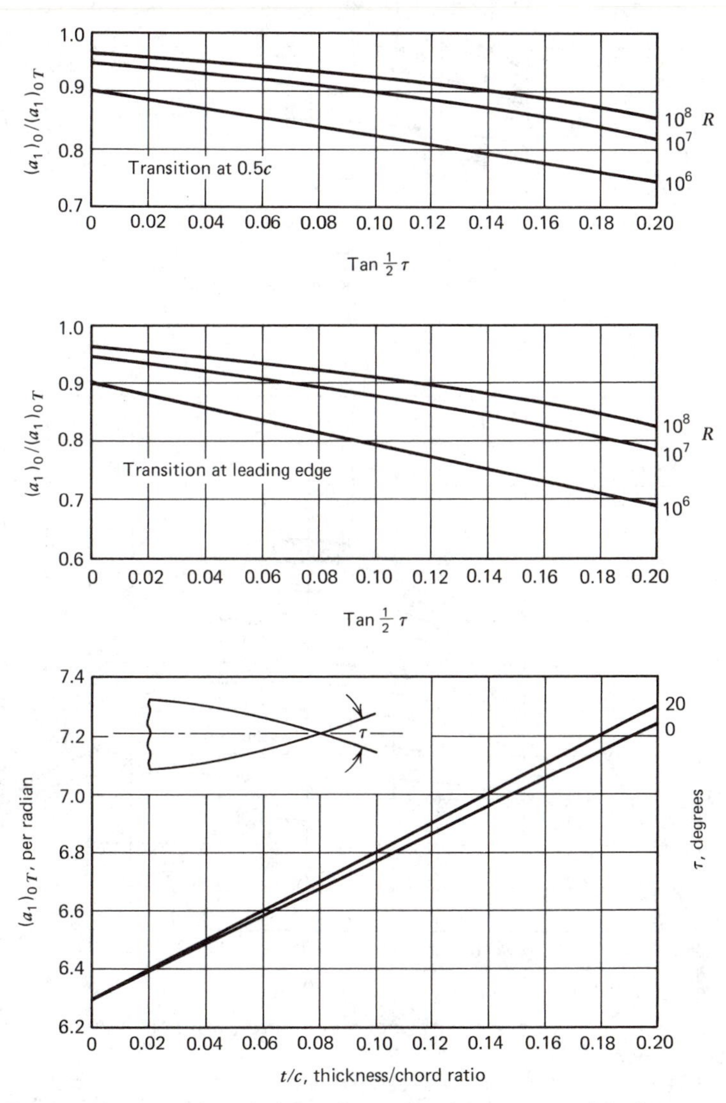

FIG. B.1,1 Lift-curve slope for two-dimensional incompressible flow. (Reproduced from Royal Aeronautical Society Data Sheet Wings 01. 01. 05)

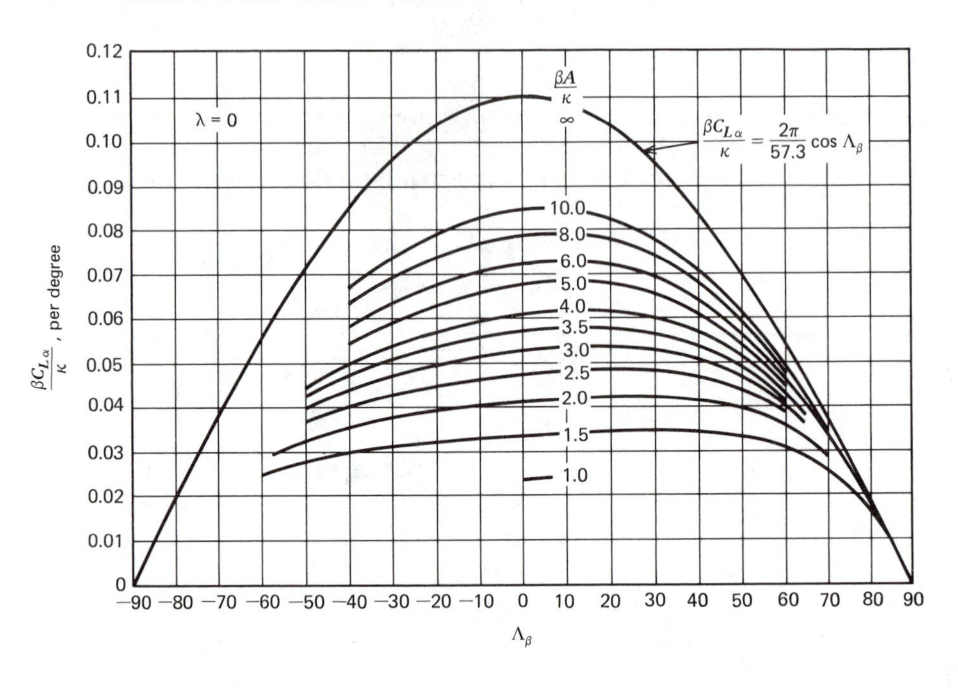

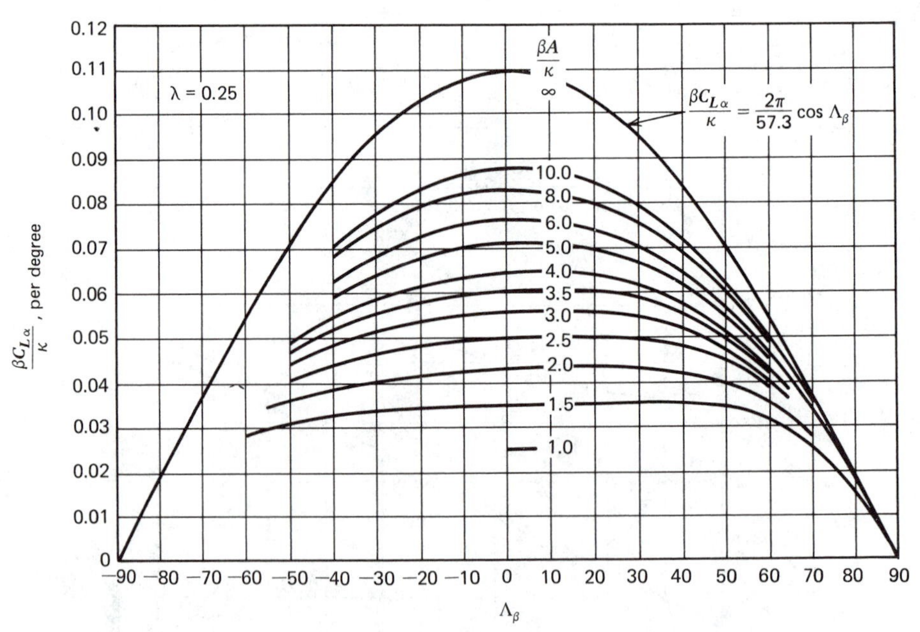

FIG. B.1,2 Lift-curve slope for swept and tapered wings at speeds below critical Mach number.
(Reproduced from *NACA Rept. 921* by J. DeYoung)

FIG. B.1,2 (continued) →

Figure B.1,2: C_{L_α} $(M < M_{cr})$. The data are based on Weissinger's theory for flat-plate wings in incompressible flow with straight quarter chord lines over the semispan, arbitrary sweep, aspect ratio, and continuous twist. The curves apply to high angles of attack if there is no separation of flow over the wing. The effects of compressibility are included by application of the Prandtl–Glauert rule. To utilize this data, the experimental section lift-curve slope of the wing under consideration must be known or assumed.

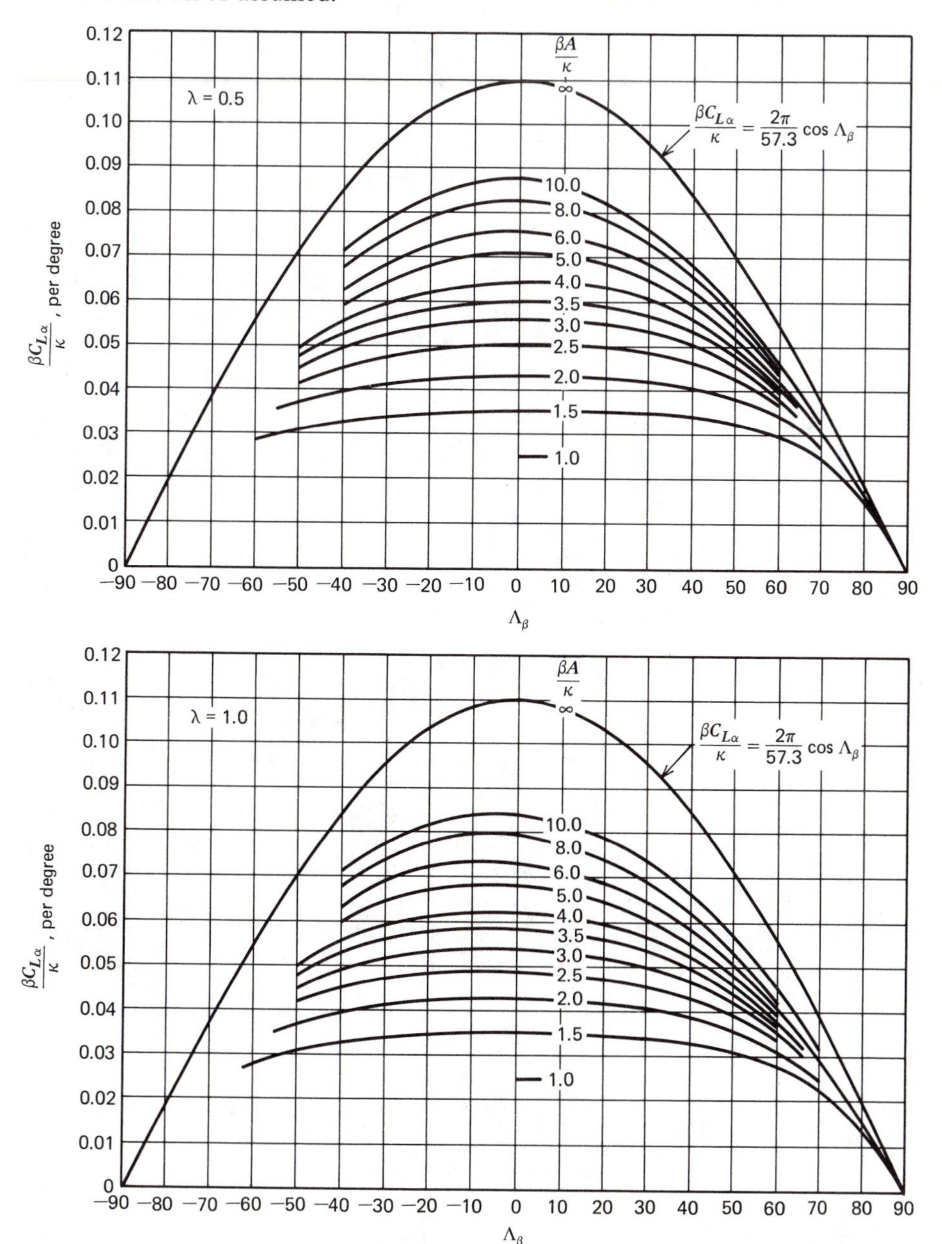

Figure B.1,3: C_{L_α} (Supersonic Speeds). The curves were based on linearized supersonic lifting-surface theory. Slender wing theory was used to obtain C_{L_α} for $A\sqrt{M^2 - 1} = 0$. The theories are based on inviscid flow and apply to flat plates at small incidence. The values of C_{L_α} for small values of $A\sqrt{M^2 - 1}$ (dotted lines) are questionable, and the kinks shown in the curves are not likely to occur in practice. The curves are accurate to within $\pm 10\%$.

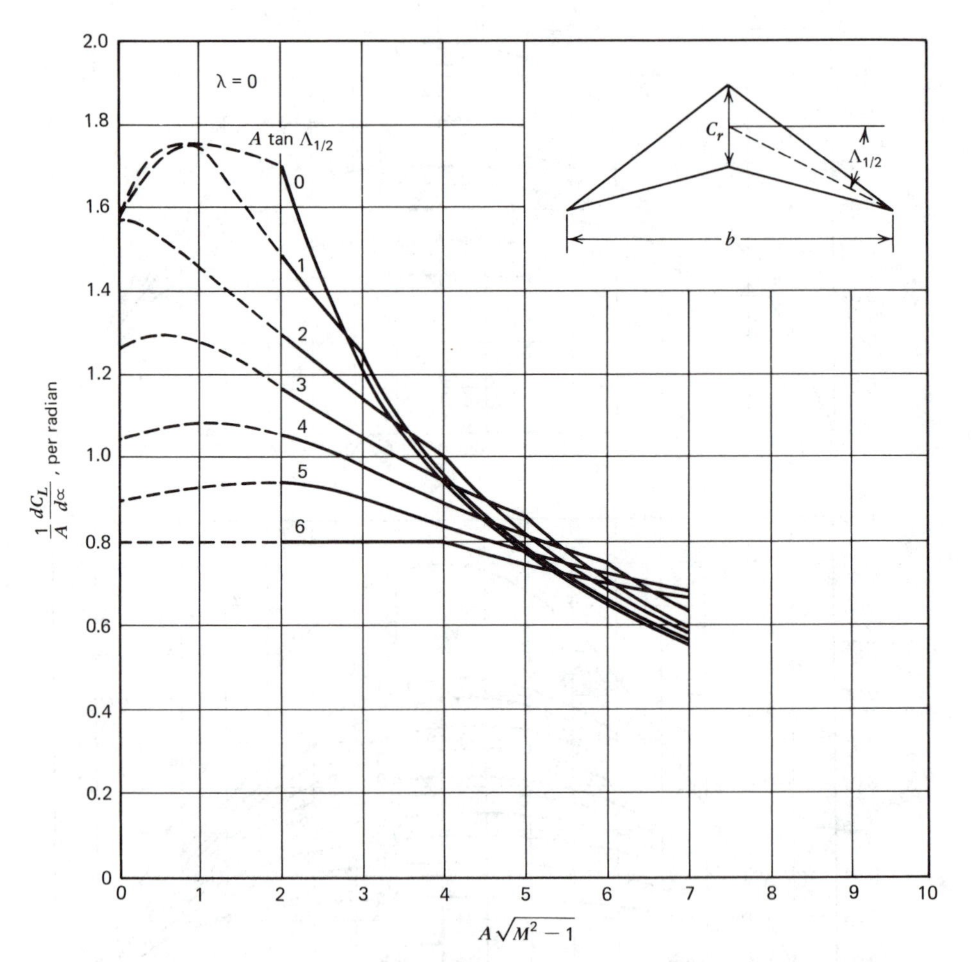

FIG. B.1,3 Theoretical lift-curve slope for swept and tapered wings at supersonic speeds.
(Reproduced from Royal Aeronautical Society Data Sheets Wings S01. 03. 03, 04, 05, 06)

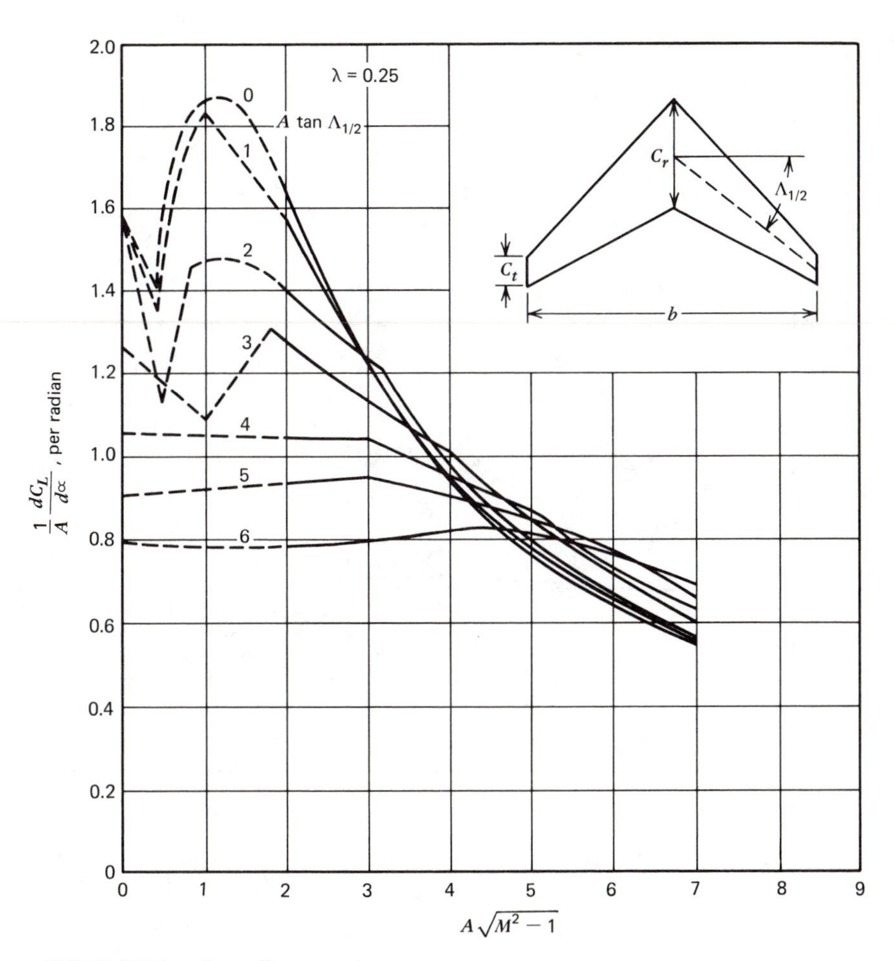

FIG. B.1,3 (continued)

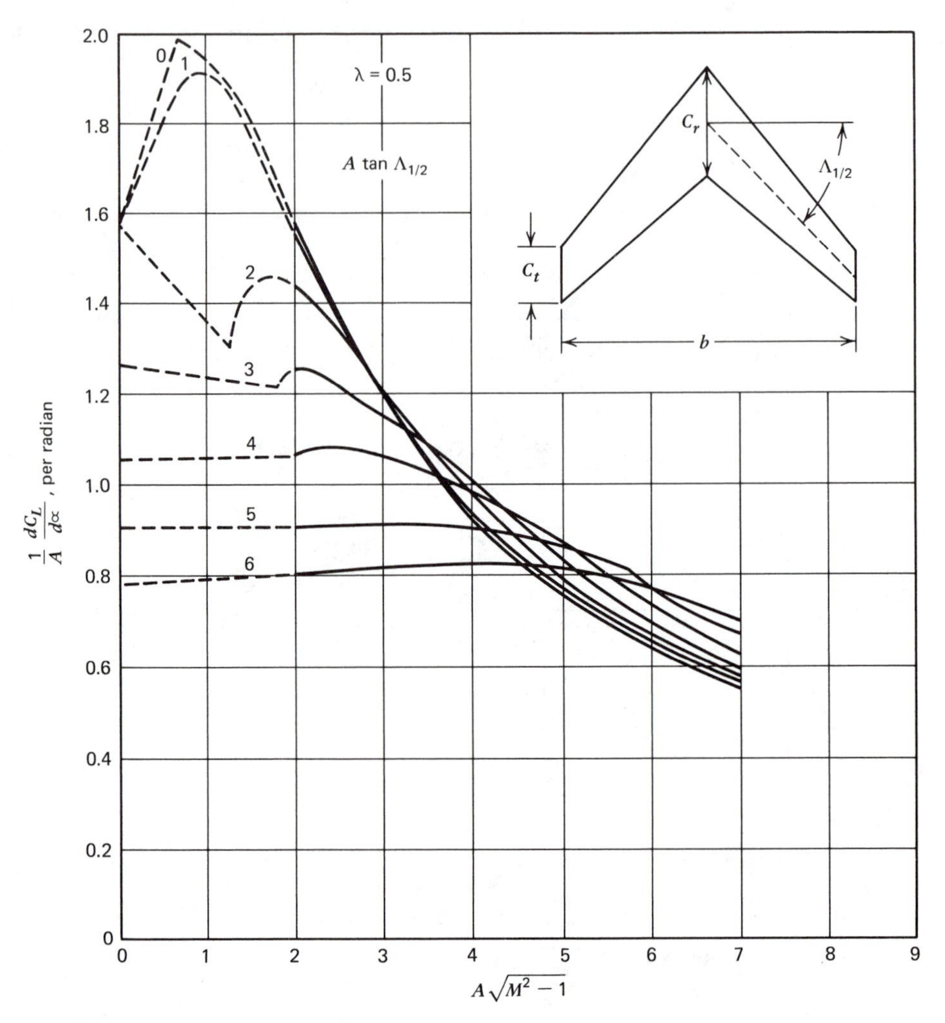

FIG. B.1,3 (continued)

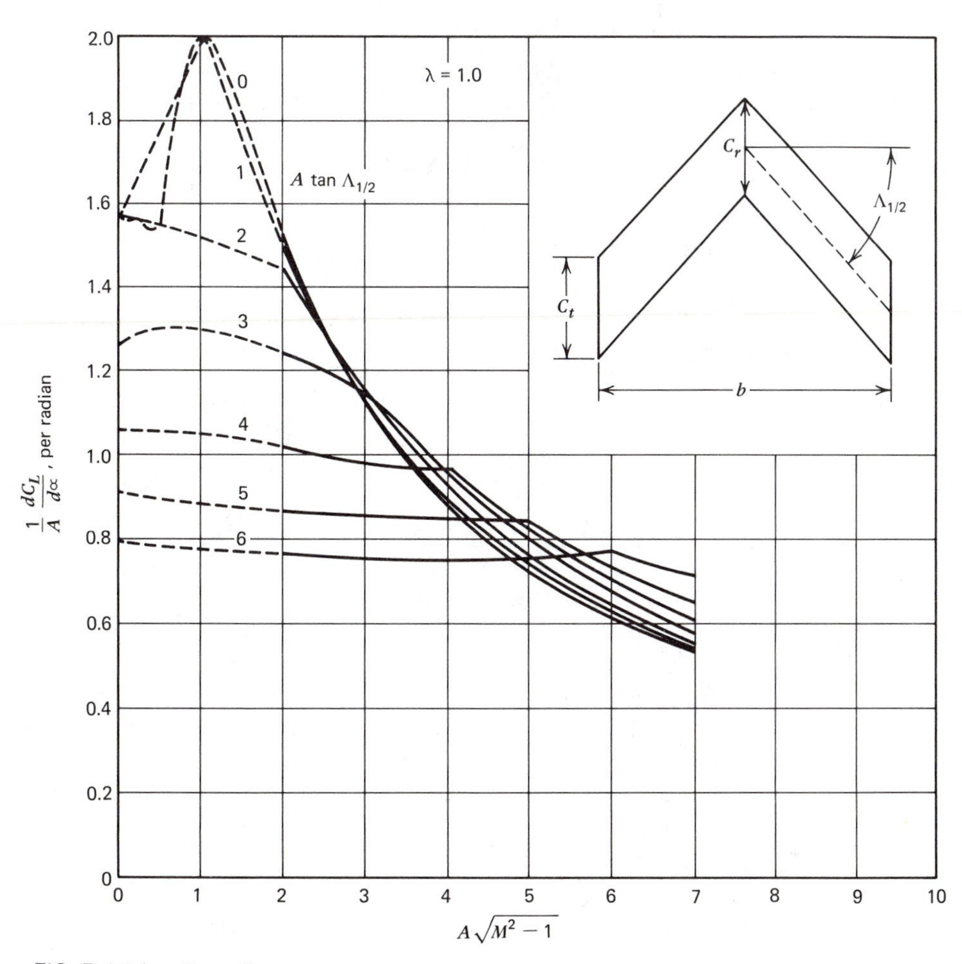

FIG. B.1,3 (continued)

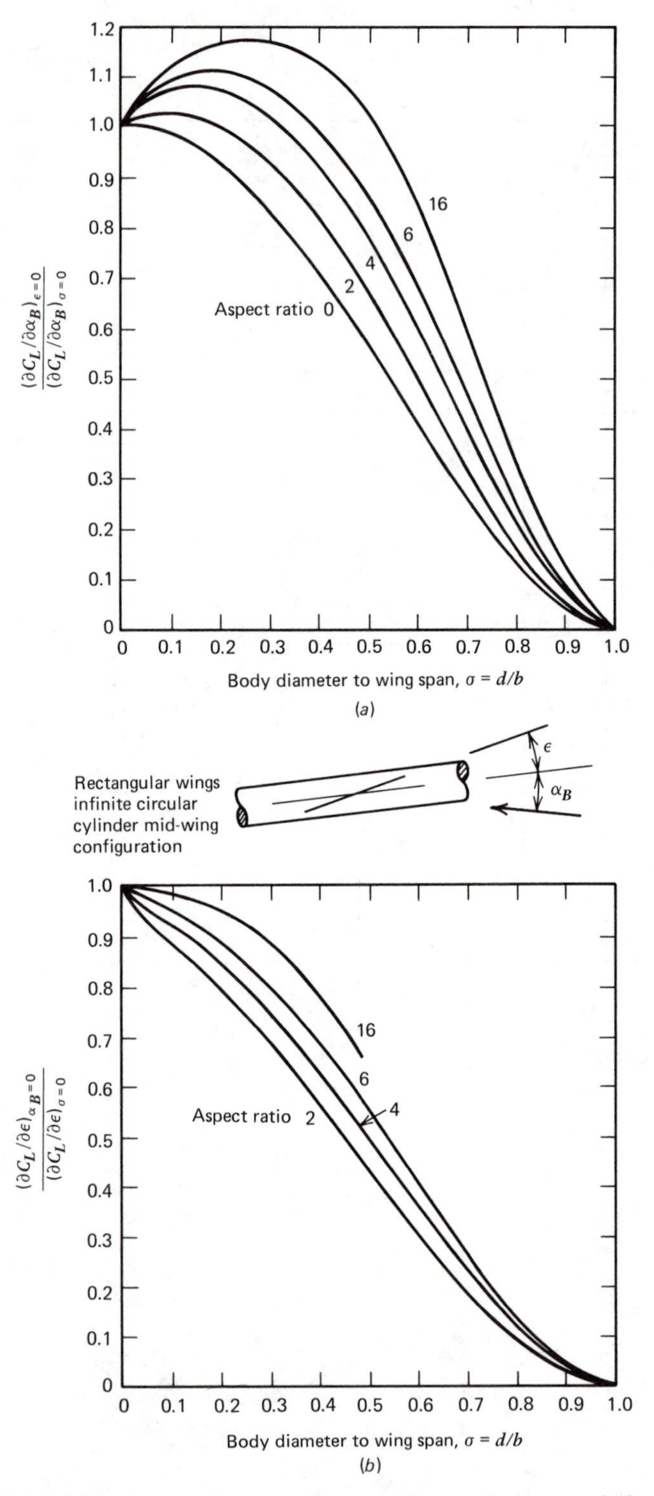

FIG. B. 1,4 Body effect on lift-curve slope expressed as a ratio of lift of wing-body combination to lift of wing alone.

(Reproduced from "Lift and Lift Distribution of Wings in Combination with Slender Bodies of Revolution" by H. J. Luckert, *Can. Aero. J.*, December 1955.)

Figure B.1,4: C_{L_α} (Body Effect). These curves apply only to unswept wings in mid-wing combination with an infinite circular cylinder body. For values of $A < 1$, the theory also applies to delta wings with pointed tips.

In Fig. B.1,4*a* the wing incidence is the same as that of the fuselage; i.e., $\varepsilon = 0$. In this case the lift of the wing-body combination increases to a maximum value, then decreases with increasing body diameter. Where there is a wing setting, i.e., $\varepsilon \neq 0$, and $\alpha_B = 0$ (Fig. B.1,4*b*), the lift of the combination decreases with increasing σ.

Figure B.1,5: $(a_F)_{BT}$. The left-hand curve shows the effect of the body on the fin lift-curve slope and gives $(a_F)_B/a_1$ in terms of D/h. This curve was based on wind-tunnel data and includes the loading induced on the body by the fin. The value of a_1 for the fin and rudder may be determined from Fig. B.1,2 using the geometric aspect ratio of the fin and rudder. The effect of the addition of a tail

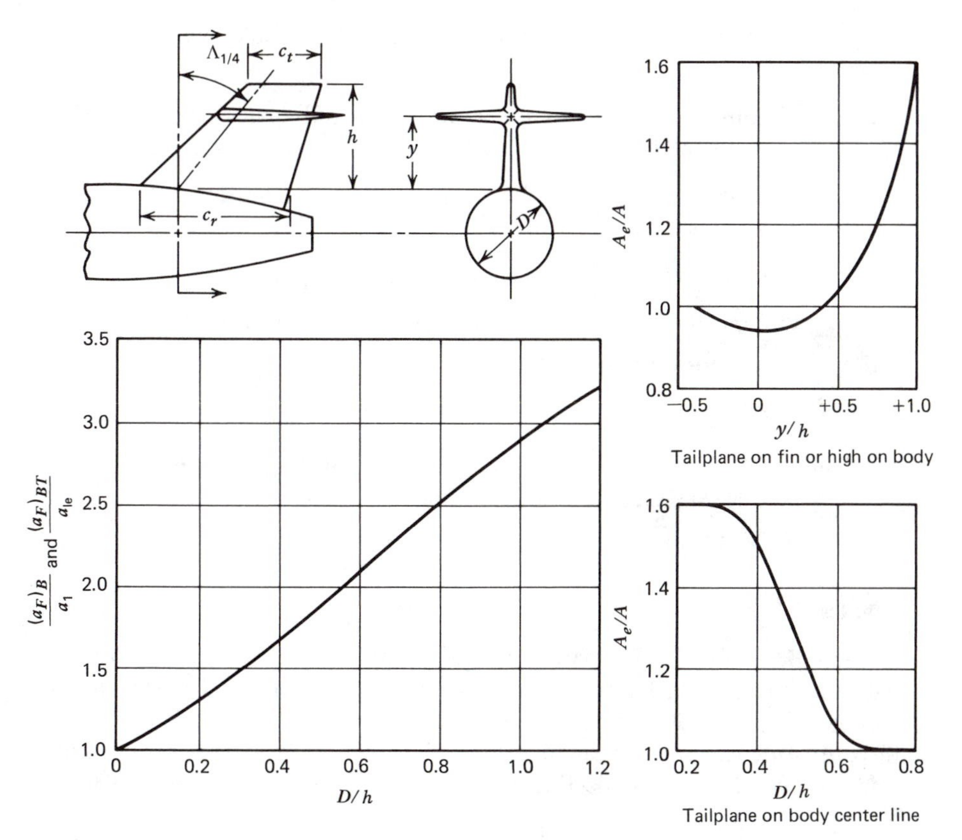

FIG. B.1,5 Lift-curve slope for single fin and rudder on a body of circular cross section.

(Reproduced from Royal Aeronautical Society Data Sheet Controls 01. 01. 05.)

plane was expressed as a change in fin aspect ratio; that is, A_e was so chosen that $(a_F)_{BT}/a_{1_e}$ is the same function of D/h as $(a_F)_B/a_1$. The lift-curve slope a_{1_e} is calculated from Fig. B.1,2 using the equivalent aspect ratio A_e.

The data were based on tests in which there was no wing, or in which the wing chord was on the body center line; with the wing at the bottom of the body the increased sidewash may increase a_F by a factor of 1.4, and decreased sidewash with the wing at the top of the body may reduce a_F by as much as 0.7′ The accuracy of the data is within $\pm 10\%$, except for a tail plane on the body center line, for which the accuracy was estimated to be $\pm 15\%$.

B.2 CONTROL EFFECTIVENESS

Notation

$(a_2)_{0T}$ theoretical rate of change of lift coefficient with control deflection for incompressible, inviscid two-dimensional flow

$(a_2)_0$ actual control effectiveness for incompressible two-dimensional flow

$(a_2)_M$ control effectiveness for a wing with full span controls at Mach number M

$(a_2)'_M$ control effectiveness for a wing with part span controls at Mach number M

$(a_1)_G$ lift-curve slope for a wing with control gap unsealed

$(a_2)_G$ control effectiveness for a wing with full-span control gap unsealed

f a factor defined by

$$(a_2)_G = f \cdot a_2[(a_1)_G/a_1]$$

Balance ratio of control-surface area forward of the hinge line to control-surface area aft of the hinge line

Notes

Figure B.1,2: $(a_2)_0$. The data presented in these curves apply only within the linear range of the wing-lift–incidence curve for controls with sealed gaps and for the range of control deflections where the increment in lift coefficients is linear with control deflection. For small values of incidence the curves should apply within a range of control deflection of $\pm 15°$. The accuracy of the data is within $\pm 5\%$.

Values of $(a_1)_0/(a_1)_{0T}$ for a particular airfoil may be determined from Fig. B.1,1.

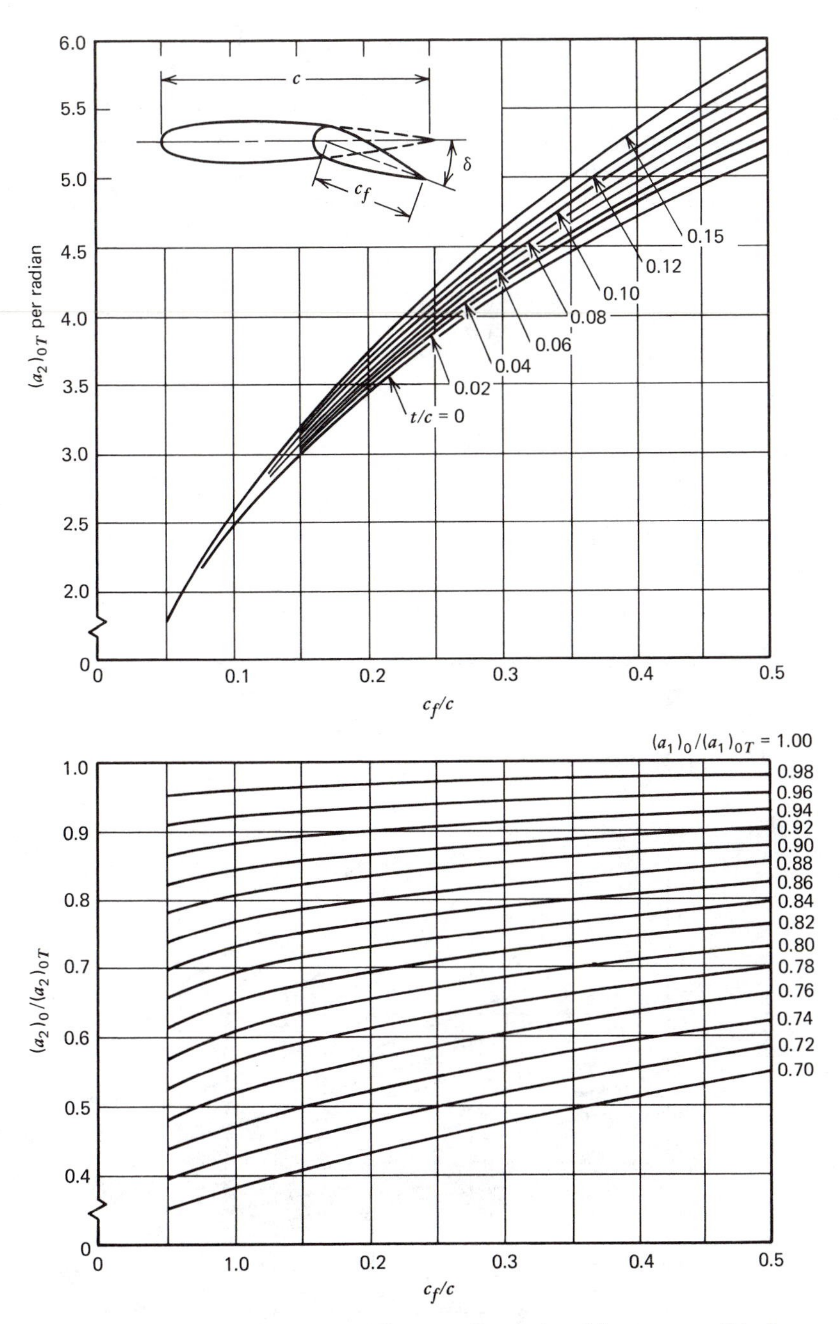

FIG. B.2,1 Control effectiveness for two-dimensional incompressible flow. (Reproduced from Royal Aeronautical Society Data Sheet Controls 01. 01. 03.)

Figure B.2,2: $(a_2)_M$. The curves are based on calculations (Multhopp's method) modified by experimental data on wings with fullspan flaps and apply to straight tapered wings with streamwise tips and constant c_f/c. The linearity limitations given for $(a_2)_0$ also apply to $(a_2)_M$.

The data can be used for Mach numbers less than the critical Mach number of the wing section. Values of $(a_2)_M$ calculated from the curves are accurate to within $\pm 10\%$.

Values of $(a_1)_M/(a_1)_0$ can be obtained from Sec. B.1.

If the control chord varies spanwise, the control effectiveness can be calculated by splitting the control into a number of part-span sections of assumed constant chord-ratio. The effectiveness is the sum of the effectiveness for each part-span section determined by the method of Fig. B.2,3.

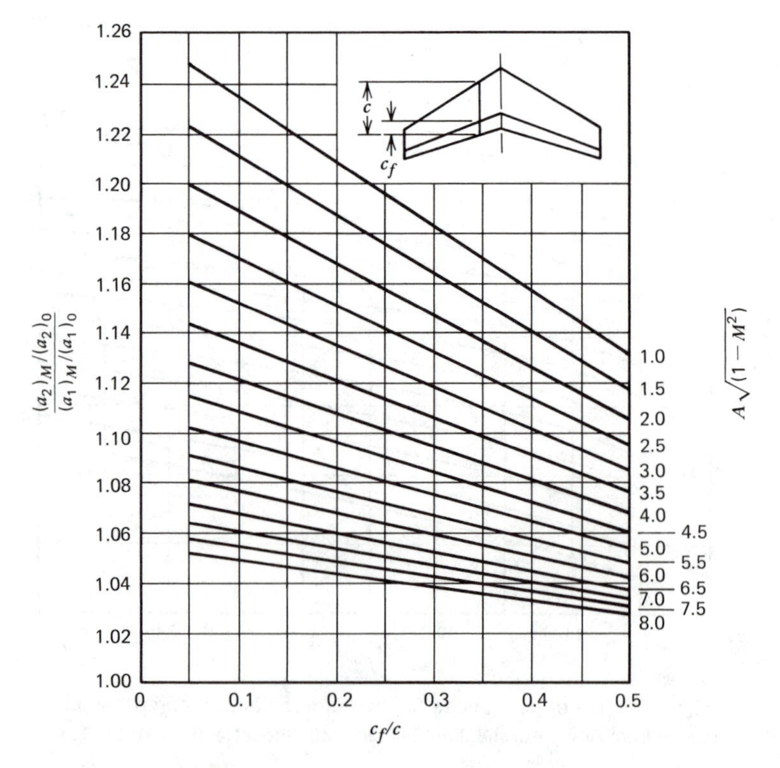

FIG. B.2,2 Control effectiveness for full-span controls on swept and tapered wings. (Reproduced from Royal Aeronautical Society Data Sheets Controls 01. 01. 06.)

Figure B.2,3: $(a_2)'_M$. The control effectiveness for part-span controls (or flaps) extending to the wing center line may be estimated from the $(a_2)_M$ for the corresponding full-span controls by means of an empirical relation derived from Weissinger's method:

$$(a_2)'_M = K(a_2)_M$$

where

$$K = k_1\{1 + k_2(A\sqrt{(1 - M^2)} - 6) + k_3 \sin[\tan^{-1}(\tan \Lambda_{1/2}/\sqrt{1 - M^2})]\}$$

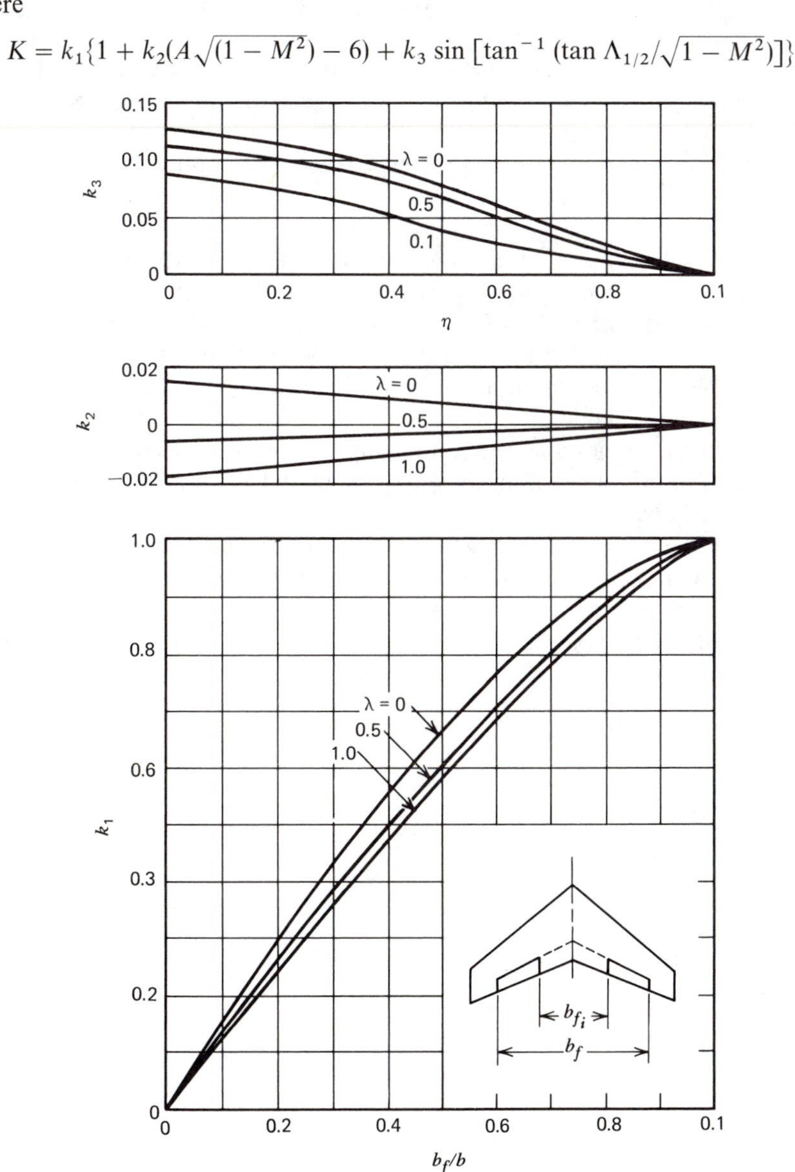

FIG. B.2,3 Conversion factor to obtain control effectiveness for part-span controls. (Reproduced from Royal Aeronautical Society Data Sheet Flaps 01. 01. 07.)

The factors k_1, k_2, k_3 given on the curves apply to straight tapered wings with streamwise tips and aspect ratios between 1.5 and 10. The ratio c_f/c should be constant across the span of the control. For unswept wings the difference between values calculated from the curves and the Weissinger calculations is less than $\pm 2\%$, and for swept wings this difference is less than $\pm 5\%$. Test data have shown that the curves are as reliable as the Weissinger calculations.

For cases where the inner ends of the control are not at the wing center line, $(a_2)'_M$ is equal to the difference between the control effectiveness for the two sets of controls whose spans (b_f and b_{fi}) are the distances between the outer and inner ends of the actual control.

A method for estimating control effectiveness of controls with supersonic leading and trailing edges is given in ref. 2.14.

Figure B.2,4: $(a_2)_G$. The gap effect on a_2 depends on the factor f and the ratio $(a_1)_G/a_1$. The values of $(a_1)_G/a_1$ given in the curves were based on wind-tunnel data and apply only when there is no flow separation over the airfoil. The effect of gap in two-dimensional flow may be used for calculating the effectiveness of controls on wings. The effect of gap at aspect ratio three can be utilized for the average tailplane or fin and rudder.

The factor f applies only within the range of control deflections for which a_2 is constant. For this data a range of $\pm 10^\circ$ was used. Unsealing the gap has only a small effect on this range.

The values of f were determined from tests where the c_f/c ratios varied from 0.25 to 0.40. Variations in f will occur for values of c_f/c outside of this range. The factor f applies to gaps greater than $0.004c$; for smaller gaps the magnitude of $(1 - f)$ should be reduced linearly with gap size.

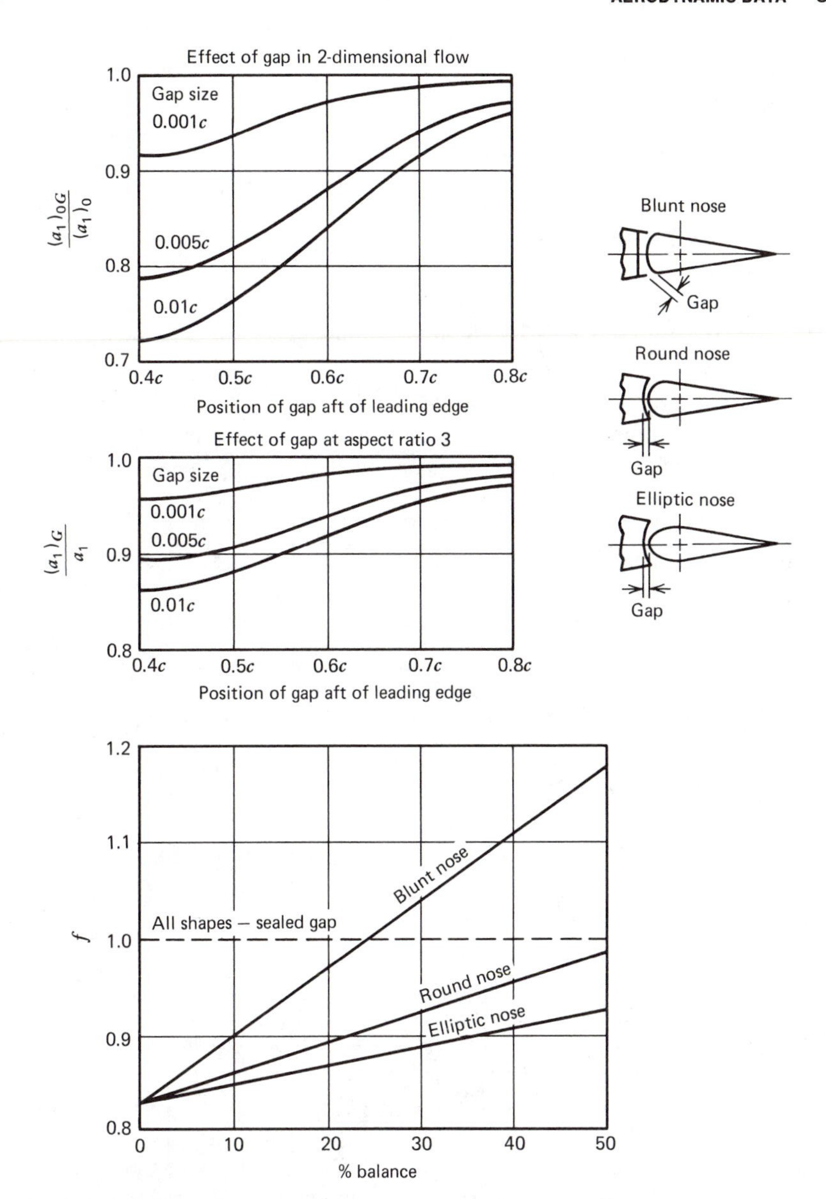

FIG. B.2,4 Effect of gap on lift-curve slope and control effectiveness.
(Reproduced from Royal Aeronautical Society Data Sheet Controls 01. 01. 04.)

B.3 FLAP EFFECTS

Notation

$\Delta\alpha_0$ change in zero-lift angle

$\Delta\delta$ flap deflection

E ratio flap chord ÷ airfoil chord

$\Delta C_{m_{ac_f}}$ increment in section $C_{m_{ac}}$ due to flaps

Figures B.3,1 and B.3,2 present data applicable to two-dimensional airfoils in low-speed flow. Figure B.3,1 shows the change in zero-lift angle for simple flaps with sealed gaps, and Fig. B.3,2 gives the change in C_{m_0} for several types of flap. Additional data on the effects of sweep-back and part-span flaps may be found in the Data Sheets of the Royal Aeronautical Society, and in Appendix B.2.

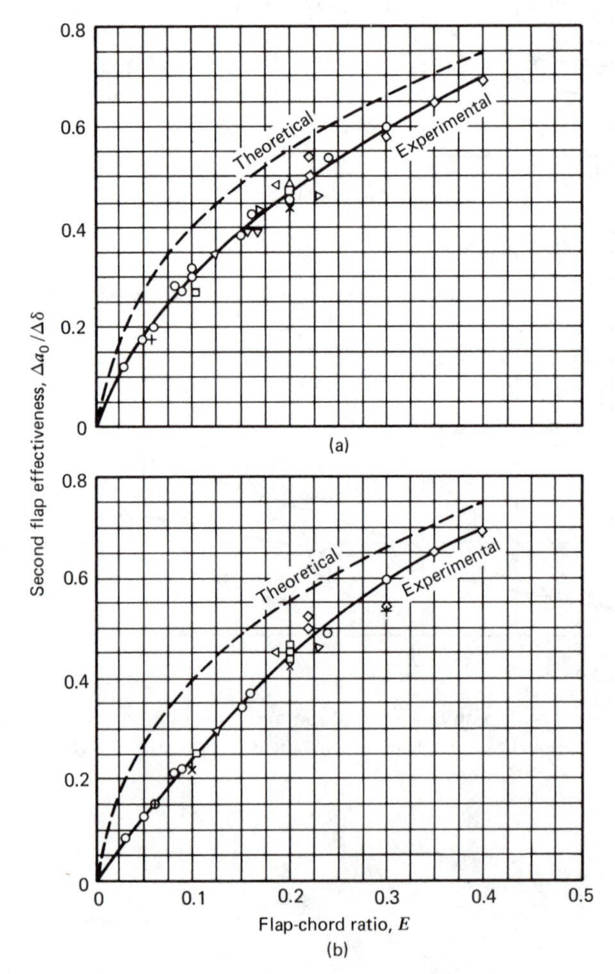

FIG. B.3,1 Change of zero-lift angle due to flap. (*a*) δ range from 0° to 10°. (*b*) δ range from 0° to 20°.

(By permission from *Theory of Wing Sections* by Abbott and Van Doenhoff, Fig. 96, copyright 9149, McGraw-Hill Book Co.)

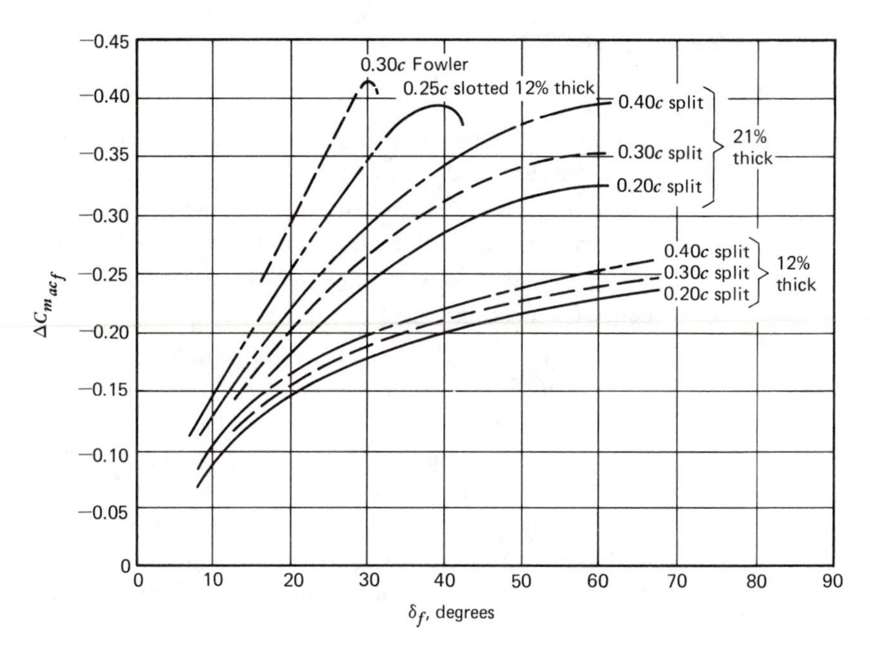

FIG. B.3,2 Effect of flap deflection on section pitching moments.
(Reproduced from *Airplane Performance, Stability, and Control* by Perkins and Hage, John Wiley & Sons, 1949, Fig. 5-40.)

B.4 HINGE MOMENTS

Notation

τ	trailing-edge angle defined by the tangents to the upper and lower surfaces at the trailing edge
$(b_1)_{0T}$	theoretical rate of change of hinge-moment coefficient with incidence for incompressible inviscid two-dimensional flow
$(b_1)_0$	actual rate of change of hinge-moment coefficient with incidence for incompressible two-dimensional flow
$(b_2)_{0T}$	theoretical rate of change of hinge-moment coefficient with control deflection for incompressible inviscid two-dimensional flow
$(b_2)_0$	actual rate of change of hinge-moment coefficient with control deflection for incompressible two-dimensional flow
$(b_1)_{0_{\text{bal}}}, (b_2)_{0_{\text{bal}}}$	rates of change of control hinge-moment coefficients with incidence and control-surface deflection, respectively, in two-dimensional flow for control surfaces with sealed gap and nose balance
$F_1, \alpha_i/\delta$	induced angle-of-incidence correction to $(b_1)_0$ and $(b_2)_0$, respectively, where F_1 is the value of $(\alpha_i/\delta)[(a_1)_0/(a_2)_0]$ when $c_f = c$
$F_2, \Delta(b_2)$	stream-line curvature correction to $(b_1)_0$ and $(b_2)_0$, respectively, where F_2 is the value of $\Delta(b_2)$ when $c_f = c$
F_3	factor to F_2 and $\Delta(b_2)$ allowing for nose balance
Balance	ratio of control-surface area forward of hinge line to control-surface area behind hinge line

Notes

Figures B.4,1 and B.4,2. The curves of Fig. B.4,1 were derived for a standard series of airfoils with plain controls for which $\tan(\frac{1}{2})\tau = t/c$ (referred to by an asterisk). To correct for airfoils with $\tan(\frac{1}{2})\tau$ different from t/c, values of $(b_1)^*_{0T}$, $(a_1)^*_{0T}$, and $(a_1)^*_0$ are calculated for the given t/c ratio; then $(b_1)_0$ is calculated from

$$(b_1)_0 = (b_1)^*_0 + 2[(a_1)^*_{0T} - (a_1)^*_0](\tan(\tfrac{1}{2}\tau - t/c). \qquad \text{(B.4,1)}$$

Values of $(a_1)^*_0/(a_1)^*_{0T}$ may be obtained from Fig. B.1,1.

The curves apply for values of incidence and control deflection for which there is no flow separation over the airfoil; for these conditions $(b_1)_0$ can be estimated to within ± 0.05. The data refer to sealed gaps but may be used if the gap is not greater than 0.002c. For larger gaps, a correction to the hinge-moment coefficient can be obtained from ref. 2.10.

The above discussion also applies to the data given in Fig. B.4,2 for $(b_2)_0$. The subscript 1 in Eq. B.4,1 becomes a subscript 2, and values of $(a_2)_0^*/(a_2)_{0T}^*$ may be obtained from Fig. B.2,1.

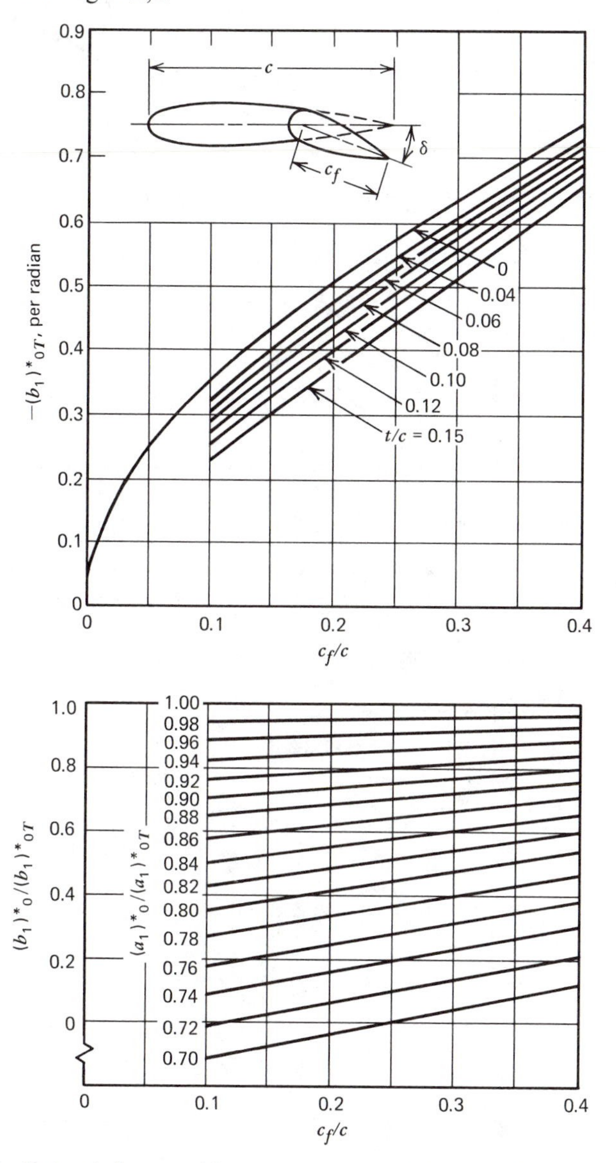

FIG. B.4,1 Rate of change of hinge-moment coefficient with incidence for a plain control in incompressible two-dimensional flow.
(Reproduced from Royal Aeronautical Society Data Sheet Controls 04. 01. 01.)

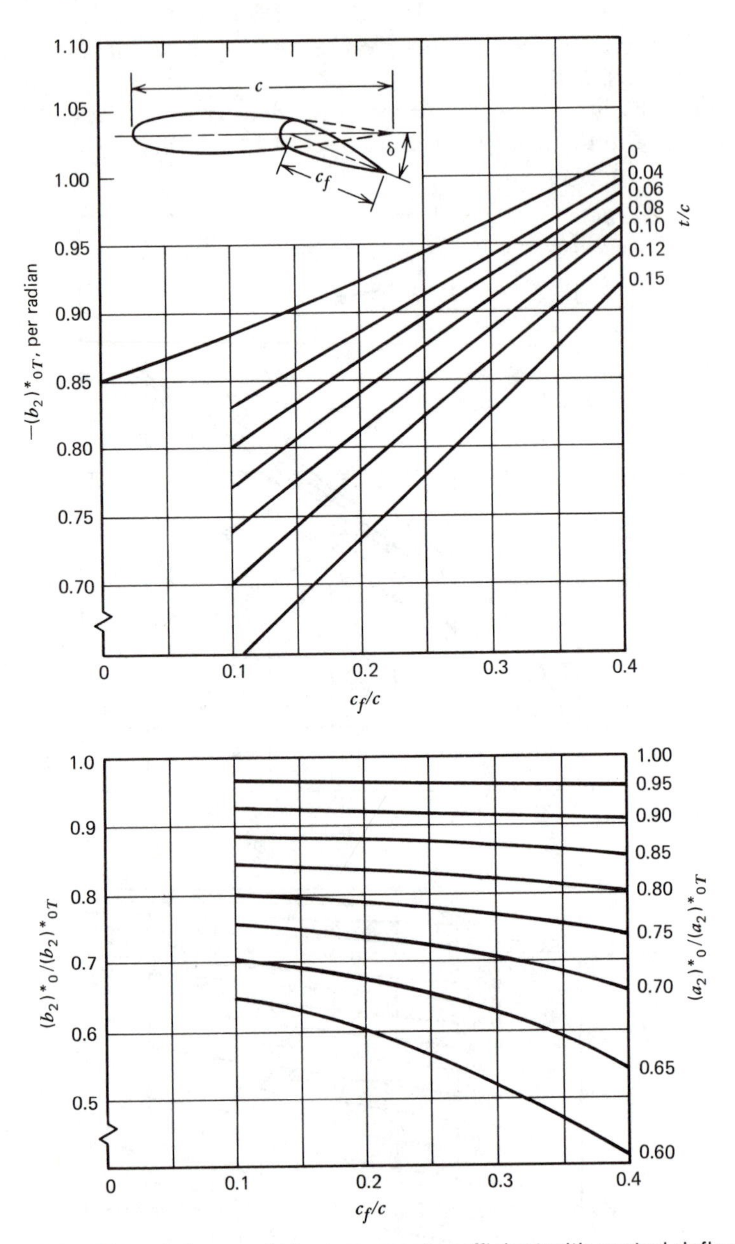

FIG. B.4,2 Rate of change of hinge-moment coefficient with control deflection for a plain in two-dimensional flow.

(Reproduced from Royal Aeronautical Society Data Sheet Controls 04. 01. 02.)

Figure B.4,3. The effect of nose balance on $(b_1)_0$ and $(b_2)_0$ can be estimated from the curves given on this figure. The data were obtained from wind-tunnel tests on airfoils with control-chord/airfoil-chord ratio of 0.3. Relatively small changes in nose and trailing-edge shape, and airflow over the control surface, may have a large effect on hinge moments for balanced control surfaces, so that estimates of nose-balance effect will be fairly inaccurate. If the control-surface gap is unsealed, the hinge-moment coefficients of plain and nosebalanced controls will generally become more positive.

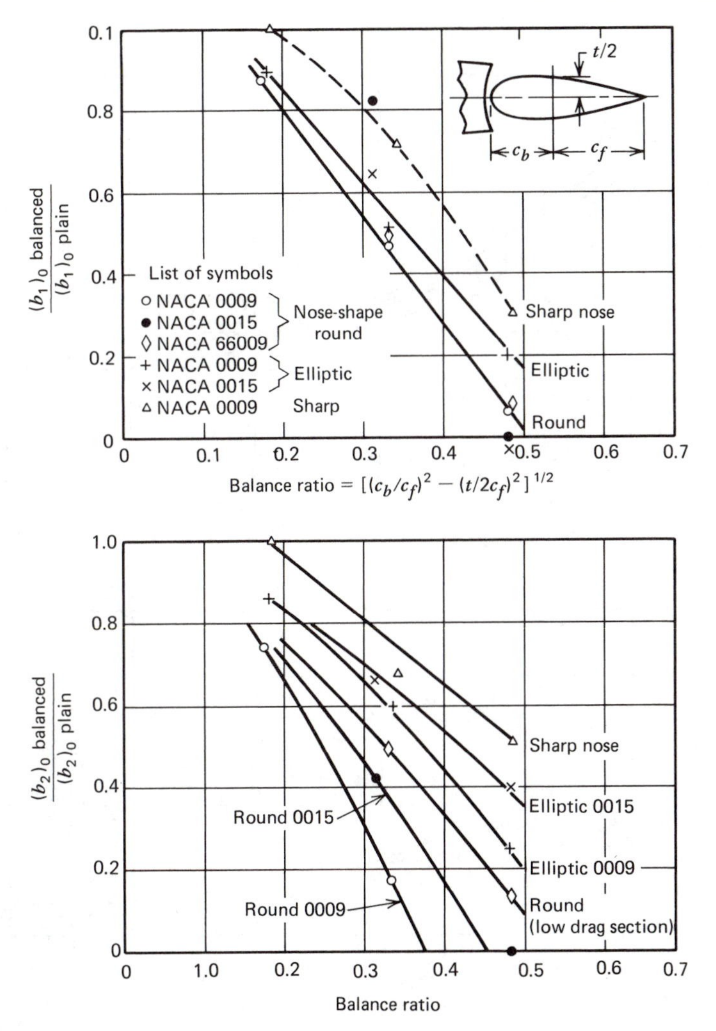

FIG. B.4,3 Effect of nose balance on two-dimensional plain-control hinge-moment coefficients.
(Reproduced from Royal Aeronautical Society Data Sheet Controls 04. 01. 03.)

Figure B.4,4. Two-dimensional hinge-moment coefficients for control surfaces with nose balance can be corrected to finite aspect ratio using the factors given in the curves and the following equations:

$$b_1 = (b_1)_0(1 - F_1) + F_2 F_3 (a_1)_0 \quad \text{(B.4,2)}$$

$$b_2 = (b_2)_0 - (\alpha_i/\delta)(b_1)_0 + \Delta(b_2) F_3 (a_2)_0 \quad \text{(B.4,3)}$$

For plain control surfaces the above equations are used with $F_3 = 1$. $(b_1)_0$ and $(b_2)_0$ can be obtained from Fig. B.4,1 and B.4,2, respectively, for plain controls.

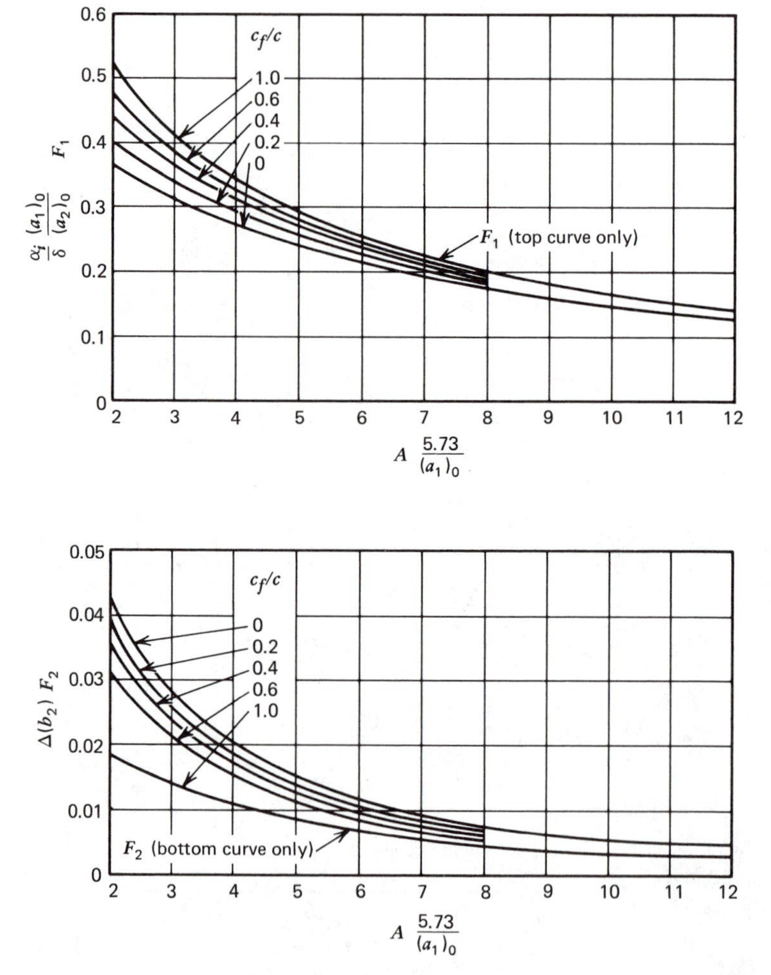

FIG. B.4,4 Finite-aspect-ratio corrections for two-dimensional plain and nose-balanced control hinge-moment coefficients.
(Reproduced from Royal Aeronautical Society Data Sheet Controls 04. 01. 05.)

For nose-balanced controls, the two-dimensional coefficients must include the effect of nose balance. Values of $(a_1)_0$ can be obtained from Fig. B.1,1, and those for $(a_2)_0$ from Fig. B.2,1.

Lifting-surface theory was applied to unswept wings with elliptic spanwise lift distribution to derive the factors. Full-span control surfaces were assumed together with constant ratios of c_f/c and constant values of $(b_1)_0$ and $(b_2)_0$ across the span. The factors apply to wings with taper ratios of 2 to 3 if c_f/c, $(b_1)_0$ and $(b_2)_0$ do not vary by more than $\pm 10\%$ from their average values. Hinge-moment coefficient corrections for variations with span of control-chord ratios and section shapes are given on Royal Aeronautical Society Data Sheet Controls 04.01.06, 1949.

A method for estimating hinge-moment parameters for controls with supersonic leading and trailing edges is given in ref. 2.14.

B.5 TAB EFFECTIVENESS

Notation

b_3	rate of change of control hinge-moment coefficient with tab angle
S'_F	area of control surface aft of control hinge line measured over span of tab
S_1	area of control surface forward of control hinge line, measured over span of tab
S_1/S'_F	local forward balance
τ	trailing-edge angle
F	correction factor for trailing-edge angle for tabs with sealed gap

Notes

The curves given in Fig. B.5 are based on experimental data for controls at incidences up to 5° and control and tab deflections within $\pm 10°$. The airfoil sections tested had maximum thickness at about $0.30c$ behind the leading edge and a forward transition point. Tabs with gap of the order of $0.001c'$ may be considered to be sealed. The accuracy of the data is within $\pm 10\%$, and variations in b_3 for gaps larger than $0.001c'$ will be within this range.

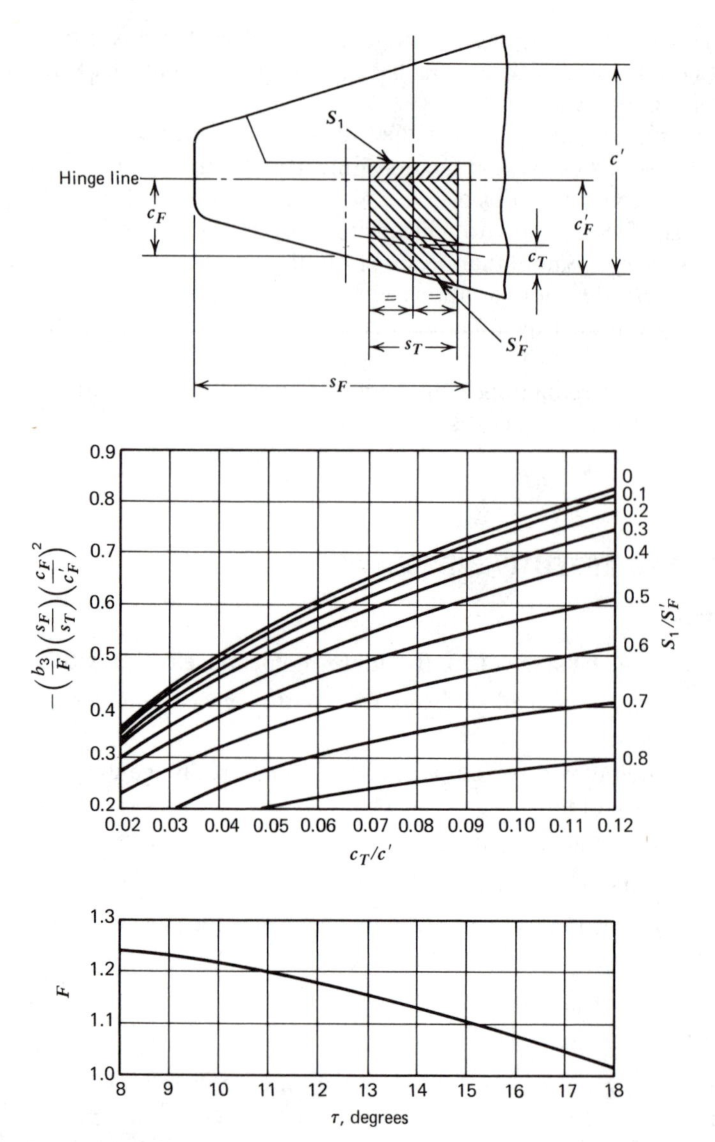

FIG. B.5 Tab effectiveness.
(Reproduced from Royal Aeronautical Society Data Sheet Controls 04. 01. 08.)

B.6 DOWNWASH

Notation

δ wing semispan

A aspect ratio

x distance behind quarter-chord point of root chord, parallel to the direction of air flow

h distance (up or down) from median plane of vortex sheet, measured normal to the direction of air flow

d distance of median plane of vortex sheet, measured normal to the direction of air flow

C_L wing lift coefficient

ε downwash angle, averaged over mid 30% of span

The graphs presented have been reproduced from the Data Sheets of the Royal Aeronautical Society. They were derived from potential flow theory, using a lifting line representation of the wing. They apply to subsonic flow only, and to wings of sweep less than 15°. For interpolation, values of ε/C_L should be cross-plotted against $1/A$.

Figure B.6,1 gives the values of the vortex-sheet deflection (d/sC_L) from which the value of d required for the determination of h is found. Figures B.6,2 and 3 give ε/C_L for various taper ratios, aspect ratios, and tail positions.

For a method applicable to swept wings see ref. 2.5, and for supersonic wings see ref. 2.13.

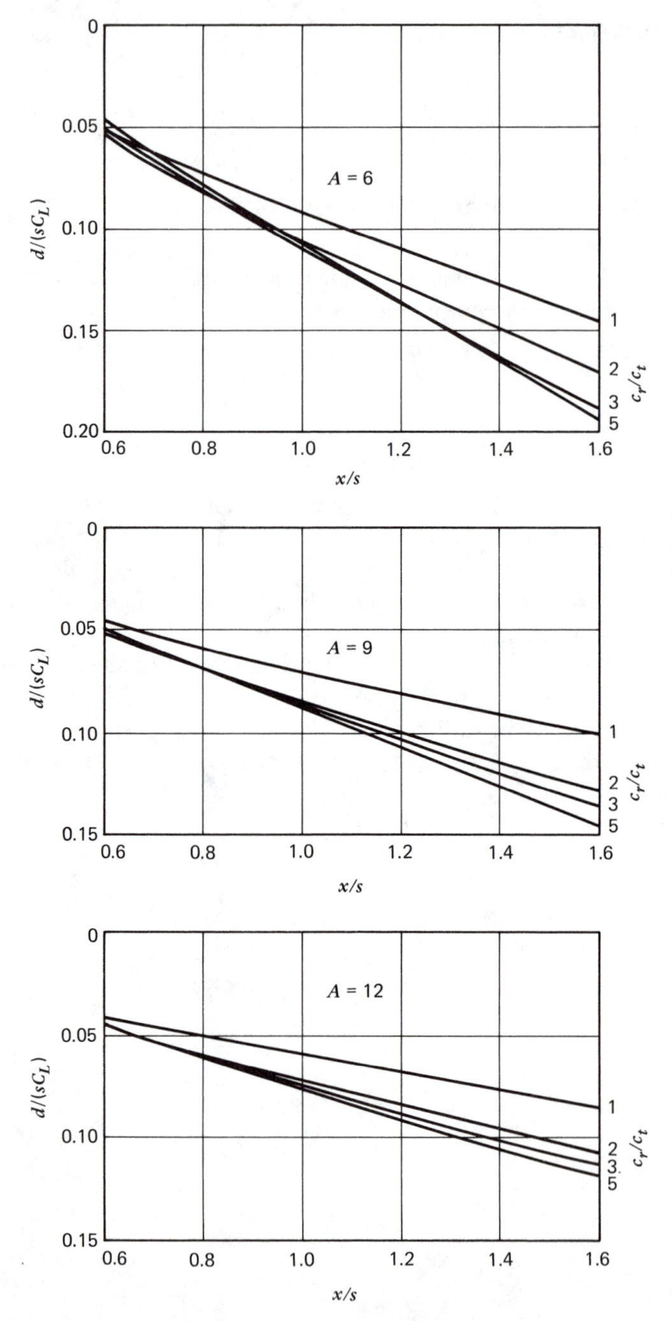

FIG. B.6,1 Displacement of median plane of vortex sheet.
(Reproduced from Royal Aeronautical Society Data Sheet Aircraft 08. 01. 04.)

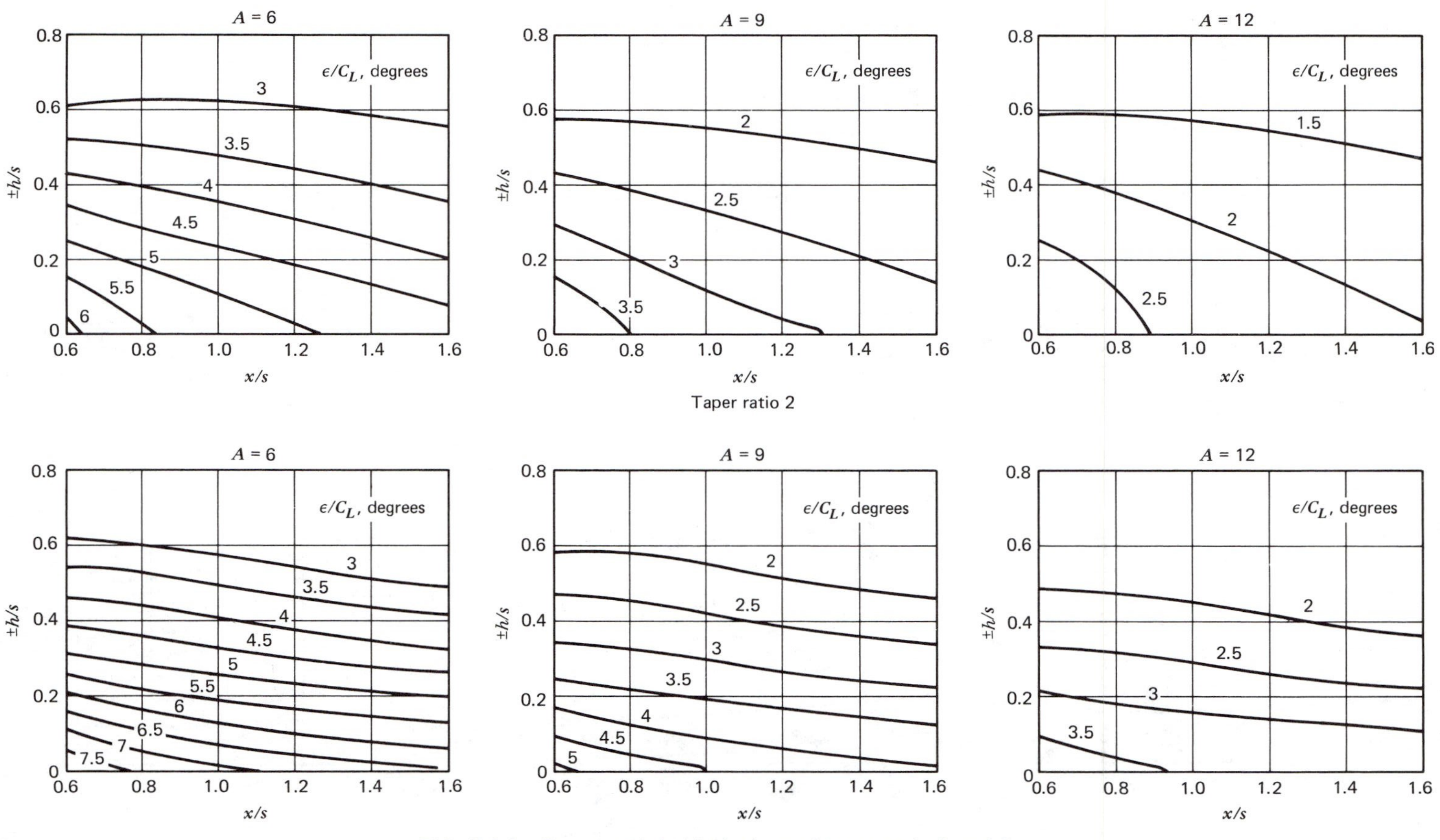

FIG. B.6,2 Downwash behind wings of taper ratio 0 and 2.
(Reproduced from Royal Aerontujical Society Data Sheet Aircraft 08. 01. 02.)

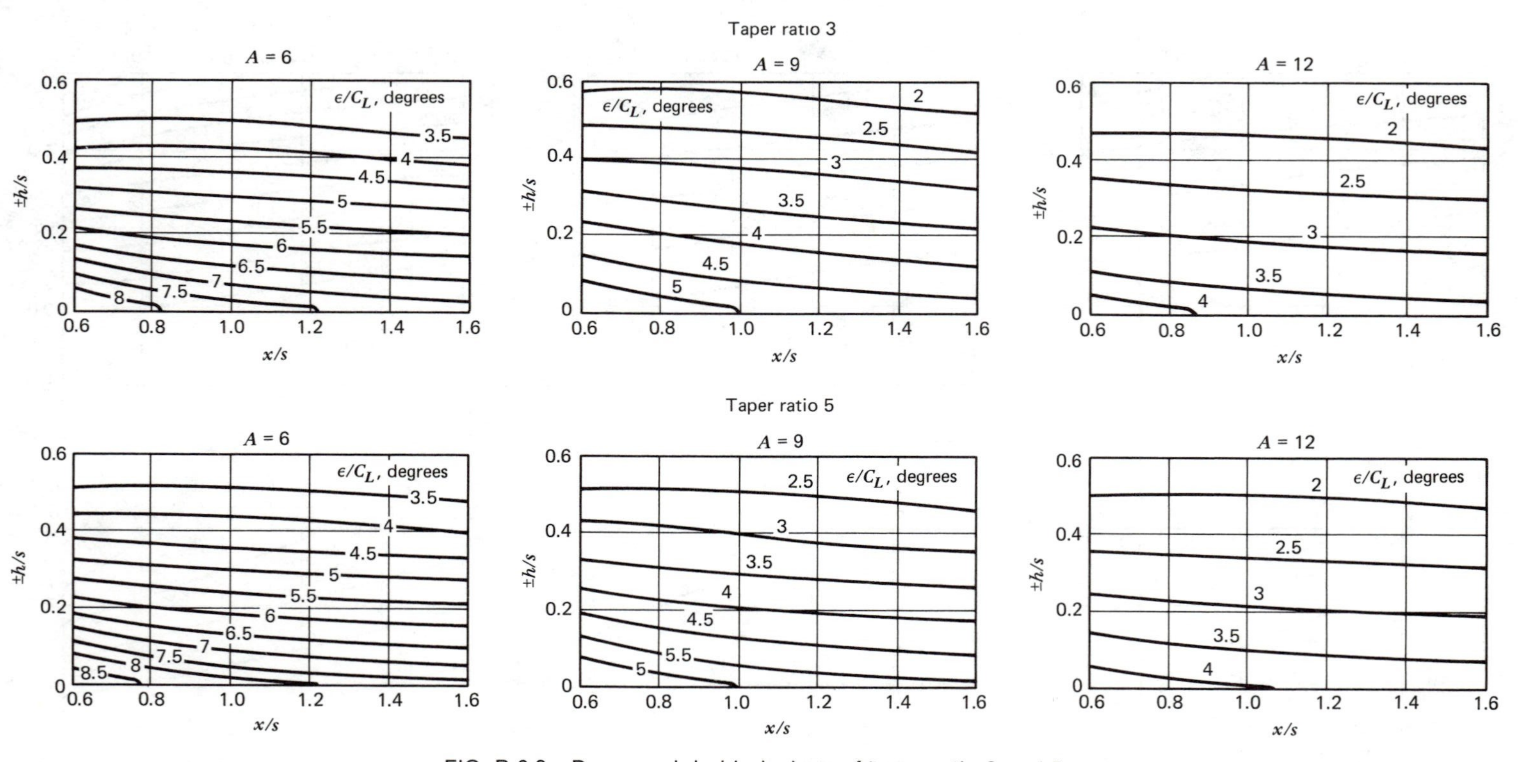

FIG. B.6,3 Downwash behind wings of taper ratio 3 and 5.
(Reproduced from Royal Aeronautical Society Data Sheet Aircraft 08. 01. 03.)

B.7 GROUND EFFECT

The principal effects of the ground on a nearby airplane are an increase in the lift-curve slope of wing and tail, and a decrease in the downwash at the tail.

The first of these is given in Fig. B.7,1, taken from ref. 3.14. These data may be used to estimate the value of C_{L_α} in the presence of the group ($C_{L_{\alpha G}}$) for both wing and tail plane.

The decrease in the downwash angle for wings with flaps deflected, and with or without slipstream present, can be estimated by the method of Owen and Hogg (ref. 3.20). It is given by

$$(\Delta\varepsilon)_g = \varepsilon \frac{b_1{}^2 + 4(h - H)^2}{b_1{}^2 + 4(h + H)^2}$$

where ε = downwash angle remote from the ground, at the same α and thrust coefficient

h = height of tailplane root quarter-chord-point above the ground

H = height of the wing root quarter-chord-point above the ground

$b_1 = (C_{Lw} + \Delta C_{L_f})/(C_{Lw}/b'_w + \Delta C_{L_f}/b'_f)$

C_{Lw} = lift coefficient of wing, flaps retracted

ΔC_{L_f} = increment in lift coefficient due to flaps

b'_w, b'_f are given in Fig. B.7,2.

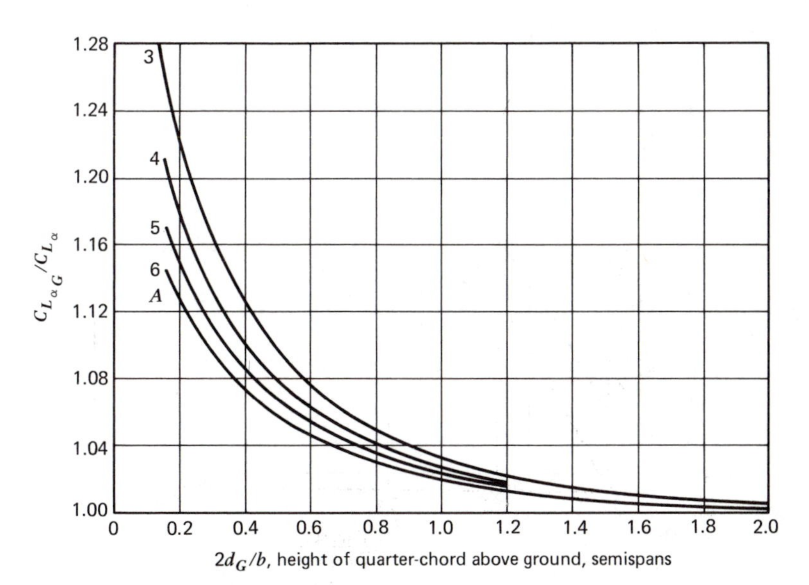

FIG. B.7,1 Ground effect on lift-curve slope.
(Reproduced from *NACA Wartime Rept. WR L95*, 1944, by R. F. Goranson.)

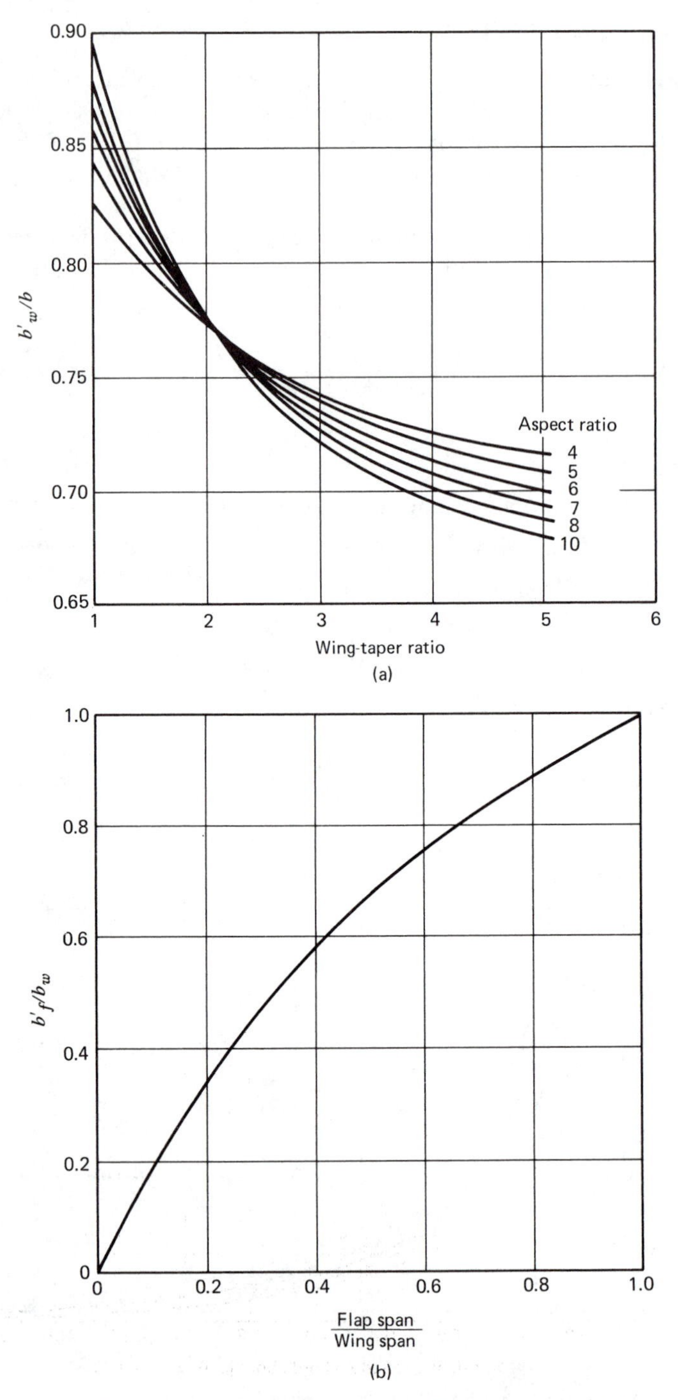

FIG. B.7,2 (*a*) Effective span of basic wing. (*b*) Effective span of flaps. (Reproduced from *ARC R&M 2449*, 1952, by P. R. Owen and H. Hogg.)

B.8 EFFECT OF BODIES ON NEUTRAL POINT AND C_m

Notation

c	local wing chord at center line of fuselage or nacelle
$\bar{c}$	mean aerodynamic chord
w	maximum width of fuselage or nacelle
S	gross wing area
Δh_n	shift of neutral point due to fuselage or nacelle as a fraction of $\bar{c}$, positive aft
S_B	area of planform of body
S_{BF}	area of planform of body, forward of $0.25\bar{c}$
c'	root chord of wing without fillets
$(C_{m_0})_B$	increment to C_{m_0} due to a body at zero lift
θ	reflex angle of fillet, i.e., angle between wing root chord and lower surface of fillet for upswept fillets, or the upper surface for downswept fillets, positive as indicated in Fig. B.8,2
λ	fillet lift-increment ratio, i.e., $(a_2)_0/(a_1)_0$, considering the fillet to be a flap of chord l_f

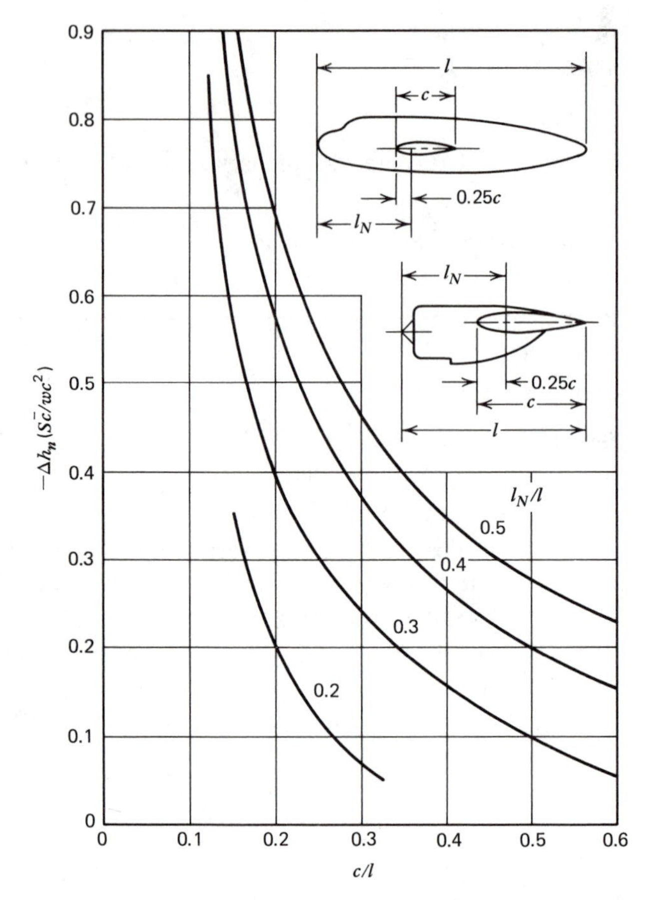

FIG. B.8,1 Effect of a fuselage or nacelle on neutral-point position. (Reproduced from Royal Aeronautical Society Data Sheet Aircraft 08. 01. 01.)

Notes

Figure B.8,1: Δh_n. The data for estimating Δh_n presented in this graph were derived from wind-tunnel tests. The forward shift in neutral point is mainly dependent on the length and width of the body forward of the wing. The values of Δh_n given by the curves are accurate to within $\pm 0.01\bar{c}$, and are about 5% higher for low-wing, and the same amount lower for high-wing configurations. The data are inapplicable if the wing is clear of the body. Separate values should be computed for fuselages and nacelles, and the results added to obtain the total neutral-point shift.

Figure B.8,2: $(C_{m_0})_B$. The curves given in this figure apply to stream-line bodies of circular or near circular cross section with midwing configurations. For high- or low-wing configurations a positive or negative $\Delta(C_{m_0})_B = 0.004$ is added,

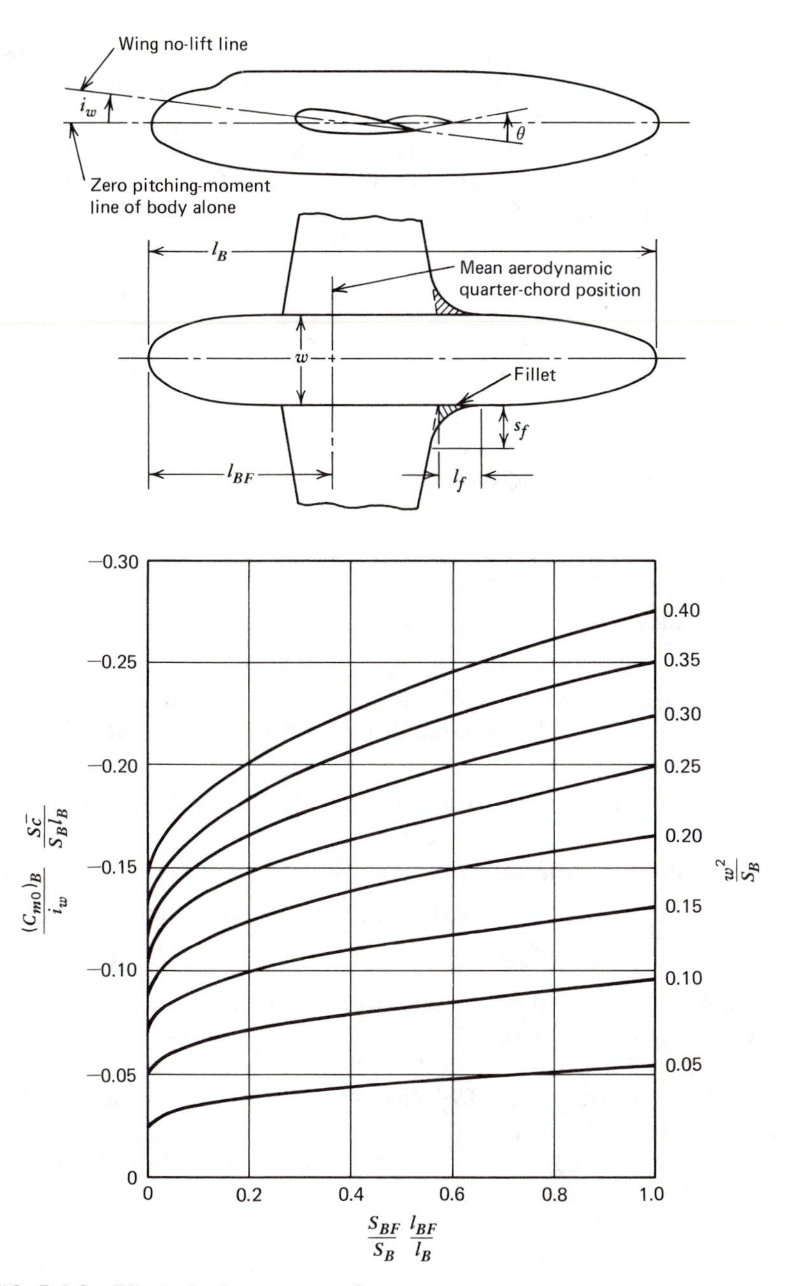

FIG. B.8,2 Effect of a fuselage on C_{m_0}.
(Reproduced from Royal Aeronautical Society Data Sheet Aircraft 08. 01. 07.)

respectively, to the value derived from the curves. The curves apply only for angles of incidence up to about 15° for stream-line bodies where the pitching moment of the body varies linearly with incidence.

In the wind-tunnel tests from which the data were derived, the wings had straight trailing edges at the wing-body junction. Fillets have a large effect on C_{m_0}, however, especially if θ is large. The following equation may be used to estimate the fillet effect if $0.12 < l_f/c < 0.5$ and $0.03 < S_f/b < 0.075$:

$$C_{m_0} \text{ due to fillets} = [0.046 + 0.08(dC_m/dC_L)_w \lambda\theta - 0.2(c + l_f)/c](w + S_f)/b$$

The value obtained from the curves, the fillet effect, and the effect due to wing position are added to determine $(C_{m_0})_B$.

B.9 JET-INDUCED DOWNWASH

Notation

- r radial distance from jet axis
- S wing area
- T'_c thrust coefficient, $T/(\frac{1}{2})\rho V^2 S$
- x axial distance from point at which jet, in accordance with the law of spreading that holds at large distances from the exit, would have zero cross section
- α_e angle of attack of thrust axis relative to average flow between jet and tail
- ε jet-induced flow inclination
- $\bar{\varepsilon}$ mean jet-induced downwash angle over horizontal tail
- θ local inclination of jet axis to general flow

The downwash induced on the tail by one or more jets may be estimated by the method of Ribner (ref. 3.2). In order to use the data given, it may be assumed that the theoretical origin of the jet (from which x is measured) is 2.3 diameters upstream of the jet exit. Figure B.9,1 gives curves from which ε can be estimated for any distance r from the center of the jet. The location of the jet center line relative to the tail can be found from the airplane geometry, the angular deviation derived from Fig. B.9,2, and the relation

$$\Delta r = -\frac{\alpha_e}{57.3}(x - x_j)(1 - \theta/\alpha_e)_{\text{avg}}$$

Here $(x - x_j)$ is the distance from the jet exit to the tail, and $(1 - \theta/\alpha_e)_{\text{avg}}$ is the average value of $(1 - \theta/\alpha_e)$ between the jet exit and the elevator hinge line, minus the value at the jet orifice.

Finally the average effect of the jet across the tailplane is obtained from Fig. B.9,3.

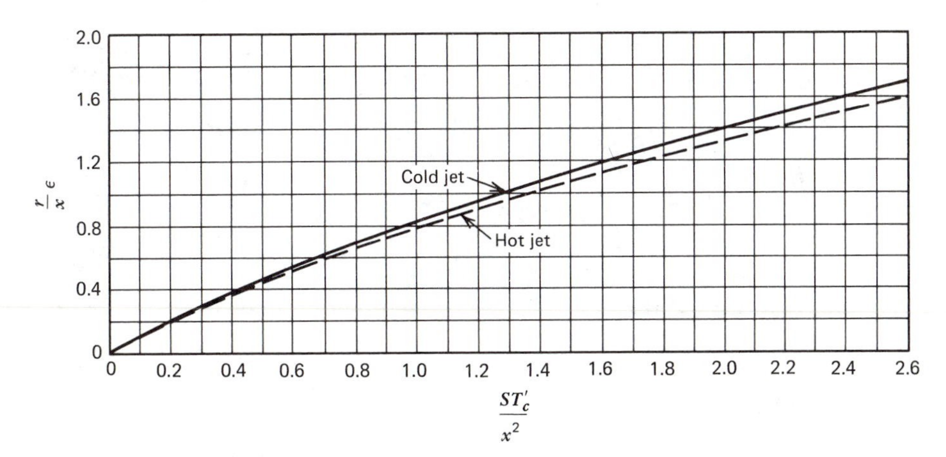

FIG. B.9,1 Flow inclination outside a jet.
(Reproduced from *NACA Wartime Rept. L-213*, 1946, by H. S. Ribner.)

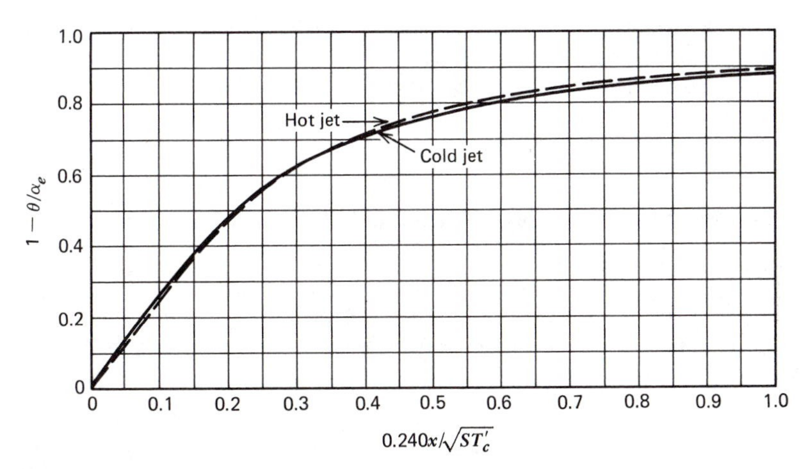

FIG. B.9,2 Angular deviation of jet due to angle of attack.
(Reproduced from *NACA Wartime Rept. L-213*, 1946, by H. S. Ribner.)

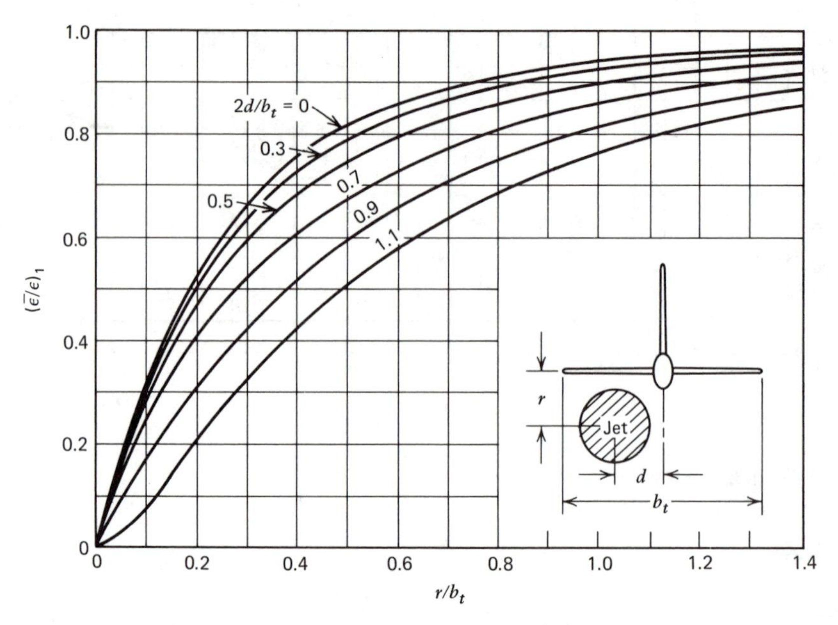

FIG. B.9,3 Ratio of the effective mean downwash $\bar{\varepsilon}$ induced by the jet over the tail plane to the flow inclination ε at radius *r*.
(Reproduced from *NACA Wartime Rept. L-213*, 1946, by H. S. Ribner.)

B.10 PROPELLER AND SLIPSTREAM EFFECTS

Propeller Normal Force

The following method of estimating the propeller normal force is due to Ribner (ref. 3.3). The normal force is expressed in terms of the derivative $\partial C_{N_p}/\partial \alpha_p$ (see Sec. 3.5), which is given by

$$\partial C_{N_p}/\partial \alpha_p = f C_{Y'_{\psi_0}}$$

The factor f is the same for all propellers, and is given in Fig. B.10,1 as a function of $T_c = T/\rho V^2 d^2$. The value of $C_{Y'_{\psi_0}}$ varies with the propeller and its operating condition. The values for a particular propeller family are given in Fig. B.10,2. Extrapolation to other propellers can be made by means of Fig. B.10,3, on the basis of the "side-force-factor," SFF. This is a geometrical propeller parameter, given approximately by

$$\text{SFF} = 525[(b/D)_{0.3} + (b/D)_{0.6}] + 270(b/D)_{0.9}$$

where (b/D) is the ratio of blade width to propeller diameter, and the subscript is the relative radius at which this ratio is measured.

Also given in ref. 3.3 are some curves which are useful for estimating the upwash or downwash at the propeller plane. These are reproduced in Fig. B.10,4.

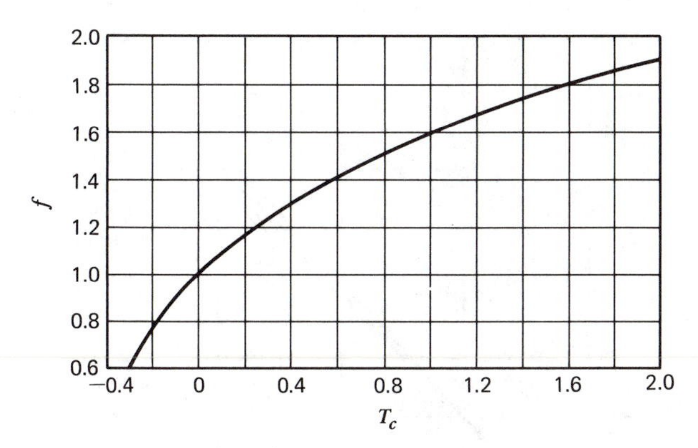

FIG. B.10,1 Variation of f with T_c.
(Reproduced from *NACA Wartime Rept. L-25*, 1944, by H. S. Ribner.)

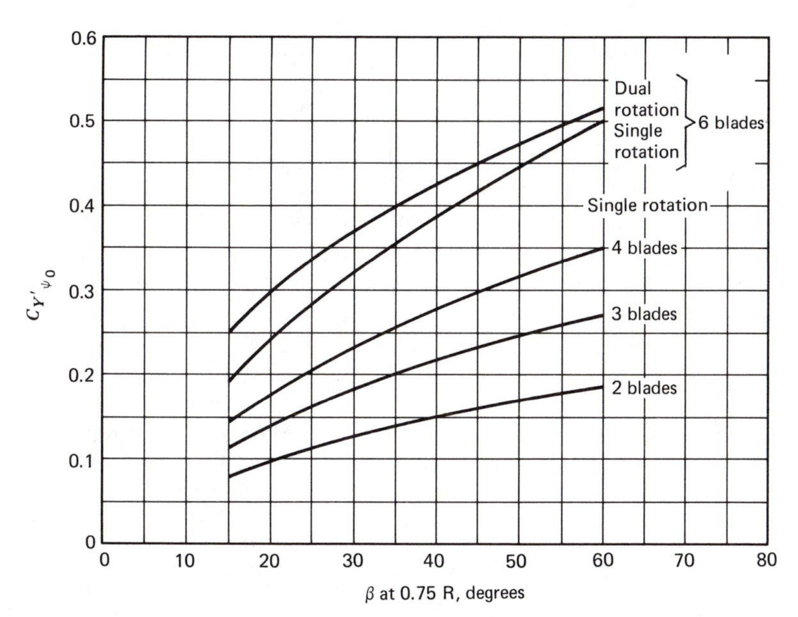

FIG. B.10,2 Variation of $C_{Y'_{\psi 0}}$ with blade angle.
(Reproduced from *NACA Wartime Rept. L-25*, 1944, by H. S. Ribner.)

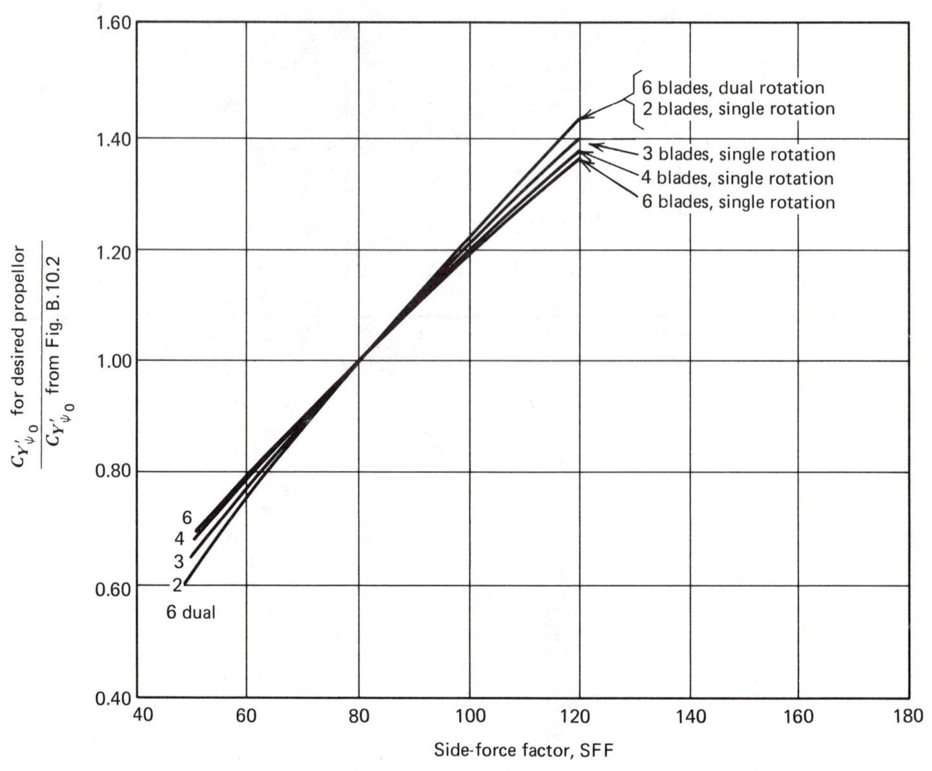

FIG. B.10,3 Ratio of normal force derivatives.
(Reproduced from *NACA Wartime Rept. L-25*, 1944, by H. S. Ribner.)

Lift Due to Slipstream

The method of Smelt and Davies (ref. 3.5) can be used to estimate the added wing lift due to the slipstream. It is given by

$$\Delta C_L = \frac{D_1 c}{S} s(\lambda C_{L_0} - 0.6a_0\theta)$$

where D_1 = diameter of slipstream at the wing C.P.

$= D[(1 + a)/(1 + s)]^{1/2}$

c = wing chord on center line of slipstream

S = wing area

$s = a + ax/(D^2/4 + x^2)^{1/2}$

D = propeller diameter

$a = -\frac{1}{2} + \frac{1}{2}(1 + 8T_c/\pi)^{1/2}$

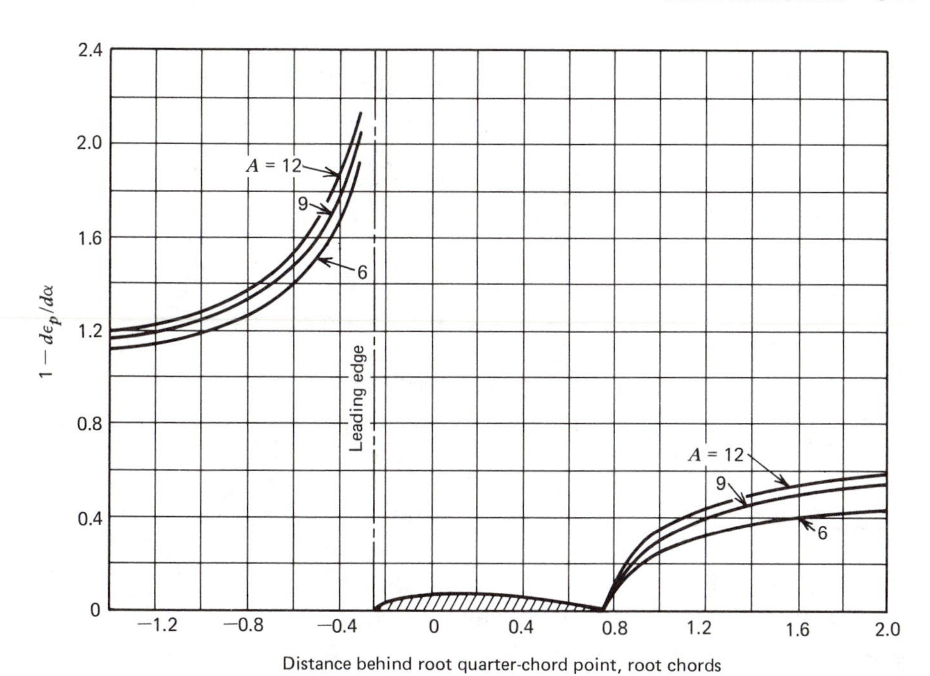

FIG. B.10,4 Value of $1 - d\varepsilon/d\alpha$ on longitudinal axis of elliptic wing for aspect ratios 6,9, and 12.
(Reproduced from *NACA Wartime Rept. L-25*, 1944, by H. S. Ribner.)

x = distance of wing C.P. behind propeller

C_{L0} = lift coefficient at section on slipstream center line, in absence of the slipstream

a_0 = two-dimensional lift-curve slope of wing section

θ = angle of downwash of slipstream at wing C.P. calculated from the equation

$$1/\theta^{0.8} = 0.016x/D + 1/\theta_0{}^{0.8}$$

where $\theta_0 = a\phi/(1 + a)$

ϕ = angle between propellor axis and direction of motion.

λ is an empirical constant given in Fig. B.10,5.

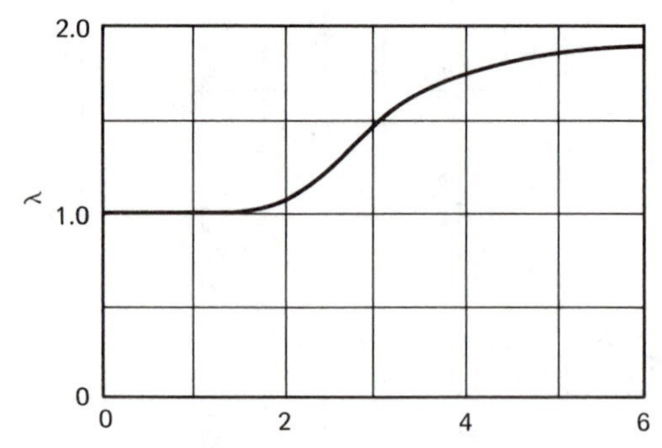

FIG. B.10,5 Empirical factor λ.
(Reproduced from *ARC R&M 1788*, 1937, by R. Smelt and H. Davies.)

B.11 ROLLING-MOMENT DERIVATIVES

C_{l_p}

Theoretical values of C_{l_p} for straight tapered wings are given in Fig. B.11,1. They are for subsonic flow, and apply to Mach numbers below the critical. The following notation applies:

β $\sqrt{1 - M^2}$

κ ratio of actual section lift-curve slope to theoretical value

Λ_e $\tan^{-1}[(1/\beta)\tan\Lambda_{1/4}]$ degrees

λ c_t/c_r

For sources of data on supersonic wings, see Table II of ref. 5.22.

C_{l_β}

Values of the contribution of the wing planform to C_{l_β} are given in Fig. B.11,2. They are for subsonic flow, and are based on test results and theory. To these values must be added the increments due to dihedral and wing-body interference. The former may be estimated from Fig. B.11,3, and the latter from the formula given by Campbell and McKinney (ref. 5.22); i.e.,

$$\Delta C_{l_\beta} = 1.2\sqrt{A}\,\frac{z_w}{b}\cdot\frac{h + w}{b}$$

where A = aspect ratio

z_w = vertical distance of wing-root $\frac{1}{4}$ chord point below fuselage center line

b = wing span

h = average fuselage height at wing root

w = average fuselage width at wing root

For sources of information C_{l_β} of supersonic wings, see Table II of ref. 5.22.

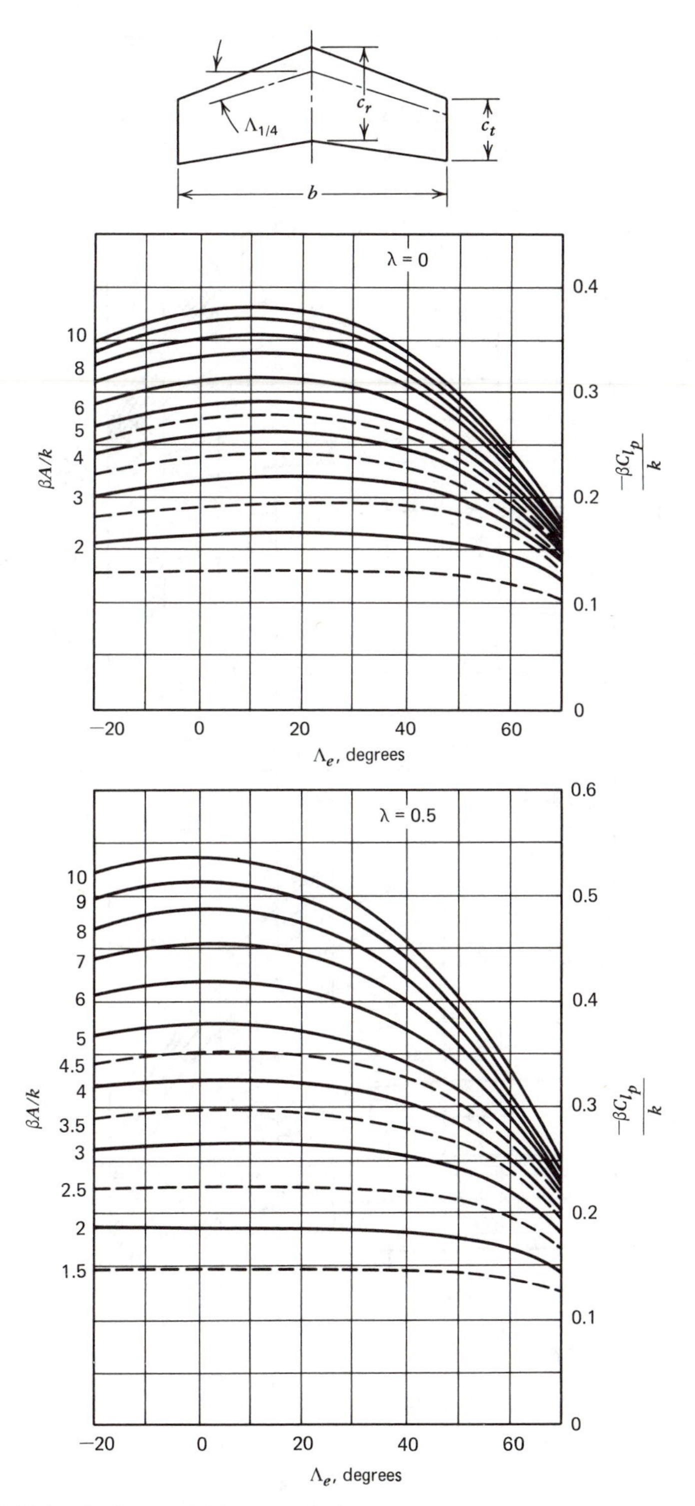

FIG. B.11,1 C_{l_p} for straight-tapered wings.
(Reproduced from Royal Aeronautical Society Data Sheet Aircraft 06. 01. 01.)

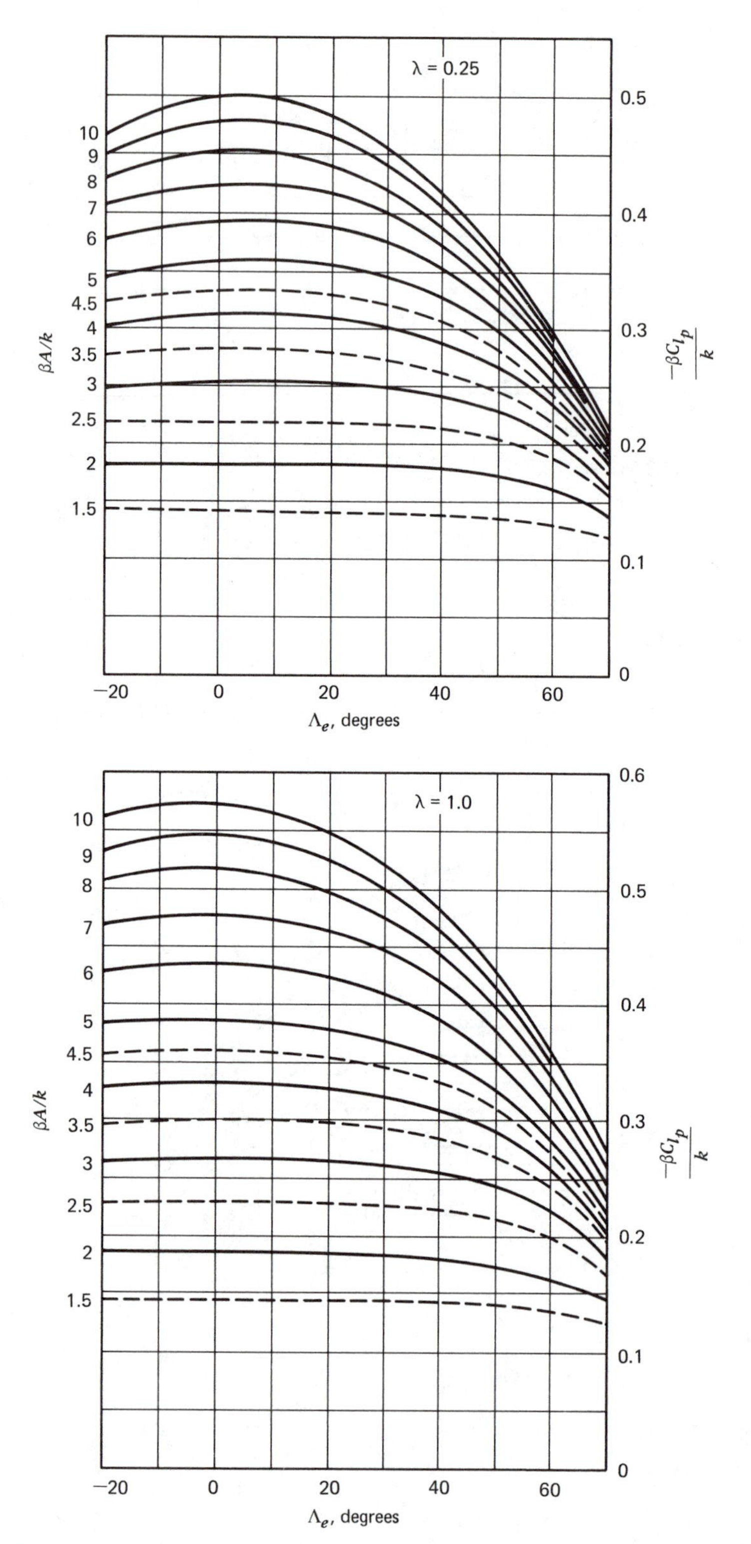

FIG. B.11,1 (concluded)

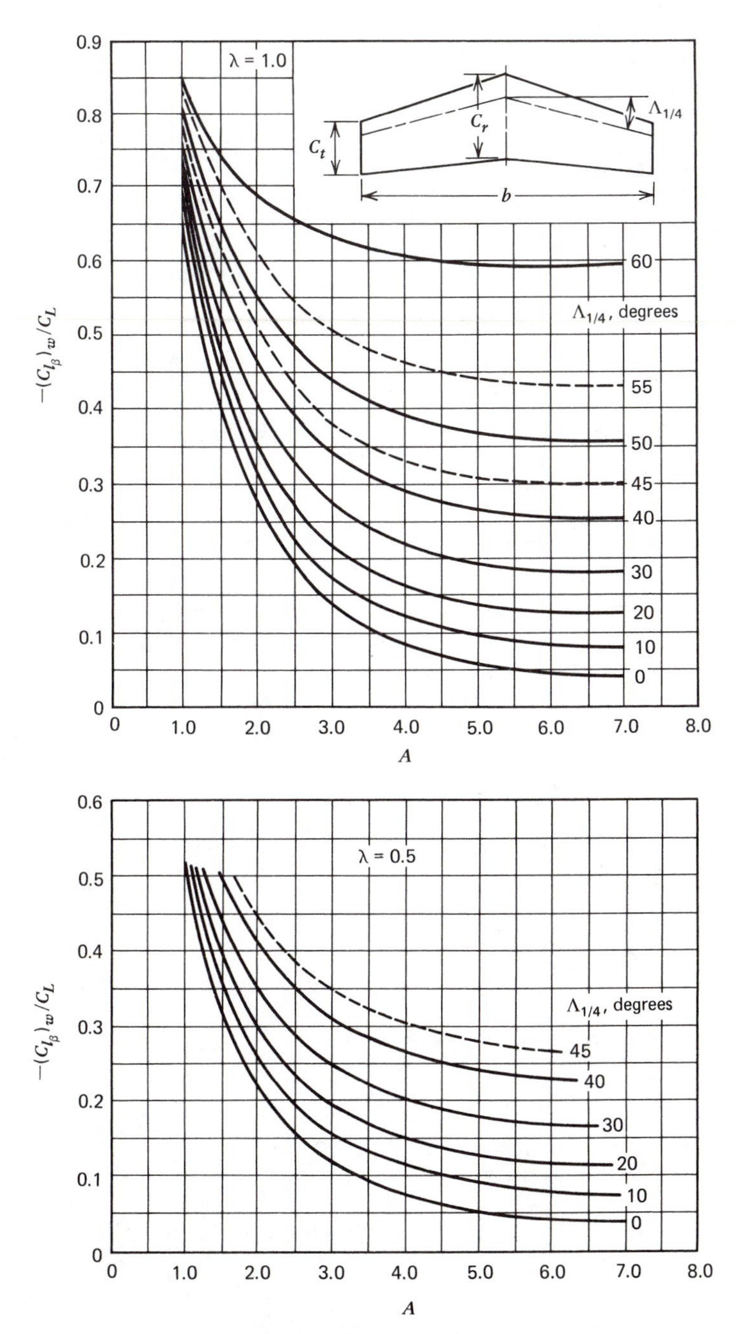

FIG. B.11,2 C_{l_β} of straight-tapered wings, zero dihedral.
(Reproduced from Royal Aeronautical Society Data Sheet Aircraft 06. 01. 04.)

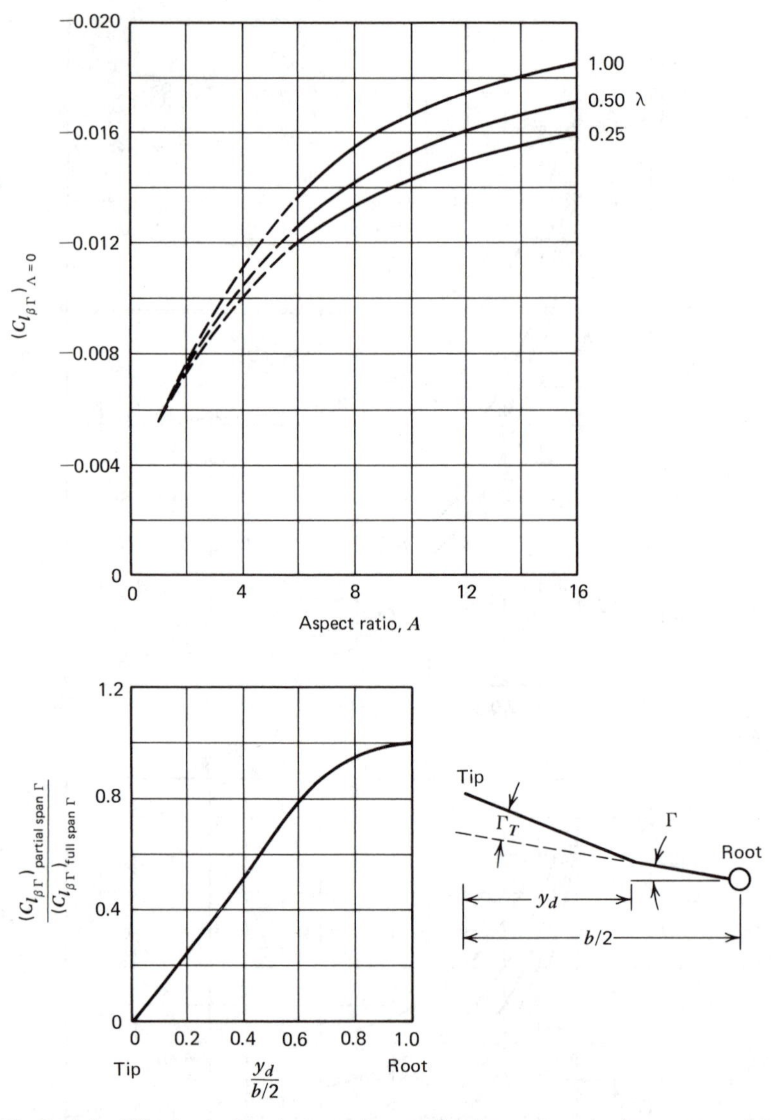

FIG. B.11,3 Effect of dihedral angle on C_{l_β} for subsonic incompressible flow (C_{l_β} in radians, Γ in degrees).

$$C_{l_\beta} = C_{l_{\beta_\Gamma}}\left[\Gamma + \frac{(C_{l_{\beta_\Gamma}})_{\text{partial span }\Gamma}}{(C_{l_{\beta_\Gamma}})_{\text{full span }\Gamma}}\Gamma_T\right]$$

where

$$C_{l_{\beta_\Gamma}} = \frac{A + 4\cos\Lambda}{(A+4)\cos\Lambda}(C_{l_{\beta_\Gamma}})_{\Lambda=0}$$

(Reproduced from *NACA Rept. 1098*, 1952, by J. P. Campbell and M. O. McKinney.)

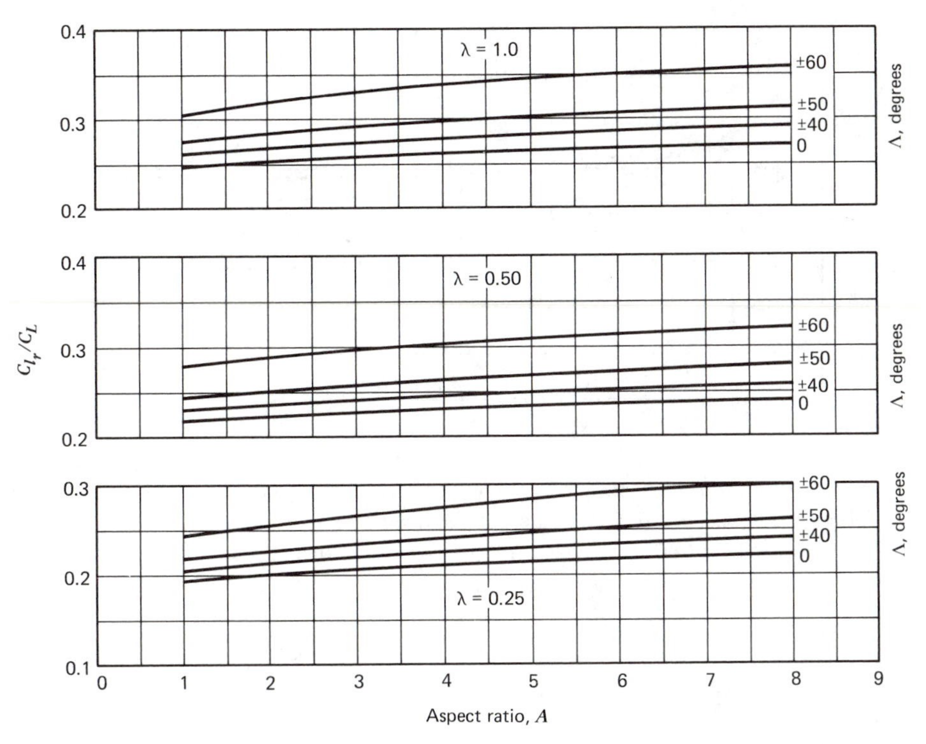

FIG. B.11,4 Charts for estimating C_{l_r} for subsonic incompressible flow. λ = taper ratio, Λ = sweepback of $\frac{1}{4}$ chord line.
(Reproduced from *NACA Rept. 1098*, 1952, by J. P. Campbell and M. O. McKinney.)

C_{l_r}

Theoretical values of C_{l_r} for swept, straight-tapered wings in incompressible flow are given in Fig. B.11,4. These values may be used in combination with experimental values of C_{l_β} to obtain an improved estimate as follows:

$$C_{l_{r_{\text{wing}}}} = C_L\left(\frac{C_{l_r}}{C_L}\right)_{\text{theory}} + C_L\left(\frac{C_{l_\beta}}{C_L}\right)_{\text{theory}} - C_{l_{\beta_{\text{exp}}}}$$

For sources on information on C_{l_r} of supersonic wings, see Table II of ref. 5.22.

B.12 YAWING-MOMENT DERIVATIVES

C_{n_β}

Theoretical values of C_{n_β} for swept untapered wings are given in Fig. B.12,1. The effect of taper ratio may be neglected. Campbell and McKinney (ref. 5.22) give an empirical formula for obtaining the fuselage contribution from measurements

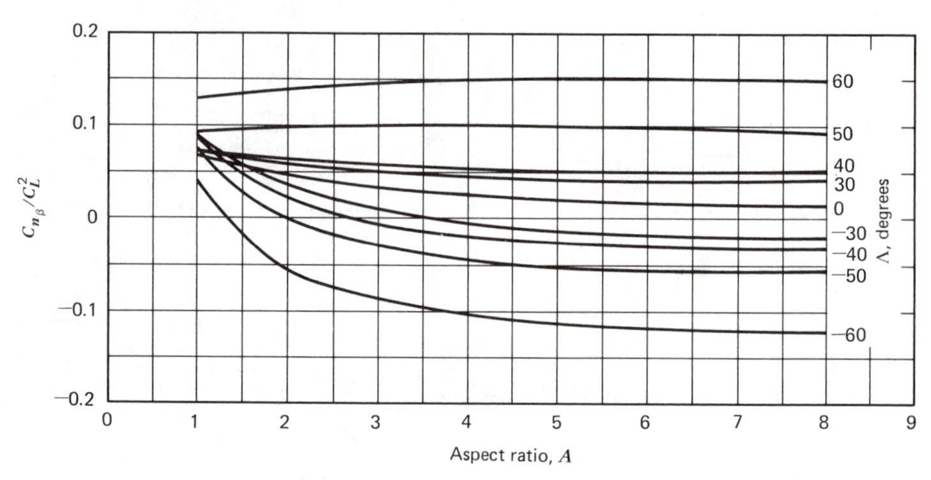

FIG. B.12,1 Variation of C_{n_β}/C_L^2 with aspect ratio and sweep for subsonic incompressible flow. $\lambda = 1.0$. C.G. coincident with wing m.a. ctr.
(Reproduced from *NACA Rept. 1098*, 1952, by J. P. Campbell and M. O. McKinney.)

on a similar configuration. It may, however, be used to obtain an approximation to the fuselage contribution directly. The formula is

$$C_{n_{\beta_{\text{fus}}}} = -1.3 \frac{\text{fuselage volume}}{Sb} \frac{h}{w}$$

where h and w are as defined in Sec. B.11.

For sources of information on C_{n_β} of supersonic wings, see Table II of ref. 5.22.

C_{n_p}

Figure B.12,2 gives theoretical values of C_{n_p} for swept untapered wings. The data may be used, however, for tapered wings. The quantity $(C_{D_0})_\alpha = (\partial/\partial\alpha)(C_D - C_L^2/\pi A)$; i.e., it is approximately the derivative of the profile drag.

For sources of information on C_{n_p} of supersonic wings, see Table II of ref. 5.22.

C_{n_r}

Theoretical values of C_{n_r} are given in Fig. B.12,3 for swept straight-tapered wings in incompressible flow.

For sources of information on C_{n_r} of supersonic wings, see Table II of ref. 5.22.

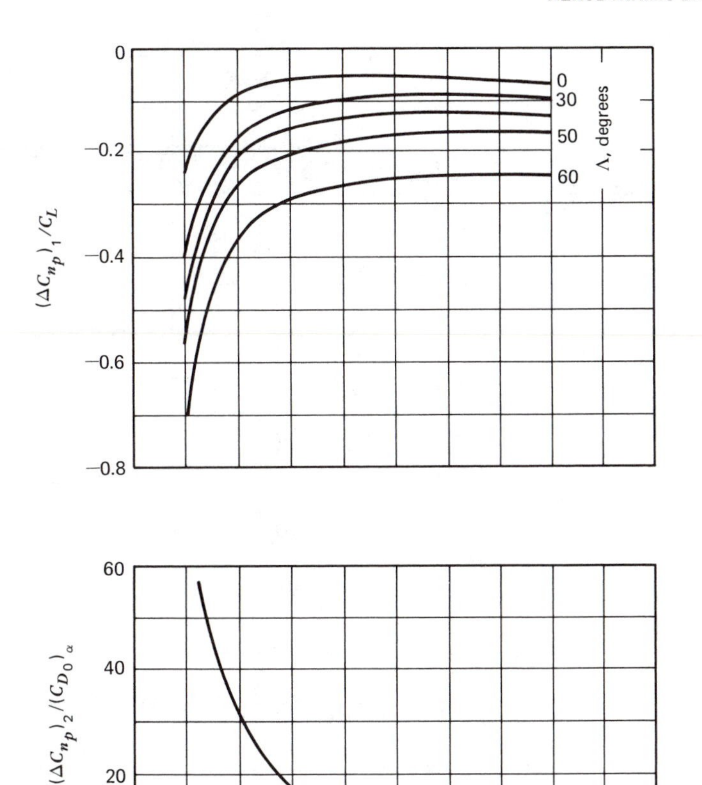

FIG. B.12,2 Variation of $(\Delta C_{n_p})_1/C_L$ and $(\Delta C_{n_p})_2/(C_{D_0})_\alpha$ with aspect ratio for subsonic incompressible flow. $\lambda = 1.0$.

$$C_{n_p} = \frac{(\Delta C_{n_p})_1}{C_L} C_L + \frac{(\Delta C_{n_p})_2}{(C_{D_0})_\alpha} (C_{D_0})_\alpha$$

(Reproduced from *NACA Rept. 1098*, 1952, by J. P. Campbell and M. O. McKinney.)

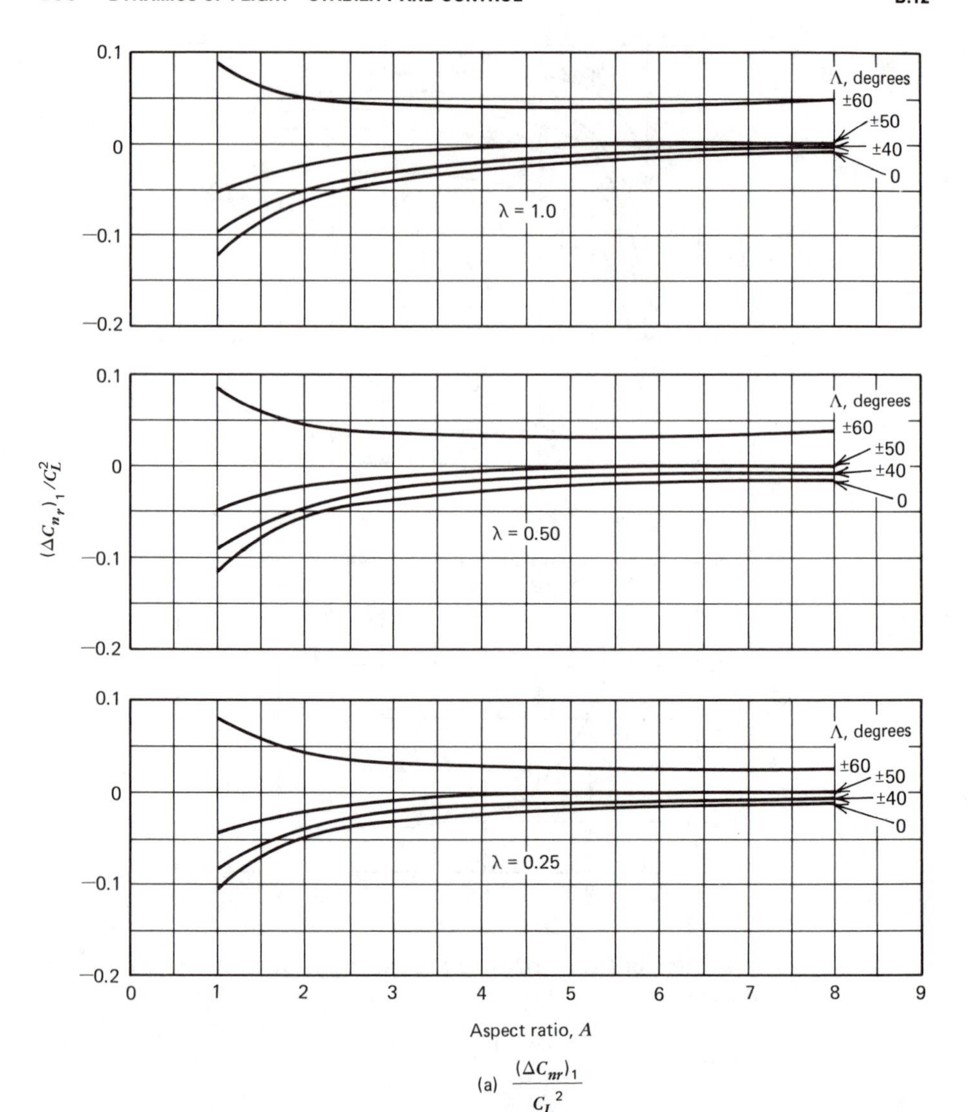

(a) $\dfrac{(\Delta C_{nr})_1}{C_L{}^2}$

FIG. B.12,3 Charts and formula for estimating C_{n_r} for subsonic incompressible flow.

$$C_{n_r} = C_L{}^2 \frac{(\Delta C_{n_r})_1}{C_L{}^2} + C_{D_0} \frac{(\Delta C_{n_r})_2}{C_{D_0}}$$

(Reproduced from *NACA Rept. 1098*, 1952, by J. P. Campbell and M. O. McKinney.)

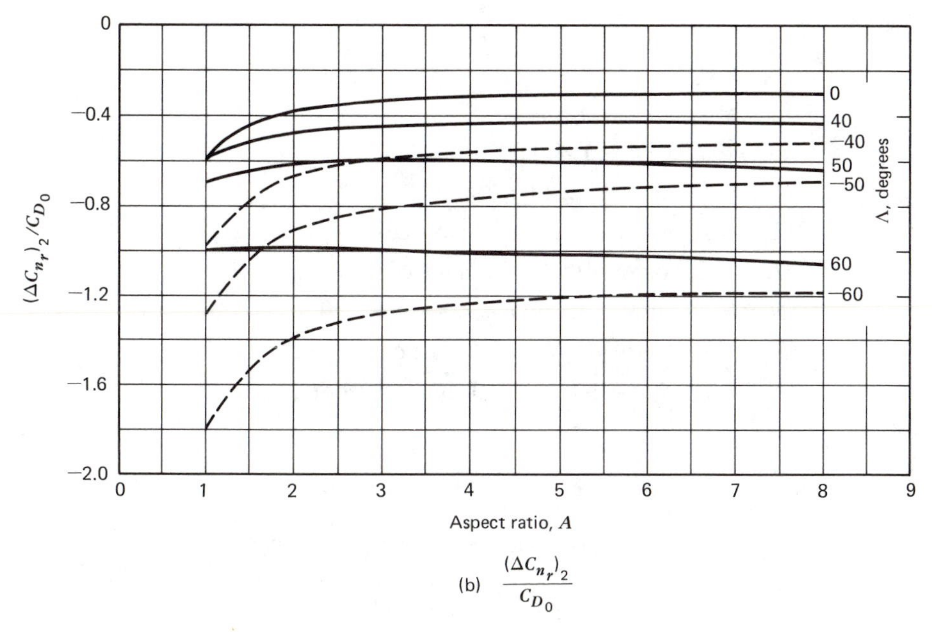

(b) $\dfrac{(\Delta C_{n_r})_2}{C_{D_0}}$

FIG. B.12,3 (concluded)

B.13 CHANGE OF STABILITY DERIVATIVES WITH ROTATION OF AXES

As an aid in transforming the stability derivatives from one set of body axes to another, the following transformation equations are reproduced (with slight change in notation) from the Royal Aeronautical Society Data Sheet Aircraft 00.00.06. The unprimed symbols refer to axes $Cxyz$; the primed symbols to axes $Cx'y'z'$. The axis Cx' makes an angle ε with Cx, as in Fig. 4.3.

Longitudinal

$$(X_u)' = X_u \cos^2 \varepsilon - (X_w + Z_u) \sin \varepsilon \cos \varepsilon + Z_w \sin^2 \varepsilon$$
$$(X_w)' = X_w \cos^2 \varepsilon + (X_u - Z_w) \sin \varepsilon \cos \varepsilon - Z_u \sin^2 \varepsilon$$
$$(X_q)' = X_q \cos \varepsilon - Z_q \sin \varepsilon$$
$$(Z_u)' = Z_u \cos^2 \varepsilon - (Z_w - X_u) \sin \varepsilon \cos \varepsilon - X_w \sin^2 \varepsilon$$
$$(Z_w)' = Z_w \cos^2 \varepsilon + (Z_u + X_w) \sin \varepsilon \cos \varepsilon + X_u \sin^2 \varepsilon$$
$$(Z_q)' = Z_q \cos \varepsilon - X_q \sin \varepsilon$$
$$(M_u)' = M_u \cos \varepsilon - M_w \sin \varepsilon$$
$$(M_w)' = M_w \cos \varepsilon + M_u \sin \varepsilon$$
$$(M_q)' = M_q$$
$$(M_{\dot{w}})' = M_{\dot{w}} \cos \varepsilon + M_{\dot{u}} \sin \varepsilon^1$$

[1] The term $M_{\dot{u}}$ is usually neglected.

Lateral

$$
\begin{aligned}
(Y_v)' &= Y_v \\
(Y_p)' &= Y_p \cos \varepsilon - Y_r \sin \varepsilon \\
(Y_r)' &= Y_r \cos \varepsilon + Y_p \sin \varepsilon \\
(L_v)' &= L_v \cos \varepsilon - N_v \sin \varepsilon \\
(L_p)' &= L_p \cos^2 \varepsilon - (L_r + N_p) \sin \varepsilon \cos \varepsilon + N_r \sin^2 \varepsilon \\
(L_r)' &= L_r \cos^2 \varepsilon - (N_r - L_p) \sin \varepsilon \cos \varepsilon - N_p \sin^2 \varepsilon \\
(N_v)' &= N_v \cos \varepsilon + L_v \sin \varepsilon \\
(N_p)' &= N_p \cos^2 \varepsilon - (N_r - L_p) \sin \varepsilon \cos \varepsilon - L_r \sin^2 \varepsilon \\
(N_r)' &= N_r \cos^2 \varepsilon + (L_r + N_p) \sin \varepsilon \cos \varepsilon + L_p \sin^2 \varepsilon
\end{aligned}
$$

MEAN AERODYNAMIC CHORD, MEAN AERODYNAMIC CENTER, AND C_{m0_w}

G. K. DIMOCK

APPENDIX C

C.1 BASIC DEFINITIONS

In the normal flight range, the resultant aerodynamic forces acting on any lifting surface can be represented as a lift and drag acting at the mean aerodynamic center ($\bar{x}$, $\bar{y}$, $\bar{z}$), together with a pitching couple C_{m0_w} which is independent of angle of attack (see Fig. 2.6).

The pitching moment of a wing is nondimensionalized by the use of the mean aerodynamic chord $\bar{c}$.

Both the m.a. center and the m.a. chord lie in the plane of symmetry of the wing. However, in determining them it is convenient to work with the half-wing.

These quantities are defined by (Fig. C.1)

$$\bar{c} = \frac{2}{S}\int_0^{b/2} c^2\, dy \tag{C.1,1}$$

$$\bar{x} = \frac{2}{C_L S}\int_0^{b/2} C_{l_a} cx\, dy \tag{C.1,2}$$

$$\bar{y} = \frac{2}{C_L S}\int_0^{b/2} C_{l_a} cy\, dy = \eta_{cp}\frac{b}{2} \tag{C.1,3}$$

$$\bar{z} = \frac{2}{C_L S}\int_0^{b/2} C_{l_a} cz\, dy \tag{C.1,4}$$

where b = wing span

c = local chord

C_L = total lift coefficient

C_{l_a} = local additional lift coefficient

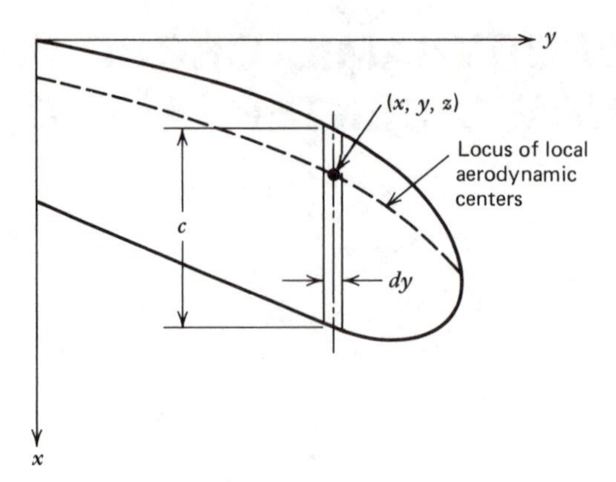

FIG. C.1 Local aerodynamic center coordinates.

S = wing area

y = spanwise coordinate of local aerodynamic center measured from axis of symmetry

x = chordwise coordinate of local aerodynamic center measured aft of wing apex

z = vertical coordinate of local aerodynamic center measured from xy plane

η_{cp} = lateral position of the center of pressure of the additional load on the half-wing as a fraction of the semispan

The coordinates of the m.a. center depend on the additional load distribution; hence the position of the true m.a. center will vary with wing incidence if the form of the additional loading varies with incidence (as it may with swept wings). For a wing that has no aerodynamic twist, the m.a. center of the half-wing is also the center of pressure of the half-wing. If there is a basic loading (i.e., at zero overall lift, due to wing twist), then $(\bar{x}, \bar{y}, \bar{z})$ is the center of pressure of the additional loading.

The height and spanwise position of the local aerodynamic centers may be assumed known, and hence $\bar{y}$ and $\bar{z}$ for the half-wing can be calculated once the additional spanwise loading distribution is known. However, in order to calculate $\bar{x}$, the fore-and-aft position of each local aerodynamic center must be known first. If all the local aerodynamic centers are assumed to lie on the nth-chord line (assumed to be straight), then

$$\bar{x} = nc_r + \bar{y} \tan \Lambda_n \tag{C.1,5}$$

where c_r = wing root chord

Λ_n = sweepback of nth-chord line, degrees

Ideal two-dimensional flow theory gives $n = \frac{1}{4}$ for subsonic speeds and $n = \frac{1}{2}$ for supersonic speeds.

The m.a. chord is located relative to the wing by the following procedure:

1. In Eq. C.1,2 replace C_{l_α} by C_L, and for x use the coordinates of the $\frac{1}{4}$-chord line.
2. The value of $\bar{x}$ so obtained (the mean quarter-chord point) is the $\frac{1}{4}$-point of the m.a. chord.

The above procedure and the definition of $\bar{c}$ (Eq. C.1,1) are used for *all* wings.

C.2 COMPARISON OF M.A. CHORD AND M.A. CENTER FOR BASIC PLANFORMS AND LOADING DISTRIBUTIONS

In Table C.1 taken from ref. C.1, values of m.a. chord and $\bar{y}$ are given for some basic planforms and loading distributions.

In the general case the additional loading distribution and the spanwise center-of-pressure position can be obtained by methods such as those of ref. C.2,3 and 4. For a trapezoidal wing with the local aerodynamic centers on the nth-cord line, the chordwise location of the aerodynamic center from the leading edge of the m.a. chord expressed as a fraction of the m.a. chord h_{n_w} is given by

$$h_{n_w} = n + \frac{3(1+\lambda)^2}{8(1+\lambda+\lambda^2)}\left[n_{cp} - \frac{1+2\lambda}{3(1+\lambda)}\right] A \tan \Lambda_n \tag{C.2,1}$$

TABLE C.1

Planform	*Additional Loading Distribution*	*M.A.C.,* $\bar{c}$	$\bar{y}$
Constant taper and sweep (trapezoidal)	Any	$\frac{2c_r}{3}\frac{1+\lambda+\lambda^2}{1+\lambda}$	$\eta_{cp}\cdot\frac{b}{2}$
Constant taper and sweep (trapezoidal)	Proportional to wing chord (uniform C_{l_α})	$\frac{2c_r}{3}\frac{1+\lambda+\lambda^2}{1+\lambda}$	$\frac{b}{2}\cdot\frac{1+2\lambda}{3(1+\lambda)}$
Constant taper and sweep (trapezoidal)	Elliptic	$\frac{2c_r}{3}\frac{1+\lambda+\lambda^2}{1+\lambda}$	$\frac{b}{2}\cdot\frac{4}{3\pi}$
Elliptic (with straight sweep of line of local a.c.)	Any	$\frac{c_r}{3}\cdot\frac{8}{\pi}$	$\eta_{cp}\cdot\frac{b}{2}$
Elliptic (with straight sweep of line of local a.c.)	Elliptic (uniform C_{l_α})	$\frac{c_r}{3}\cdot\frac{8}{\pi}$	$\frac{b}{2}\cdot\frac{4}{3\pi}$
Any (with straight sweep of line of local a.c.)	Elliptic	$\frac{2}{S}\int_0^{b/2} c^2\,dy$	$\frac{b}{2}\cdot\frac{4}{3\pi}$

where A = aspect ratio, b^2/S

λ = taper ratio, c_t/c_r

c_t = wing-tip chord

The length of the chord through the centroid of area of a trapezoidal half-wing is equal to $\bar{c}$. For the same wing with uniform spanwise lift distribution (i.e., C_{l_a} = const) and local aerodynamic centers on the nth-chord line, the m.a. center also lies on the chord through the centroid of area. The chord through the centroid of area of a wing having an elliptic planform is not the same as $\bar{c}$, but the m.a. center for elliptic loading and the centroid of area both lie on the same chord (see ref. C.1).

C.3 M.A. CHORD AND M.A. CENTER FOR SWEPT AND TAPERED WINGS

The ratio $\bar{c}/c_r$ is plotted against λ in Fig. C.2 for straight tapered wings with streamwise tips. The spanwise position of the m.a. center of the half-wing (or the center of pressure of the additional load) for uniform spanwise loading is also given in Fig. C.2. These functions are given in Table C.1.

The m.a. chord is located by means of the distance x of the leading edge of the m.a. chord aft of the wing apex, given in Fig. C.3:

$$x = \frac{b}{2} \cdot \frac{1}{3} \frac{1 + 2\lambda}{1 + \lambda} \tan \Lambda_0$$

$$= \frac{1 + 2\lambda}{12} c_r A \tan \Lambda_0 \tag{C.3,1}$$

where Λ_0 = sweepback of wing leading edge, degrees.

The sweepback of the leading edge is related to the sweep of the nth-chord line Λ_n by the relation

$$A \tan \Lambda_0 = A \tan \Lambda_n + 4n \frac{1 - \lambda}{1 + \lambda} \tag{C.3,2}$$

Using Eq. C.3,2 and the expression for $\bar{c}/c_r$, x can be obtained in terms of $\bar{c}$ and Λ_n from

$$\frac{x}{\bar{c}} = \frac{(1 + 2\lambda)(1 + \lambda)}{8(1 + \lambda + \lambda^2)} \left[A \tan \Lambda_n + 4n \frac{1 - \lambda}{1 + \lambda} \right] \tag{C.3,3}$$

The fractional distance of the m.a. center aft of the leading edge of the m.a. chord, h_{n_w}, is given for swept and tapered wings at low speeds and small incidences in Fig. C.4. The dotted lines show the aerodynamic-center position for wings with unswept trailing edges. The curves have been obtained from theoretical and experimental data (ref. C.5). The curves apply only within the linear range of the curve of wing lift against pitching moment, provided that the flow is subsonic over the entire wing. The probable error of h_{n_w} given by the curves is within 3%.

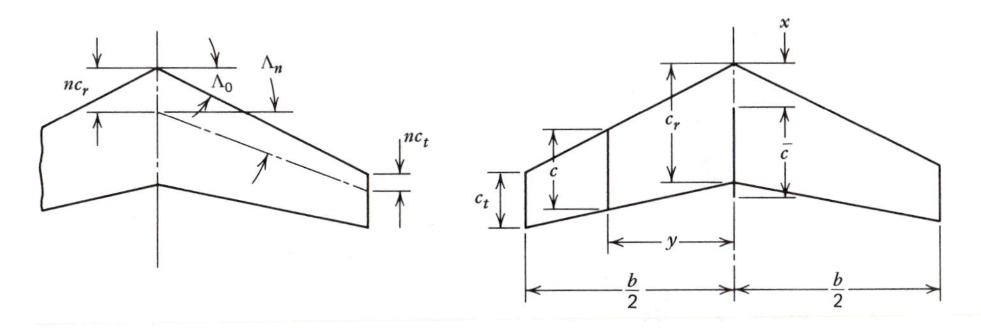

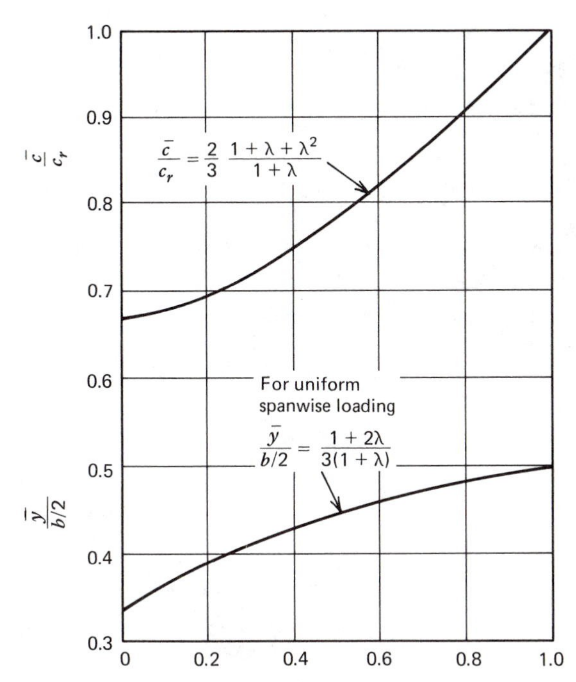

FIG. C.2 Mean aerodynamic chord for straight tapered wings; and spanwise position of mean aerodynamic center for uniform spanwise loading (i.e., constant C_{L_α}).
(Reproduced from "Notes on the Mean Aerodynamic Chord and the Mean Aerodynamic Center of a Wing" by A. H. Yates, *J. Roy. Aero. Soc.*, June 1952.)

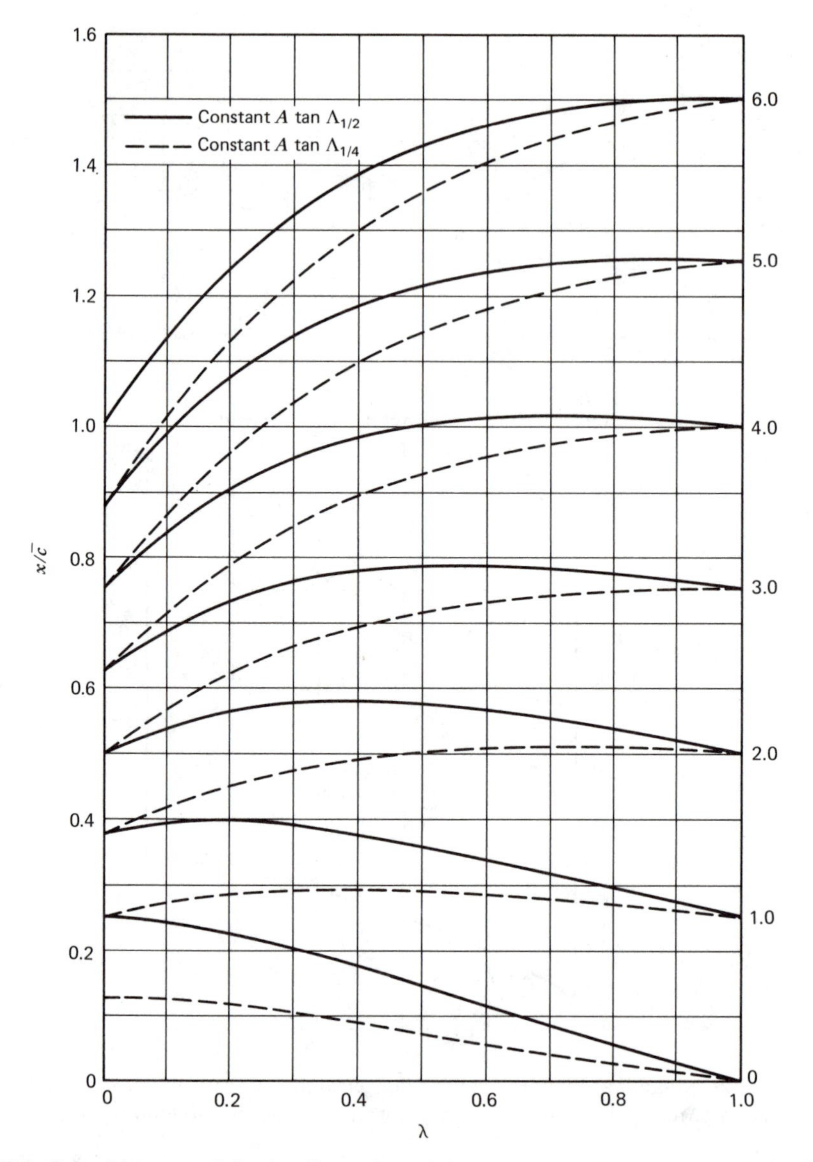

FIG. C.3 Distance of the leading edge of the mean aerodynamic chord aft of the wing apex for swept and tapered wings.

(Reproduced from Royal Aeronautical Society Data Sheet Wings 00. 01. 01.)

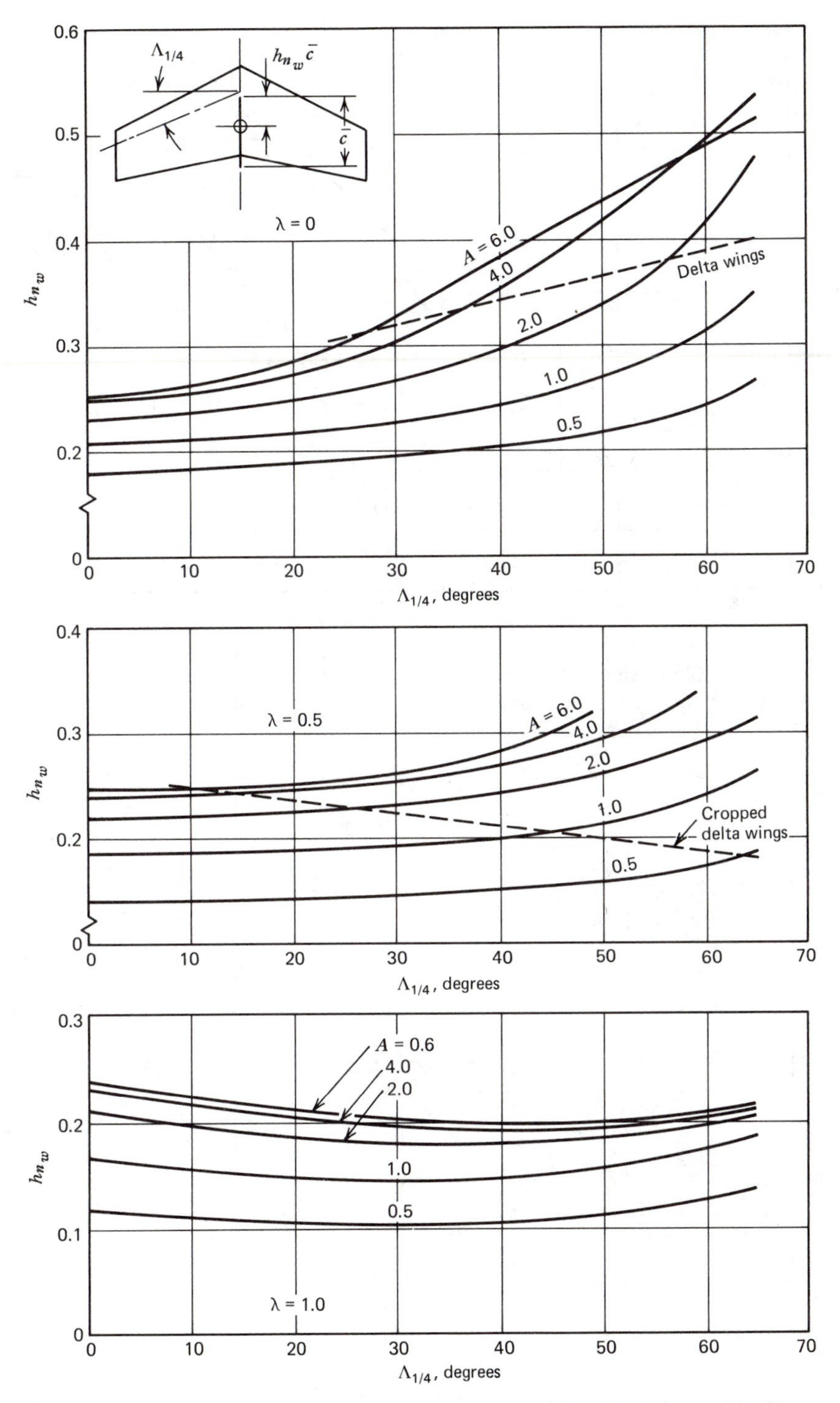

FIG. C.4 Chordwise position of the mean aerodynamic center of swept and tapered wings at low speeds expressed as a fraction of the mean aerodynamic chord. (Reproduced from Royal Aeronautical Data Sheet Wings 08. 01. 01.)

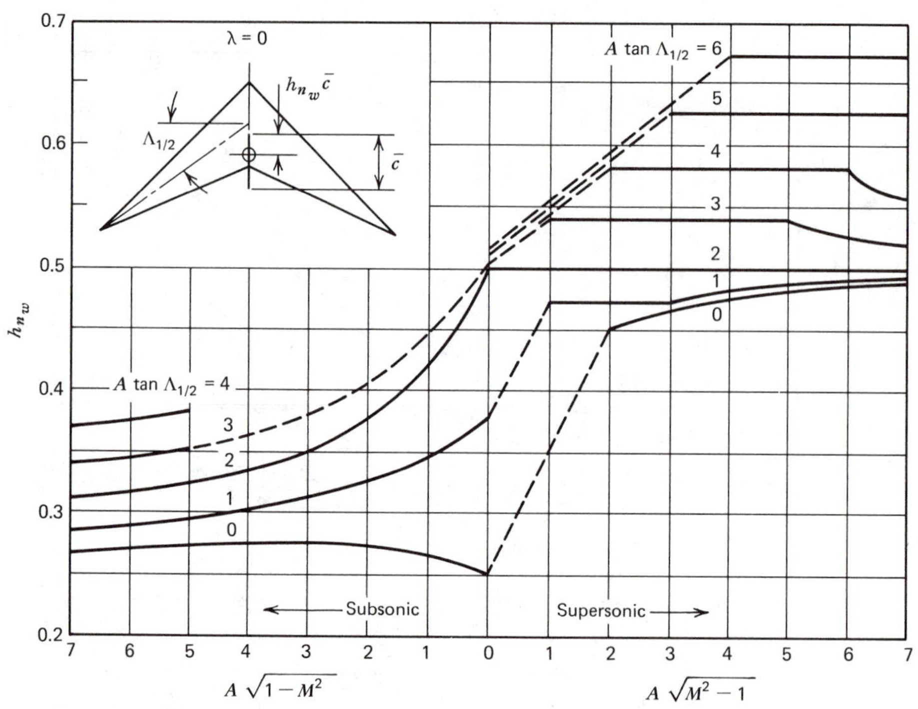

FIG. C.5 Chordwise position of the mean aerodynamic center of swept and tapered wings at high speeds expressed as a fraction of the mean aerodynamic chord. (Reproduced from Royal Aeronautical Society Data Sheet Wings S. 08. 01. 02.)

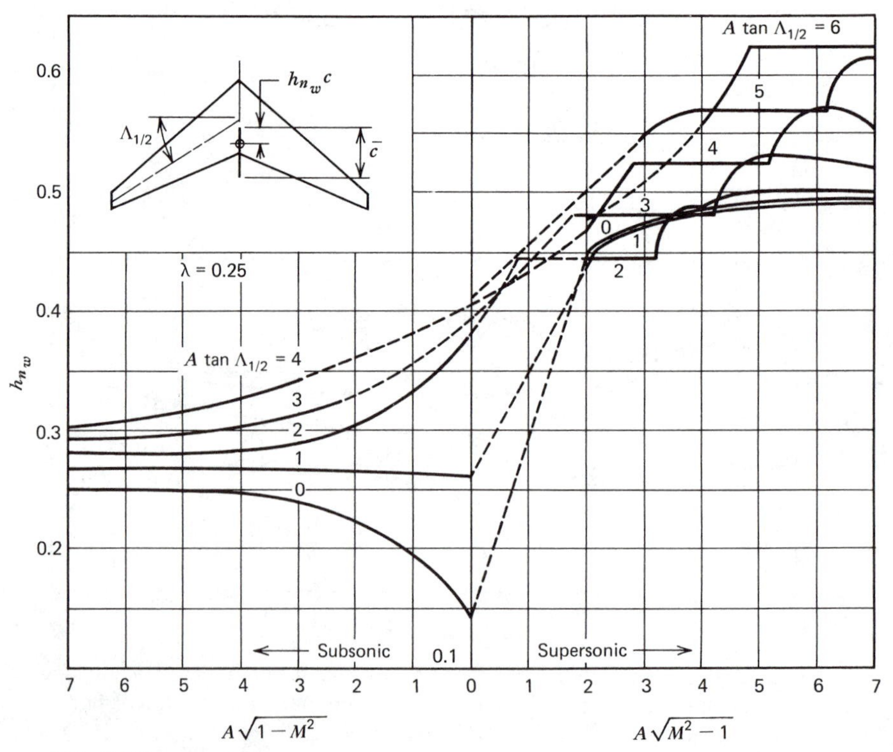

FIG. C.5 (continued)

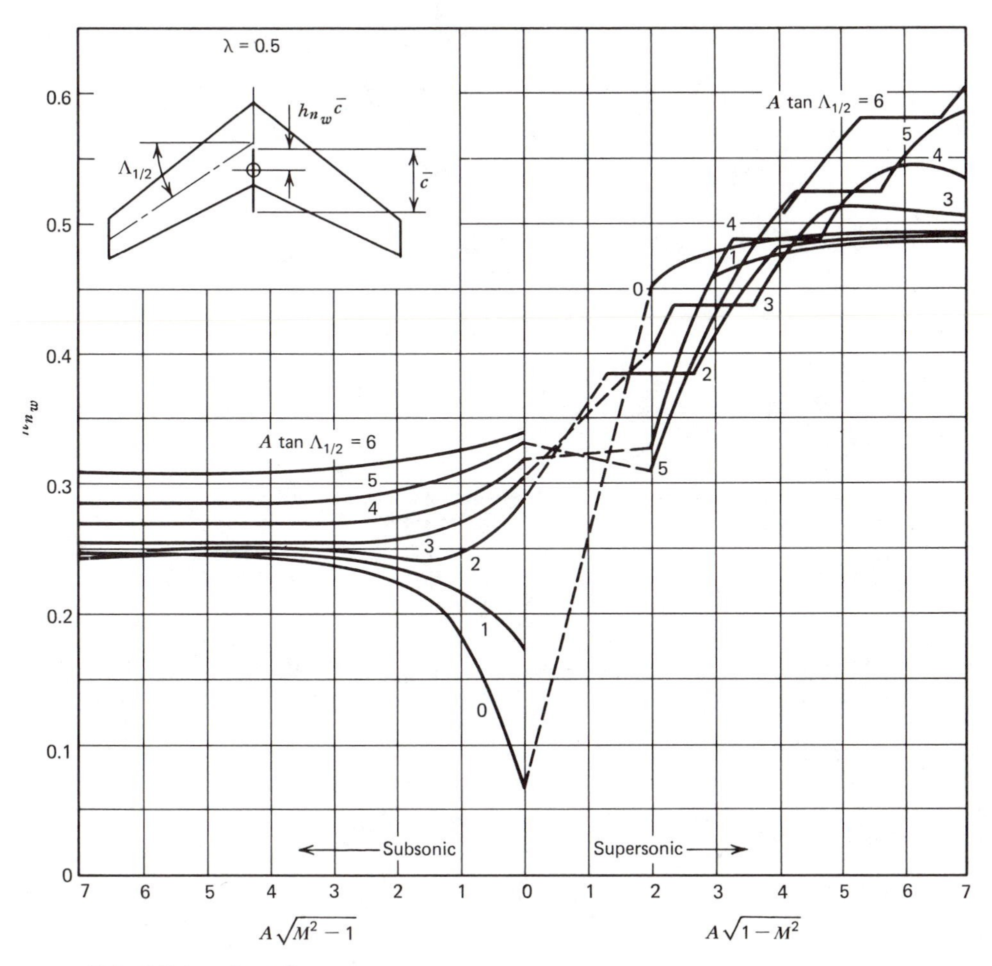

FIG. C.5 (continued)

For higher Mach numbers and in particular for supersonic speeds, h_{n_w} may be obtained from the theoretical data given in Fig. C.5. The curves for subsonic flow were obtained by applying the similarity laws of linearized perturbation theory to Weissinger's method, whereas those for supersonic flow were based on linearized supersonic lifting-surface theory. Slender-wing theory was used to give h_{n_w} for $A\sqrt{M^2-1}=0$. In regions where the theoretical solutions were inadequate, the curves are shown dotted.

The theories are based on inviscid flow and apply to flat plates at small incidence. For small values of $A\sqrt{M^2-1}$ their practical validity is doubtful. The kinks shown in the curves are not likely to occur in practice. The curves are expected to be of acceptable accuracy where the variation of aerodynamic-center position with Mach number is smooth (ref. C.6).

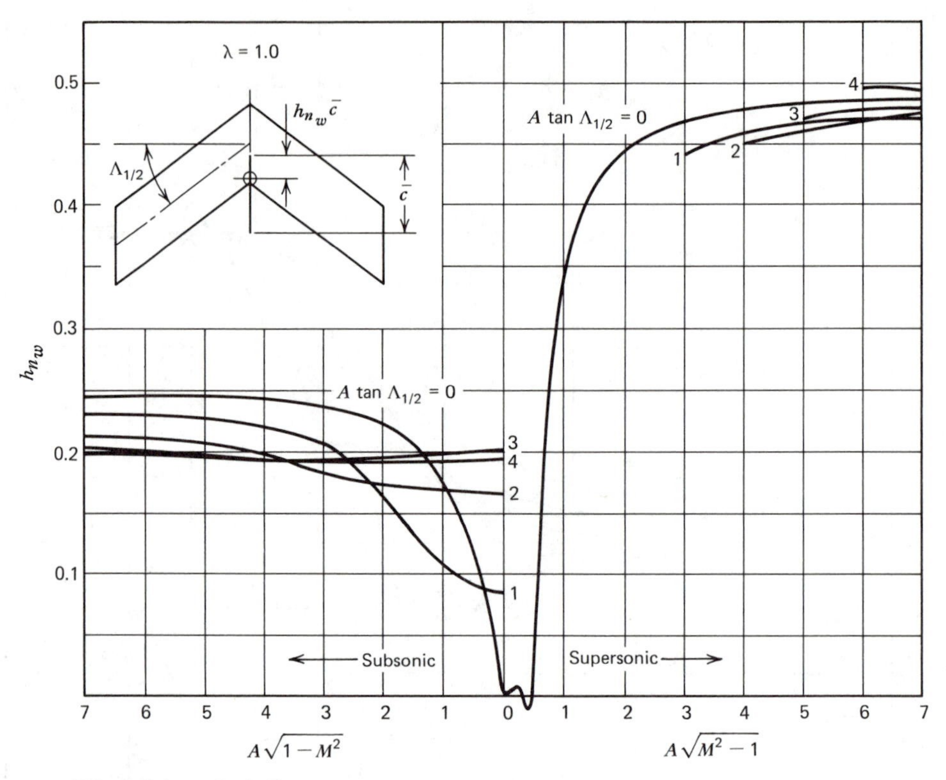

FIG. C.5 (concluded)

C.4 $C_{m_{0_w}}$

The total load on each section of a wing has three parts as illustrated by Fig. C.6*a*. The resultant of the local additional lift l_a, is the lift L_a acting through the m.a. center (Fig. C.6*b*).

The resultant of the distribution of the local basic lift l_b is a pitching couple whenever the line of aerodynamic centers is not straight and perpendicular to x. This couple is given by

$$M_1 = 2\int_0^{b/2} (x - \bar{x})l_b\,dy = 2\int_0^{b/2} xl_b\,dy$$

since the resultant of $l_b = 0$.

Then

$$C_{m_1} = \frac{2}{qS\bar{c}}\int_0^{b/2} xC_{l_b}qc\,dy$$

$$= \frac{2}{S\bar{c}}\int_0^{b/2} C_{l_b}xc\,dy \tag{C.4,1}$$

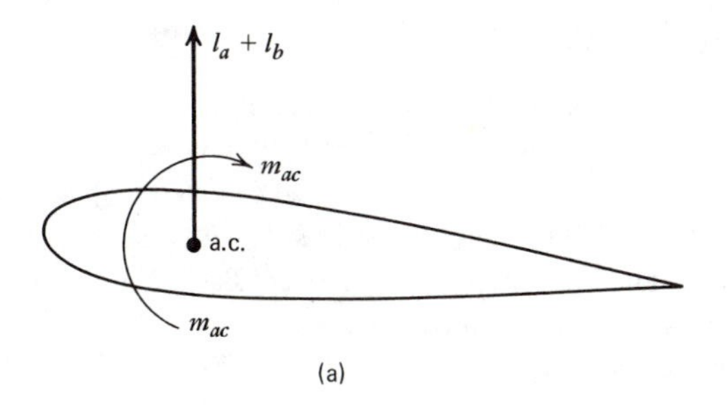

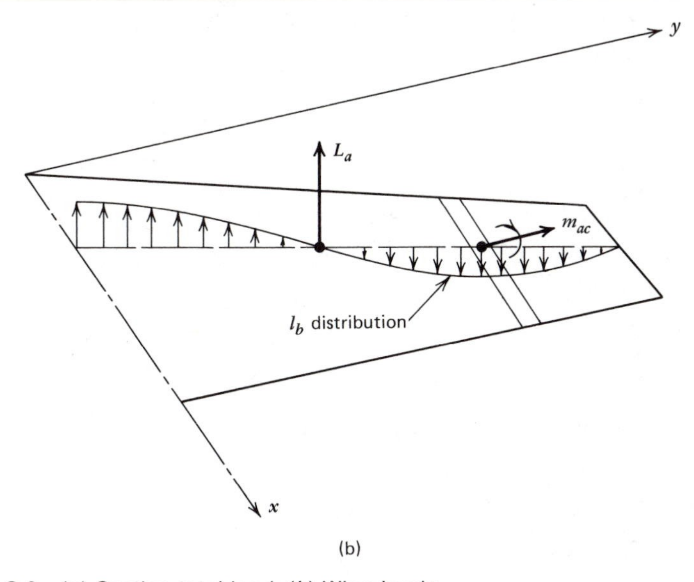

FIG. C.6 (*a*) Section total load. (*b*) Wing loads.

The resultant of the m_{ac} distribution is given by

$$C_{m_2} = \frac{2}{S\bar{c}} \int_0^{b/2} C_{m_{ac}} c^2 \, dy \tag{C.4,2}$$

The total pitching-moment coefficient about the m.a. center is then

$$C_{m_{0_w}} = C_{m_1} + C_{m_2} = \text{const} \tag{C.4,3}$$

If $C_{m_{ac}}$ is constant across the span, and equals C_{m_2}, then Eq. C.4,2 also becomes the defining equation for $\bar{c}$.

C.5 BIBLIOGRAPHY

C.1 A. H. Yates. Notes on the Mean Aerodynamic Chord and the Mean Aerodynamic Centre of a Wing. *J. Roy. Aero. Soc.*, vol. 56, p. 461, June 1952.

C.2 J. De Young and C. W. Harper. Theoretical Symmetric Span Loadings at Subsonic Speeds for Wings Having Arbitrary Planforms. *NACA Rept. 921*, 1948.

C.3 J. Weissinger. The Lift Distribution of Swept Back Wings. *NACA TM 1120*, 1947.

C.4 R. Stanton-Jones. An Empirical Method for Rapidly Determining the Loading Distributions on Swept Back Wings. *Coll. Aero.* (*Cranfield*) *Rept. 32*, 1950.

C.5 Royal Aeronautical Society. Data Sheet Wings .08.01.01, 1955.

C.6 Royal Aeronautical Society. Data Sheet Wings S.08.01.02, 1955.

INDEX